Geographic Thought

Critical Introductions to Geography

Critical Introductions to Geography is a series of textbooks for undergraduate courses covering the key geographical sub-disciplines and providing broad and introductory treatment with a critical edge. They are designed for the North American and international market and take a lively and engaging approach with a distinct geographical voice that distinguishes them from more traditional and outdated texts.

Prospective authors interested in the series should contact the series editor:

John Paul Jones III
School of Geography and Development
University of Arizona
jpjones@arizona.edu

Published

Water: A Critical Introduction
Katie Meehan, Naho Mirumachi, Alex Loftus, and Majed Akhter

Environment and Society: A Critical Introduction, 3e
Paul Robbins, John Hintz, Sarah A. Moore

Political Geography: A Critical Introduction
Sara Smith

Political Ecology: A Critical Introduction, 3e
Paul Robbins

Economic Geography: A Critical Introduction
Trevor J. Barnes, Brett Christophers

Health Geographies: A Critical Introduction
Tim Brown, Gavin J. Andrews, Steven Cummins, Beth Greenhough, Daniel Lewis, Andrew Power

Urban Geography: A Critical Introduction
Andrew E.G. Jonas, Eugene McCann, Mary Thomas

Geographic Thought: A Critical Introduction
Tim Cresswell

Mapping: A Critical Introduction to Cartography and GIS
Jeremy W. Crampton

Research Methods in Geography: A Critical Introduction
Basil Gomez, John Paul Jones III

Geographies of Media and Communication
Paul C. Adams

Social Geography: A Critical Introduction
Vincent J. Del Casino Jr.

Geographies of Globalization: A Critical Introduction
Andrew Herod

Geographic Thought: A Critical Introduction, 2e
Tim Cresswell

Forthcoming
Energy, Society, and Environment: A Critical Introduction
Dustin Mulvaney

Introduction to Cultural Geography: A Critical Approach
Jamie Winders, Declan Cullen

Geographic Thought

A Critical Introduction

Second Edition

Tim Cresswell

WILEY Blackwell

This second edition first published 2024
© 2024 John Wiley & Sons Ltd

Edition History
John Wiley & Sons Ltd (1e, 2013).

Registered Offices
John Wiley & Sons, Inc., 111 River Street, Hoboken, NJ 07030, USA
John Wiley & Sons Ltd, The Atrium, Southern Gate, Chichester, West Sussex, PO19 8SQ, UK

For details of our global editorial offices, customer services, and more information about Wiley products visit us at www.wiley.com.

Wiley also publishes its books in a variety of electronic formats and by print-on-demand. Some content that appears in standard print versions of this book may not be available in other formats.

Library of Congress Cataloging-in-Publication Data

Name: Cresswell, Tim, author. | John Wiley & Sons, publisher.
Title: Geographic thought : a critical introduction / Tim Cresswell.
Other titles: Critical introductions to geography
Description: Second edition. | Hoboken, NJ : Wiley-Blackwell, 2024. |
   Series: Critical introductions to geography | Includes bibliographical references and index.
Identifiers: LCCN 2023037268 (print) | LCCN 2023037269 (ebook) | ISBN
   9781119602828 (paperback) | ISBN 9781119602842 (adobe pdf) | ISBN
   9781119602835 (epub)
Subjects: LCSH: Human geography–Philosophy.
Classification: LCC GF21 .C74 2024 (print) | LCC GF21 (ebook) | DDC 910/.01–dc23/eng/20231025
LC record available at https://lccn.loc.gov/2023037268
LC ebook record available at https://lccn.loc.gov/2023037269

Cover Design: Wiley
Cover Image: Courtesy of 2010 Tintin Wulia

Set in 10/12pt Minion by Straive, Pondicherry, India
Printed and bound by CPI Group (UK) Ltd, Croydon CR0 4YY
C9781119602828_090124

Contents

Preface to Second Edition

As I write this, it has been 10 years since the first edition of *Geographic Thought*. A lot of thought happens in a decade. At the same time, the basic story of the development of theory in geography has not fundamentally changed. After all, the history of theory in geography up to 2013, when the first edition was published, is still the same history it was then. Strabo, Alexander von Humboldt, and Ellen Semple still did their thing. They have not been erased from history. So, writing a second edition is, in some ways, harder than writing a first edition. It is more difficult to insert myself into my own narrative of theory in geography. I said what I had to say, so why even do a second edition? As it turns out, there are a number of good reasons. First of all, clearly, a lot *has* happened in a decade. The biggest changes in this edition, nor surprisingly, happen at the end of the book, where two chapters have now become four chapters covering "more-than-physical" as well as "more-than-human" geographies, as well as chapters on postcolonial, decolonial, and anticolonial geographies and Black geographies – no longer crammed into one chapter on geography's exclusions. These latter chapters are also implicated in the second reason for a revision – a determined attempt to diversify the geographers represented in most of the other chapters. I hesitate to call this "decolonizing" my own book (and myself) for reasons that should be clear from Chapter 14 – but it is a concerted attempt to seek out and find different voices representing different positionalities – to make the account of geographic thought less white and less male. In doing this, I paid particular attention to those authors I have substantially engaged with in order to tell the story of geographic thought. I am convinced that this exercise in diversification has made the account richer and even more inspiring than the one I gave 10 years ago.

In addition to the four new chapters (replacing two of the old ones), all except two of the chapters in the book have been updated and refreshed. Inevitably, and in line with other multiple-edition books, my account has lengthened. It is relatively easy to recognize things to include that were previously absent; it is much harder to get rid of things that I previously thought worthy of inclusion. This has happened, but sparingly. As with the first edition, writing a textbook on geographic thought comes with some interesting challenges. It is clearly supposed to be accessible enough for the students who read this as part of their education in the discipline of geography, and, at the same time, I cannot help but be aware that the book becomes part of the very thing it is about – the story of geographic thought. Textbooks are important components of the assemblage that produces the history of geography. They are among the most widely read books written in an academic context, and they are implicated in the creation of master (and other) narratives of the discipline. Sometimes this seems like quite a heavy responsibility – both telling the story of the main theoretical approaches within the discipline and recognizing that this very story is one that excludes all that is not included in the pages that follow. I remain convinced that it is important for all of us, however critically inclined, to understand the main bodies of thought that have led to where we are. I also understand that there are other stories that have not been told. I urge readers of this book to seek out those stories.

I owe thanks to the continuing insistence of Justin Vaughan at Wiley for continuing to chase me up at conferences and between times to convince me to work on this edition. I am grateful to various critics and commentators who gently (or not so gently) pointed out deficiencies in the first edition. These include the late Ron Johnston, Janice Monk, Eden Kincaid, and Lauren Fritzsche. John Paul Jones III, as editor of the "Critical Introductions," was amazingly helpful both in the detailed care he took with the manuscript and with the thoughtful side notes, which gave me plenty to think about. Many thanks. I remain eternally grateful for my family of adventurers: my wife Carol and my no-longer kids, Owen, Alice, and Maddy. Particular thanks to Carol Jennings for the much-improved index!

Chapter 1

Introduction

Good evening. Welcome to Difficult Listening Hour. The spot on your dial for that relentless and impenetrable sound of Difficult Music. So sit bolt upright in that straight-backed chair, button that top button, and get set for some difficult music. (Laurie Anderson – "Difficult Listening Hour," from "Home of the Brave," 1986)

Hostility to theory usually means an opposition to other people's theories and an oblivion of one's own. (Eagleton 2008: xii)

If the scientific investigation of any subject be the proper avocation of the philosopher, Geography, the science of which we propose to treat, is certainly entitled to a high place ... (Strabo 1912 [AD 7–18]: 1)

Geography is a profound discipline. To some this statement might seem oxymoronic. Profound geography seems as likely as "military intelligence." Geography is often the butt of jokes in the United Kingdom. A school friend of mine who was about to start a degree in pure mathematics described my chosen degree as the "science of common sense." I once appeared on a public radio quiz show in the United States. When the host asked me what I did and I explained I was a geography student, he asked what geographers had left to do – surely we know where Milwaukee is already? I mumbled an apologetic answer. Taxi drivers ask me to name the second highest mountain in the world, trying to catch me out by avoiding the obvious first highest. My parents thought I was going to be a weather forecaster. So why is geography profound? Why indeed would the classical Greek/Roman scholar Strabo (more on him in Chapter 2) suggest that geography deserves a "high place" and that it constitutes "philosophy"?

Strabo presented a number of answers ranging from the fact that many "philosophers" and "poets" of repute had taken geography as central to their endeavors to the fact that geography was indispensable to proper government and statecraft. But perhaps most profoundly:

In addition to its vast importance in regard to social life, and the art of government, Geography unfolds to us the celestial phenomena, acquaints us with the occupants of the land and ocean, and the vegetation, fruits, and peculiarities of the various quarters of the earth, a knowledge of which marks him who cultivates it as a man earnest in the great problem of life and happiness. (Strabo 1912 [AD 7–18]: 1–2)

"The great problem of life and happiness." This was and is a central philosophical and theoretical problem. How do we lead a happy life? What constitutes a good life? How should people relate to the nonhuman world? How do we make our life meaningful? These are profound questions and they are also geographical questions.

In addition to being profound, geography is also everywhere. The questions we ask are profound because of, not in spite of, the everydayness of geographical concerns. This point is well made in this extended extract from an essay by the cultural geographer, Denis Cosgrove:

On Saturday mornings I am not, consciously, a geographer. I am, like so many other people of my age and lifestyle, to be found shopping with my family in my local town-sector precinct. It is not a very special place, artificially illuminated under the multi-storey car park, containing an entirely predictable collection of chain stores – W.H. Smith, Top Shop, Baxters, Boots, Safeway and others – fairly crowded with well-dressed, comfortable family consumers. The same scene could be found almost anywhere in England. Change the names of the stores and then the scene could be typical of much of western Europe and North America. Geographers might take an interest in the place because it occupies the peak rent location of the town, they might study the frontage widths or goods on offer as part of a retail study, or they might assess its impact on the pre-existing urban morphology. But I am shopping.

Then I realise other things are also happening: I'm asked to contribute to a cause I don't approve of; I turn a corner and there is an ageing, evangelical Christian distributing tracts. The main open space is occupied by a display of window panels to improve house insulation – or rather, in my opinion, to destroy the visual harmony of my street. Around the concrete base of the precinct's decorative tree a group of teenagers with vividly coloured Mohican haircuts and studded armbands cast the occasional scornful glance at middle-aged consumers. …

The precinct, then, is a highly textured place, with multiple layers of meaning. Designed for the consumer to be sure, and thus easily amenable to my retail geography study, nevertheless its geography stretches way beyond that narrow and restrictive perspective. The precinct is a symbolic place where a number of cultures meet and perhaps clash. Even on a Saturday morning I am still a geographer. Geography is everywhere. (Cosgrove 1989: 118–119)

Here, Cosgrove reflects on the way our discipline sticks close to the banal everydayness of life. It is not possible to get through an hour, let alone a day, without confronting potentially geographical questions. Shopping centers in medium-sized British towns do not seem particularly profound (when compared to the question of the origins of the universe, say), but they are. They are full of geography. But this geography is not always readily apparent. It is not just *there* like park benches or shop windows. To see it we have to have the tools to see it. We need to know about the importance of a "peak rent location" or even what a "symbolic place" is, and to know this we have to think about geography theoretically. So, geography is at the same time "profound" and everyday. Unlike theoretical physics or literary theory, it is hard to escape geography. Once you are a geographer, particularly one interested in theory, you always are a geographer. It is this confluence of the profound and the banal that gives geographical theory its special power.

This book is focused on key geographical questions. It is based on my belief that geography is profound: that the ideas geographers deal in are some of the most important ideas there are. Each of the chapters that follow may occasionally seem slightly arcane as I recount the arguments that geographers and others have with each other in the pages of journals and monographs. But at the heart are important questions. They are important both for the existential dimension of how we lead a good life and for more worldly issues of equality, justice, and our connections to the natural world. Seemingly abstract questions like how we make a home in the world and what constitutes a good life are important precursors to pressing and grounded issues of living in the third decade of the twenty-first century, including species extinction, climate change, structural racism, global pandemics, and ongoing colonialism. I am convinced that thinking through the theoretical issues of geography at least makes us more aware of ourselves, of the world, and of our relationship with the world. It also allows us insight into so-called wicked problems facing humanity in the current moment.

While geographical questions remain central to this book, I make no claims to completeness. Geographers, like practitioners of many other disciplines, are constantly arguing about ideas. Often it is the people who are supposed to be in agreement that are doing the arguing. We are used to the idea of advocates of competing ideas clashing with each other. In these arguments, large numbers of people are lumped together as "positivists" or "Marxists" for instance. But if we look closely, we find that these groups are constantly arguing with each other too, over what it means to be a positivist or a Marxist. A book like this cannot hope to recount each and every one of these arguments. Such a book would be an encyclopedia of many volumes. Here I hope to convey what, to me, are the essential questions that geographic theory helps us to answer – questions that all of us can apply to our everyday lives in order to help us make sense of the world. This will necessarily involve ignoring the vast majority of work in geography including, undoubtedly, some work that my colleagues and others may feel is central. This book reflects my own fascinations and predilections. Theory in human geography is more complicated by orders of magnitude than what I present here. To engage with these complications, I provide suggested readings along the way (indicated with an asterisk (*) in the References section at the end of each chapter). This is a road map and there are many small towns and hamlets and even some major cities that these roads do not connect. You will have to go off road occasionally to find them.

This book is likely to play an important role in a ritual. At some point, either as an undergraduate or as a postgraduate, geography students (particularly human geography students) have to do a course on theory, or geographic thought, or philosophy and geography. It is a rite of passage. For many, this is much like Laurie Anderson's "difficult listening hour" – relentless and impenetrable. For two or three hours a week, students are confronted with a dizzying array of theories and philosophies each with its own particular jargon and logic. And just when one "ism" appears to make sense the next one comes over the horizon and declares it invalid, wrong, confused, or, amazingly, too simplistic. To many of us this ritual seemed a long way from doing geography. It was a diversion that took us away from getting on with our work. To some, however, and I include myself here, it made geography come alive. It was certainly difficult, but it seemed to make other parts of the discipline make sense and make our own work more profoundly connected to currents of thought that coursed not only through geography but its sister disciplines as well.

Why Theory Matters

This ritual is important. It is important because all geographical inquiry, even that which pretends otherwise, is always shaped by theory and philosophy. To paraphrase the literary scholar, Terry Eagleton: those who say they don't like theory mean that they don't like someone else's theory and are unaware of their own. So how does theory shape geographical inquiry?

First, it is there when we make choices about *what to study*. If we choose to look at the micro spaces of the home, there is a history of feminist theory urging geographers to take private space seriously. If we choose to study the structuring of public space, there are any number of theorists who have argued about the meaning of "public" (let alone the meaning of "space"). It is true that we may be unaware of these writers, and not directly influenced by them, but theory still has played a role at a number of levels. First, these previous theorists have been instrumental in making such projects acceptable as geographical research whether we have heard of them or not. A geography of the spaces of home would probably have been dismissed out of hand as a viable research project in the vast majority of geography departments in (say) 1960. Funding bodies would probably have returned a polite rejection; many of them still would! Second, we are practicing theory ourselves when we make these decisions. We are deciding what, out of all the possible projects in an infinitely complicated world, is important to us. We are prioritizing some questions over others – promoting some parts of the world as important, as interesting. Such choices are (in part) theoretical.

The second major way in which theory shapes geographical study is in the choices we make about what to include and what to ignore in our study. Once we have decided we want to explore domestic space, we still have work to do. We have to decide what might be included in such a study. What kind of domestic space? Where? How many? Do we focus on the "things" in a space or the things people do? Is it important to explore these themes at different times of the day, week, or year? Should we look at the world of children or just the adults? Shall we link the research to the kinds of spaces the family members inhabit when they are not at home? Questions such as these are endless. They are (in part) theoretical questions.

The third major way in which theory shapes geographical study is in the choices we make about how to gather information. Theory is linked to method through methodology and **epistemology** (how we know what we know). Can we answer the questions we have set ourselves through a survey of thousands of households? Will a quantitative approach be more "scientific" and generalizable? Or do we need to live life with the inhabitants of a small number of households over a long period of time in order to get some of the depth and richness of life as it is lived? Is there archival material we could access to study these issues in the past or elsewhere? These are, of course, practical questions concerning how much money, time, expertise, and energy we have. But they are also theoretical/ philosophical questions about what it is we consider important to find out, whether we are more interested in generalizability or depth. Methods are theoretical too.

The fourth major way in which theory shapes geographical study is in the choices we make about how to represent our research to others. The answer to this might seem straightforward; a standard journal paper, a monograph, in text or graphs. But we have to ask how we are going to write a text: impressionistically or with hard certainty? What kind of maps or charts will we use? Why? What journal will we choose to publish in? How will we engage with those beyond the academy? Do we even need to? All of these are theoretical questions too.

So, theory is involved in all stages of geographical research. We may not be clear about exactly how, but it is there nonetheless. And it is my assertion that it is better to be somewhat aware of this than blissfully unaware.

Claims to have no theory (claims which are frequently made) are simply delusional. Theory is everywhere, in everything we do. Without theory, life (not just geography) would be chaos. One purpose of this book is to raise awareness about which theory or theories are implicit in geographical research – to make theory less implicit and more explicit in the practice of geography. It should be an aid in making decisions about theories you like and do not like, believe in or disbelieve. Beyond that, it will provide some ways of thinking that might stimulate self-analysis about how you and those around you lead your lives. With any luck it will make you less scared of thinking difficult thoughts.

What Is Theory?

Perhaps we have jumped the gun slightly here. Perhaps we need to define theory in order that it might make sense. The term theory can seem unduly threatening and worryingly vague. At the most general level, theory seems to refer to pretty much anything that is going on in our minds. Despite its slightly imposing implications, theory is actually a word that is used frequently in everyday speech. We say things like "Tim has a theory about that" or "In theory, that might work – but not in practice." Here, theory refers to the realm of ideas. It is opposed to "practice" which itself often appears to mean "reality." Theory is thinking and practice is doing. This opposition leads many to think of theory as impractical and unreal. Theory can often be used as a term of abuse. But most things that exist in our heads are not really "theories." Thoughts and ideas may be hopes, dreams, guesses, fears, or a host of other mental phenomena that are not strictly or wholly theoretical. Theory, in the academic sense, usually refers to organized and patterned sets of ideas rather than spur-of-the-moment thoughts. Theories are more-or-less organized ways of ordering the world which exist in our minds and which we share with others. They have a collective and enduring intellectual quality.

Clearly we perceive the world in many ways using the senses of sight, sound, taste, touch, and smell. As we move through the world, we are barraged with sensations that our body has to make some sense of. Think for a minute about the everyday activity of crossing a busy road. We can see the traffic speeding past, smell the exhaust, and see recent rain on the pavement. We can hear the surrounding people and vehicles. How do we cross the road? Is it not miraculous that we get to the other side? Why don't we stand in the middle of the road and marvel at the steady stream of perception – the roar of engines, the stream of colors? Clearly, we have to order our senses to make them make sense. The middle of the road is not a good place to stop and wonder. We did not know this as a very small child. We had to become aware of it. We make sense of the world by taking what our senses present to us and ordering it, prioritizing and assembling sensations so that we might make it to the other side. In fact we are so good at this we can do it seemingly without thinking.

This is the beginning of theory – making the complexity of the world clearer – *ordering* it and prioritizing. Avoiding death. Few would actually say that the mental processes involved in crossing the road constitute theory, but it is certainly the first step to understanding what theory can do for us.

One metaphor that is frequently used to describe theory is the "lens." Think of theory as a *lens* that helps us see some things clearly – it imposes conceptual order on messy reality – it brings an indistinct blur into focus. Theory turns the perceived and experienced world into an "*interpreted* world." How this happens is extremely varied and the subject of considerable debate among geographers. People use different lenses to see the same things differently – and then argue about it. Some might say, for the sake of argument, that we need only present "the facts." This, broadly speaking, constitutes a kind of theoretical approach (whether its advocates see it this way or not) which we might call **empiricism**. An approach that tries to stay close to the things being discussed. An approach that denies abstraction. But how could we present only "the facts"? What facts? When do we stop? Which facts are relevant to our argument and which are marginal or unnecessary? To answer this, some form of lens, or ordering, is needed. In other words, we need theory.

Theory, at its most basic, is a form of ordering the multiplicity of raw experience and "facts." It allows us to get to the other side. But there are clearly different kinds of theory, different understandings of theory, even different theories of theory.

What we mean by theory differs according to which kinds of theory we subscribe to. Human and physical geographers certainly differ in the ways they talk about theory. A theory in the natural sciences, and thus physical geography, is a much more specific thing than a theory in the social sciences or humanities. In intellectual life, at least, theory usually refers to a more systematic way of ordering the world – a set of interlinked propositions about how things in the world are connected. "Theory" (with a big T) is a word that is often used to describe a general attempt to make abstract conceptual statements about broad arenas of social life. This use of the word is more common in the humanities and the social sciences and is associated with "philosophies" – ways of thinking about questions like the meaning of existence, what it is to be human, and such like.

What theory means depends on the context in which theory is raised. The everyday use of the word theory (as in "Tim has a theory about that") suggests that I have noted a few facts and come to some conclusion about why a set of facts present themselves as they do. Say, for instance, that I have a theory about why the University of Morningside (not the real name) hired Professor Long (not a real name either). As Jonathan Culler has suggested, such a theory suggests "speculation" (Culler 1997). This is different from a mere guess, as a guess suggests that there is a correct answer that I do not know. That I have a theory suggests that I have come up with a plausible explanation which includes a certain level of complexity. Not an explanation that can be easily proved or disproved – simply a plausible one. Culler also notes that a theory often provides a counterintuitive explanation: an explanation that goes beyond the obvious. There is a difference between saying that the University of Morningside hired Professor Long because he was the best person for the job and saying that they hired him because he was about to be awarded a big grant or because he was having an affair with the registrar. The first explanation is hardly a theory at all. The latter two are both speculative and not obvious. They are kinds of theories.

When we enter the more specialized world of academic discourse, we see that theory is polysymous (has many meanings). Theory comes on many levels. **Marxism** is a theoretical approach in geography and across the social sciences and humanities. So is Marxism a theory? Well, only in a general sense. As we will see in Chapter 7, Marxism includes an array of theories that add up to a coherent philosophy. It includes a theory about how history happens (historical materialism), an economic theory about how things get value (the labor theory of value), a theory about people's relationship to commodities (commodity fetishism), and any number of other theories each with a particular arena of human life that it purports to explain. Together they add up to a potent political philosophy. These theories are quite particular and logically coherent (even when wrong). They cannot be tested in quite the same way as a theory in physical science. They cannot easily be falsified. In the history of geographical theory, there are also specific theories that are meant to explain particular aspects of the human interaction with the earth. **Spatial science** is premised on a philosophy of **positivism** (see Chapter 5) but includes a number of theories such as **central place theory**, **spatial interaction theory**, etc. Again, these are specific theories that purport to explain particular things, patterns, and processes.

The twentieth century saw the emergence of a set of ideas referred to as "social theory." **Social theory** naturally formed part of sociology. As the name indicates, it provides theory about society. But social theory quickly became interdisciplinary. Social theory has been practiced by sociologists, philosophers, anthropologists, literary theorists, and human geographers, among others. Social theory addresses the way society is structured and occasionally transformed. As we will see over the course of this book, the transformation and reproduction of social distinctions such as **class** and **gender** often, perhaps always, involve elements we could call geographical – **space**, **place**, **territory**, etc. It is not surprising, therefore, that, since the 1970s at least, geographers have been keen to embrace and practice social theory. Indeed, some geographers are at the heart of what can only retrospectively be called social theory from the nineteenth century – the theories of **anarchism** inherent in the work of Elisée Reclus and Peter Kropotkin (see Chapter 3).

Since the 1970s, at least, human geographers have begun to use the word "theory" in a new kind of way. This new approach to theory does not refer to theories of something (like the labor theory of value or spatial interaction theory), but simply "theory." This new way of using theory is not unique to human geography but imported from (and shared with) literary studies, cultural studies, continental philosophy, and all places in between. Indeed "theory" is used to refer to work that seems to have utility to thinkers across a range of fields. "Theory" challenges many of the commonsense assumptions behind thinking in a range of disciplines. Most of the time we associate this realm of theory with continental European thinkers such as Michel Foucault, Roland Barthes, Jacques Derrida, or Luce Irigaray. It is hard to say which discipline someone like Foucault belongs to. His work speaks across disciplines and is thus different from, for instance, spatial interaction theory, which speaks to a small and quite specialized group of people. "Theory" is unlikely to be about something as specific as the reasons for people's movements in space. It is, as Culler has put it, "about everything under the sun" (Culler 1997: 3).

Theory is often overtly political. Certainly, the traditions of critical social theory sought not simply to understand the world but, as Marx suggested, to change it. Clearly the various theories associated with Marx are designed first to understand why the world is like it is and then to come up with a better (by which we mean, more just) alternative. Likewise, the central message of feminist theory concerns the unequal position of women vis-à-vis men in society and argues for a transformation of that situation. The term **critical theory** is often used to refer to sets of ideas that are designed to provide a critique of the way things are and promote something better – the way things could be. The Black, feminist scholar of race, gender, and many other things, bell hooks, wrote in a powerful paper about being a Black woman using theory. She has frequently been confronted with the idea that theory is irrelevant. Or even that theory is inherently "White" or "masculine." "You can't tear down the master's house using the master's tools," she was told. Theory here is yet again contrasted with practice. In this case, political practice. In response, hooks makes a spirited argument for theory as liberatory practice, as something that enlivens and enrages, as something that challenges common sense and reveals the forms of power that stand behind it. Theory is practice, she argues, and when done well, in a way that does not deliberately exclude and obfuscate, it can change lives and become a positive force toward social transformation (hooks 1994). Theory for some, then, is about the practice of politics, about seeking a fairer and more just world than the one we currently inhabit.

All this talk of "theory" will seem strange to a physical geographer. Theory in physical geography, with a few exceptions, is quite different (see Chapter 13). A textbook on theory in physical geography (a relatively rare phenomenon when compared to the array of such books available to human geographers) defined theory as "a framework of ideas that guide what we think reality is and how to go about studying it" (Inkpen 2005: 36). More specifically, some define theories as a systematically ordered set of hypotheses interlocked by a network of deductive relationships (Von Englehardt and Zimmerman 1988). A theory here is a kind of higher-level hypothesis. A theory is a grand hypothesis that sits on top of a larger set of small-scale hypotheses that themselves predict and explain certain kinds of facts in the world. As most physical geography happens in a more-or-less positivist framework, they tend to be testable. Physical geographers, on the whole, believe that knowledge can be linked to some pre-existing reality (the physical world) that it seeks to explain. Until quite recently, they have rarely been interested in how that knowledge might play a role in constituting the reality they are describing.

Theory, Writing, and Difficulty

One of the major difficulties faced by students of theory in geography and elsewhere is the kind of writing that they encounter. Some of it is simply bad. Some of it is deliberately obfuscating. "Long words strung together in no particular order," as one physical geography colleague once put it to me. There are a number of reasons for this. Writing in this way can make the writer seem clever when in fact what they have to say is simple. The historian Patricia Limerick expressed this in the following way:

In ordinary life, when a listener cannot understand what someone has said, this is the usual exchange:

LISTENER: I cannot understand what you are saying.
SPEAKER: Let me try to say it more clearly.

But in scholarly writing in the late 20th century, other rules apply. This is the implicit exchange:

READER: I cannot understand what you are saying.
ACADEMIC WRITER: Too bad. The problem is that you are an unsophisticated and untrained reader. If you were smarter, you would understand me.

The exchange remains implicit, because no one wants to say, "This doesn't make any sense," for fear that the response, "It would, if you were smarter," might actually be true. (Limerick 1993: 3)

In her article, Limerick provides two examples from academics to illustrate her point that academics often write as poorly as their students. One of these examples is the geographer Allan Pred. This is what Limerick has to say about the words of Allan Pred quoted in the first paragraph:

> If what is at stake is an understanding of geographical and historical variations in the sexual division of productive and reproductive labor, of contemporary local and regional variations in female wage labor and women's work outside the formal economy, of on-the-ground variations in the everyday content of women's lives, inside and outside of their families, then it must be recognized that, at some nontrivial level, none of the corporal practices associated with these variations can be severed from spatially and temporally specific linguistic practices, from languages that not only enable the conveyance of instructions, commands, role depictions and operating rules, but that also regulate and control, that normalize and spell out the limits of the permissible through the conveyance of disapproval, ridicule and reproach.

> In this example, 124 words, along with many ideas, find themselves crammed into one sentence. In their company, one starts to get panicky. "Throw open the windows; bring in the oxygen tanks!" one wants to shout. "These words and ideas are nearly suffocated. Get them air!" And yet the condition of this desperately packed and crowded sentence is a perfectly familiar one to readers of academic writing, readers who have simply learned to suppress the panic. (Limerick 1993: 3)

Ideally this would not be the case. Writing, at its best, is an exercise in democracy. It is about sharing ideas. If the idea is not clearly expressed, it cannot be shared.

There is, however, another side to this argument. Some ideas are simply difficult. No matter how clear the writing, the idea will remain difficult. Consider a scenario in which a mathematician or a physicist presents a new theorem. The equations are likely to be difficult (but ultimately explainable and even aesthetically pleasing). A trained scientist might have trouble grasping it and a novice student would find it totally incomprehensible. I would not know where to start. Yet I cannot imagine anyone asking the inventor of the theorem to make it simpler or easier to understand. Scientists have to live with the fact that their science is difficult. You need to be trained to understand it. You have to struggle with it before it becomes clear. So why should geography be any different? Perhaps because many people believe that human geography exists within a realm of common sense. But just as the novice physicist needs to be trained to comprehend complicated science, so the novice geographer needs a theory course to get to grips with theory. This involves reading difficult stuff. Some of this difficult stuff is, indeed, badly written. Some, however, is just difficult.

There is also some unfairness in Limerick's discussion of the passage from Pred. Allan Pred was a geographer who worked with writing his whole life – and his life was full of ideas (we will come across some of them in the pages that follow). He continually tried to invent new kinds of writing to better represent what he was trying to say (most often a style known as "montage" borrowed from the cultural theorist Walter Benjamin). Such experimentation undoubtedly provokes failure on occasions. I have often been frustrated by his writing and have given up on it. Recently, however, I read his book *The Past Is Not Dead* (Pred 2004) from beginning to end. While it would be easy to pick out a sentence or paragraph for ridicule, the effect of the whole book (which includes a strategy of seemingly endless repetition of key ideas) was extremely powerful and left me convinced of the value of experimentation.

As well as style there is the issue of jargon. It is an easy put-down of another writer to refer to the writing as full of jargon. Jargon is most often a pejorative term. To refer to writing as jargon often simply means that the reader does not understand it. But jargon merely refers to specialized language. In this sense the word "drumlin" is jargon because it is a specialized term for a smooth rounded lump in the landscape formed in a glacial environment. Most people who have not taken elementary physical geography will not know this. Why not just say "small hill"? The answer, of course, is that there are all kinds of small hill, and not all of them are produced by glaciers in a particular way. The same applies in human geography.

Writing in the realm of theory often involves unfamiliar words. Sometimes these are neologisms – or new words. Consider, for instance, the following geographical text:

> I suggest the term *spant*, an acronym for *SPace ANd Time* unit. The size of a spant could be noted as appropriately needed by subscripts referring precisely to longitudes, latitudes, dates and times of the day … History is the study of spants. When a parent tells a youngster "this is not the time or place to behave like that" the child rearing effort has been focused on a spant. (M. Melbin quoted in Billinge 1983: 409)

Why use new words when it would be much simpler, and gentler on the reader, to simply use "plain English"? Often neologisms are unnecessary and obfuscating. Certainly, Mark Billinge was upset by this particular neologism along with a whole array of writing in geography emerging in the 1980s. His response to the spant is as follows:

> This kind of manufactured jargon is really quite unnecessary. The assertion of the last sentence of the second passage is highly questionable despite the certainty implied, whilst taken at face value the whole exercise is quite absurd. Historians might be interested to know that they are really spantologists, but the amusement would soon wear off. Equally, the human race has handled its understanding of time and space thus far without recourse to spants, and it is unlikely that whatever mysteries remain will be uncovered by the incorporation of spant into the vocabulary. In practice this kind of jargon is worthless since it adds nothing to our ability to express ideas and consequently it will not endure. It will have limited "spant." (Billinge 1983: 409)

Writing 40 years later, I can confirm that the spant did not endure. On the other hand, there are good reasons to be inventive in this way. The problem with words we use every day is that, for the most part, we tend to think we already know what they mean. Words like "culture" and "nature" for instance are fairly commonplace. We have a vague idea of what they mean and, in everyday life, we don't spend too much time questioning them. In fact these two words have been described by the literary theorist Raymond Williams as two of the most complicated words in the English language and yet we think they are obvious. Williams wrote several books which attempted to understand and explain the meaning of culture and in doing so he invented terms such as "structure of feeling" and used terms such as "hegemony" (Williams 1977). Another cultural theorist, Pierre Bourdieu, invented or adopted a whole slew of new terms with which to think about culture – "doxa," "habitus," "disposition" (Bourdieu 1990). What happens when we encounter these terms? Most obviously we do not immediately think we know what the writer is talking about. We have to move out of our everyday attitude and think about what these writers mean by these terms. We are forced to reflect. Neologisms, used well, will make us think and have the power to produce new insights. Consider another example of supposedly "bad writing" given by Billinge:

> In a supportive physical environment time-space routines and *body-ballets* of the individual may fuse into a larger whole, creating a space-environment dynamic called *place-ballet*. (David Seamon quoted in Billinge 1983: 408)

Billinge described this as "the coining of new and generally superfluous terms" (Billinge 1983: 408). Unlike the ill-fated "spants," however, the notions of time–space routines, bodyballets, and place-ballets have all endured. Indeed, they feature in Chapter 6. Anyone reading the whole paper from which this quote is taken should be able to understand them easily enough and the terms help to develop new and different understandings of place which have been developed by a number of geographers in productive ways.

And human geographers are not the only ones who use jargon and resort to neologisms. I have already described how "drumlin" can be considered jargon. Consider the remarkable career of William Morris Davis, one of the two or three most important people in the development of physical geography:

> Davis's geographical language was enriched by his constant invention of new terms. He coined more than a hundred and fifty technical terms. … Many are anatomical, such as elbow of capture, eyebrow scarp, or beheading; some are typal locations such as morvan or monadnock, They heightened the universal appeal of Davisian methodology, although occasionally a certain term gave a foothold for disapproval. … A few of the terms were stillborn; but most survived and some diffused into the general language, even into modern poetry. Davis would have been flattered to read in W.H. Auden's "Age of Anxiety": "O stiffly stand, a staid monadnock/On her peneplain." (Beckinsale 1976: 455)

Theory and the History of Geography

One way of writing or reading a book about geographical theory is as a history of the discipline. Indeed the "difficult listening" course that forms a core part of most geography degrees often doubles as an introduction to theory and as a survey of the discipline's history. This is, at least in part, because accounts of theory often proceed chronologically. The passage of time imposes a kind of narrative on ideas that makes it easier to follow. When this happens, it may appear as an account of progress, with one set of ideas being challenged and replaced by another set of ideas, and so on. Eventually we get to the present where our ideas, now, are better, more correct, more subtle, cleverer than the dusty old, simplistic, inferior ideas of the past. It is certainly true that to understand theory in geography, we have to understand important elements of the history of the discipline – or geography as a body of knowledge. But they are not the same thing. Geography is a lot more than its theories. A history of geography includes the development of its national institutions, the biographies of key players, the development of techniques, the relationships between geography and the state, and a host of other, equally interesting factors.

It would also be a mistake to think that the story of geographical theory is a story of simple progress. There are plenty of instances of geographers forgetting their past and coming up with "new" ideas that are simply new versions of old ones. Similarly, most of the key theoretical contributions to the discipline did not simply disappear when challenged by new ones. Even a set of theories as widely challenged as **environmental determinism** (the idea that the natural environment determines human life and culture) still has its advocates.

It is important to recognize that the development of theory in geography exists within wider histories. When re-reading the first edition of this book, over a decade after I wrote it, I was struck by how my knowledge of theory in geography has been limited by my standpoint as a White, British, cisgendered man educated in the UK and USA. While I attempted to acknowledge these limitations in the final chapter of the book, it was insufficient. In the last several decades, Geography (and sister disciplines) has seen a concerted effort to engage with the viewpoint and theories of scholars from beyond the usual suspects – people who share elements of my upbringing and education including the privileges that come with my Whiteness, masculinity and generally Western/Global North education. My point is not to apologize for this but to work on doing something about it. In part, this means "decolonizing" my own textbook. In addition to recognizing the ways geographic thought has been complicit with all manner of injustices – something I think the first edition did reasonably well – an account of geographic thought has to engage with (not just cite) the voices and ideas of a more diverse caste of theorists. Some might argue that it is not even necessary to learn about the ideas of the figures who have become canonical in geographic thought and that a text such as this could be written that said nothing about Carl Sauer, Ellen Semple, Alexander von Humboldt, or David Harvey. They might argue that it is even dangerous to study their writings as doing so would lead us straight into conventional modes of thinking that would simply reproduce forms of privilege. I still believe it is necessary to understand the work of these people (and many more like them) in order to understand the discipline of geography in the present moment. What I have tried to do in the chapters that follow is make a concerted attempt to engage with a wider range of voices than those that were heard in the first edition. Decolonial theorists, drawing on the thoughts of the Zapatista movement, write of a conceptual **pluriverse** – recognizing the co-existence of many worlds that call into the question the claims to universality made by academia in the Global North (EZLN 1996). It is important to recognize that much of the history of geographic thought in this book involves the constitution of that universe as part of a colonial project. I hope it makes clear that this lineage of geographic thought is in fact, like other lineages elsewhere, one that is specific to particular places that have been at the heart of colonial projects.

The development of human consciousness is reflected in the history of geographical theory. How to relate this history presents me with some problems. There are many ways to write a book on geographical theory and there are many excellent books already in existence. This book could be written biographically as an account of the ideas of key figures in the development of geographical thought. Ritter met Reclus, Harvey supervised Smith, Santos collaborated with Chomsky. A family tree of geographers would not be without interest. We will see in the early chapters of this book just how influential particular geographers were in the development of ideas in the late nineteenth and early twentieth centuries. There is some of this kind of account in the pages that follow. It could be written through places where theories were developed: German geography in the late nineteenth and early twentieth centuries, radical theories in Clark University, Massachusetts, spatial science at the University of Washington, or even new cultural geography in Lampeter, Wales. Places will be referred to. It could be approached through the concept of paradigms. Here, one set of ideas holds dominance for a period of time before being challenged and essentially replaced by another set of ideas: regional geography by spatial science, spatial science by humanism and Marxism, and so on. This is not a paradigmatic account but introduces bodies of thought in more or less the order they emerged. I have no intention, however, of suggesting that sets of theories replaced each other. They are all ongoing, living traditions of thought with fierce advocates and detractors. I could also tell this story contextually, describing the development of ideas in relation to other things going on in the world beyond the discipline: historical events, social contexts – forces exerted from outside the discipline (imperialism, religion, war, etc.). I will keep an eye on these. But at the heart of this book are key questions for geographers: key ideas that geographers have puzzled over and argued about. Everything else is secondary to them. We have a lot to offer the world. We are a profound discipline.

References

Beckinsale, R. P. (1976) The international influence of William Morris Davis. *Geographical Review*, 66, 448–466.

Billinge, M. (1983) The Mandarin dialect: an essay on style in contemporary geographical writing. *Transactions of the Institute of British Geographers*, 8, 400–420.

Bourdieu, P. (1990) *The Logic of Practice*, Stanford University Press, Stanford, CA.

Cosgrove, D. E. (1989) Geography is everywhere: culture and symbolism in human landscapes, in *Horizons in Human Geography* (eds D. Gregory and R. Walford), Barnes and Noble, Totowa, NJ, pp. 118–135.

Culler, J. D. (1997) *Literary Theory: A Very Short Introduction*, Oxford University Press, New York.

Eagleton, T. (2008) *Literary Theory: An Introduction*, University of Minnesota Press, Minneapolis, MN.

EZLN (1996) *Fourth Declaration of the Lacandona Jungle*. Ejército Zapatista de Liberación Nacional/Zapatista Army of National Liberation. https://schoolsforchiapas.org/library/fourth-declaration-lacandona-jungle/ (accessed June 11, 2022) hooks, b. (1994) *Teaching to Transgress: Education as the Practice of Freedom*, Routledge, New York.

Inkpen, R. (2005) *Science, Philosophy and Physical Geography*, Routledge, London.

Limerick, P. N. (1993) Dancing with professors: the trouble with academic prose. *New York Times Book Review*, October 31.

Pred, A. R. (2004) *The Past Is Not Dead: Facts, Fictions, and Enduring Racial Stereotypes*, University of Minnesota Press, Minneapolis, MN.

Strabo (1912 [AD 7–18]) *The Geography*, G. Bell and Sons, London.

Von Englehardt, W. and Zimmerman, J. (1988) *Theory of Earth Science*, Cambridge University Press, Cambridge.

Williams, R. (1977) *Marxism and Literature*, Oxford University Press, Oxford.

Chapter 2

Early Geographies

Long ago no divisions existed between humans, animals and spirits. All things of the earth, sky, and, water were connected and all beings could pass freely between them. The Raven was a trickster full of supernatural power. He stole the sun from his grandfather Nasshahkeeyalhl and made the moon and stars from it. The Raven created lakes, rivers and filled the lands with trees. He divided night and day, then pulled the tides into a rhythm. He filled the streams with fresh water, scattered the eggs of salmon and trout, and placed animals in the forests. The first human was hiding in a giant clamshell and Raven released them onto the beaches and gave humans fire. Raven disappeared and took with him the power of the spirit world to communicate and connect with humans. (Traditional Haida Creation Story https://web.archive.org/web/20120611074256/http://www.ucalgary.ca/applied_history/tutor/firstnations/haida.html accessed June 12, 2022)

The account above is a version of the creation story of the Haida people of Haida Gwai off the west coast of British Columbia in the land that is now called Canada. It is an account of how the world came into being. It is also an example of geographic thought that has been handed down in an oral tradition for hundreds of years. Accounts like this, creation stories of Indigenous cultures, exist around the world in a myriad of different forms. Humans exist in places and in ways that relate to other living and non-living things. As long as humans have been thinking, they have been thinking geographically. Despite this, accounts of geographic thought (including this one) tend to focus on thought from within the discipline of geography. I wanted to start with this account in order to make it clear that this book is not about all geographic thought. It is about geographic thought from within the discipline of geography – a creation of the last several hundred years in Europe and North America that, thanks to various forms of imperialism and colonialism, has become the dominant way of thinking geographically. This is not just a recognition of the fact of different kinds or traditions of geographic thought – it is an acknowledgement that the ideas contained in this book have been active agents in the marginalization and, sometimes, attempted eradication, of other ways of thinking geographically. Recent work from within the discipline of geography has begun to take the existence of other modes of geographic thought seriously and we will see how this has been done in the final chapters of the book. Despite the role of disciplinary geography in the erasure of other ways

Geographic Thought: A Critical Introduction, Second Edition. Tim Cresswell.
© 2024 John Wiley & Sons Ltd. Published 2024 by John Wiley & Sons Ltd.

of knowing, however, I continue to tell its story. I do so both to show how its ideas are complicit with largely White, western, Global North forms of oppression and domination and to show how, despite this, it still contains a host of ideas that are illuminating and profound and can be and have been used for good as well as insidious ends. Martin Heidegger, a philosopher we will meet in Chapter 6, was a member of the Nazi party – but critical theorists still find value in some of his ideas. Alexander von Humboldt (who we meet in the next chapter) was clearly involved in a colonial enterprise as he traveled the world to form his theories but, as we will see, he also produced anti-colonial ideas along the way. This is not an account of good and bad in black and white. It is more nuanced and contextual. At the outset, however, I want to recognize that the discipline of geography is both the home of useful and profound ideas *and* an enterprise thoroughly implicated in the attempted erasure of other ideas from beyond its euro-American homelands. My purpose in writing this book is to trace how both have been true. It is worth remembering, however, that many forms of geographic thought existed before the formal discipline of geography emerged in the nineteenth century in Europe. As this is an account of geographic thought within the discipline of geography, the starting point, as with most academic disciplines, is the ancient Mediterranean world.

In a paper called "What time human geography?" Rhys Jones charts how human geography has, over time, become increasingly fixed in the present (Jones 2004). Geographers in the early twenty-first century have tended to focus on the world since about 1800. This has not always been the case. In the 1950s, for instance, geographers would often explore the period before 1800. In the period 1956–1960, 31% of human geography papers in *Transactions of the Institute of British Geographers* concerned time periods before 1800. In the period 1996–2000, the figure was 11%. While it is always harder to get information about the distant past, it is unlikely that it is harder now than it was 60 years ago. One of the reasons Jones gives for human geography's temporal myopia is that historical geographers (those geographers who should be most interested in the past) have increasingly turned to the history of the discipline of geography itself. As a discipline, geography has only existed since the nineteenth century. Whatever the reason, Jones argues that this foreshortening of the time periods of human geography has impoverished the discipline. The past is important geographically because present geography will one day be past and because past geography was once present. To properly understand the geography of the present, we need to know what came before. We need to know how we got here. While Jones's arguments are primarily about what geographers study (states, empires, landscape, etc.), they are also true of geographical theory. They are true in two main ways. Books about geographical theory have increasingly taken the period since 1945 (and more often 1960) as their starting point (Peet 1998; Hubbard 2002; Aitken and Valentine 2006). Like human geography itself, the study of geographical theory is being foreshortened. And this leads to the second similarity. Just as we cannot properly understand a contemporary place without some understanding of its history, so we cannot understand modern geographical theory without some understanding of where it came from. So while this chapter deals with the very distant past, the questions at the heart of it are broadly translatable into questions geographers are still asking today. These include: How is human life related to the natural world? What are the significant differences between places? How is the particular related to the general and universal? While not being quite at the level of "what is the meaning of life?" these are nonetheless profound questions that demand answers that are equally geographical and philosophical. They are questions that link all of the chapters that follow.

Knowing something about the deep past of our discipline stops us from reinventing the wheel. There is very little engagement between geographers today and geography over 20 years ago. Some of these geographers might be surprised at what they might find. Consider a paper I read while writing the first edition of this book. I will not say when it was written or who wrote it – I will simply summarize parts of its argument. At the end, of course, I will give the author due credit.

The author is arguing for a reformulation of the central interests of human geographers. He (I will give that much away) argues that human knowledge is hampered by the way disciplines have been constructed over the past century or so. The division of faculties of arts and sciences, he argues, has left geography feeling a little uncomfortable, a little unsure of itself, with no obvious place to locate. Modern geography, he argues, is too focused on the material world, on the world of things. In so doing, he continues, geography has emphasized the shape of things, the morphology of objects "that nobody had previously thought worthy of study." He focuses on those relatively fixed geographical ideas that are familiar to all of us – particularly the idea of the **region**, but also **territory** and **landscape**. He complains that these ideas are far too object-like – too brittle and immobile. "Regions are not fossilised," he writes, "they are active and growing entities, since the men who organise them are moving, working and thinking beings." He argues, then, for a dynamic view of geography that puts its emphasis not on boundaries and fixities, not even on patterns or networks, but on "men and things moving." In many ways these observations could have been written very recently (if you exclude the frequent references to "men"). Doreen Massey has asked us to consider places as constantly dynamic things produced through constant flows within place and between places. The recent mobility turn in human geography has urged us to focus on "men and things moving" (although women are also included now, of course) (Cresswell and Merriman 2010). But this was written in 1938 by the Scottish geographer P. R. Crowe (Crowe 1938). He is not a figure familiar to many of us and yet his arguments would have to be repeated several times over before we arrived where we are now. Old geographies can, indeed, be very useful to those of us practicing in the present day.

It is impossible to say where and when geography, and more precisely geographical theory, began. In a sense, we are all geographical theorists in so far as we make decisions about things like where to live, what to eat, where to avoid late at night, and where to go on holiday. In 1947, the geographer John Kirkland Wright coined the term "geosophy" to describe what we might call the geographical imagination or geographical knowledge (Wright 1947). At the time, he was making the argument that geographers could benefit from exploring the geographical knowledges of non-academic, everyday, folk – fisherman, lorry drivers, farmers, nurses – in order to understand how their ways of knowing the world influenced their everyday lives. What he was saying was that we are all geographers, all theorists – we all make sense of the randomness of the world in geographical ways. To limit an account of geographical theory to geographers, or even academics, is therefore slightly wrongheaded. If we all have geographical theories, then it seems just as certain that the earliest humans were engaging in geographical theory – perhaps concerning where food was most abundant or where they were safest from predators.

Veronica della Dora makes the continuity of geographical theory clear in her consideration of the conceptual metaphor of the earth's "mantle" over time, showing how this geographic concept of the surface of the earth appears across many geographical imaginations.

> The Babylonians envisaged the sky as a mantle. Egyptians devised a "living mantle" in the shape of the goddess Nut, whose elongated body formed an arch, literally wrapping and sheltering the earth. Indian mythology similarly refers to a hill goddess who lifted herself from the ground, acting as a protective mantle for the people and animals of her region when it was attacked by the rage of Indra, the divinity of the thunderbolt. Myths from Central Africa describe the earth being "rolled out like a mat," while the Quran likens it to a carpet spread out by Allah and held in place by "firm mountains" that serve as weights or pegs. In the Māori tradition, the Nga Uri cloak is symbolic of care and protection of the earth, while Native American mythology ascribed the birth of summer to a "new cloak of green" unfolded by the Shining One upon Mother Earth, "beautiful with all her flowers and birds," and to a soft cloak of dark blue subsequently spread over the sky, in which many a star sparkled and twinkled. (della Dora 2021: 18–19)

Here, della Dora traces the generative metaphor of the idea of the mantle across space and time reminding us that geographic thought happens continuously and in many sites that exist outside of what we normally think of as the discipline of geography.

Classical Geographical Theory

Herodotus and Eratosthenes

The earliest written accounts we know of, which are clearly geographies, were written by Greek philosophers and historians who laid down some of the foundations for geography as an intellectual enterprise. The Greeks were responsible for the production of elaborate topographical descriptions of places in the known world. These descriptions covered both the natural conditions (climate, soil fertility, etc.) and culture and way of life. The "father of history" Herodotus of Halicarnassus (485–425 bc), for instance, described the flow of the Nile and suggested that its source might be melting ice on Mount Kilimanjaro (this proved to be a mistake but a good theory nonetheless). Indeed, his account of Egypt, based on extensive travels and what we might now recognize as "interviews" of local priests and librarians, is full of observations ranging from the natural world to the customs and beliefs of the people. The following extract, for instance, starts with the climate and goes on to observe gender differences in everyday appearance and behavior: It is hardly surprising that Herodotus is claimed as the father of both "history" and "anthropology," but he surely has some claims on geography too.

> The Egyptians in agreement with their climate, which is unlike any other, and with the river, which shows a nature different from all other rivers, established for themselves manners and customs in a way opposite to other men in almost all matters: for among them the women frequent the market and carry on trade, while the men remain at home and weave; and whereas others weave pushing the woof upwards, the Egyptians push it downwards: the men carry their burdens upon their heads and the women upon their shoulders: the women make water standing up and the men crouching down: they ease themselves in their houses and they eat without in the streets, alleging as reason for this that it is right to do secretly the things that are unseemly though necessary, but those which are not unseemly, in public: no woman is a minister either of male or female divinity, but men of all, both male and female: to support their parents the sons are in no way compelled, if they do not desire to do so, but the daughters are forced to do so, be they never so unwilling. The priests of the gods in other lands wear long hair, but in Egypt they shave their heads: among other men the custom is that in mourning those whom the matter concerns most nearly have their hair cut short, but the Egyptians, when deaths occur, let their hair grow long, both that on the head and that on the chin, having before been close shaven. (Herodotus 2007 [450 bc]: npn)

Herodotus is known for a nine-volume account of the Persian wars of 490–479 bc (recently recalled to us in the film *300*). To discover the cause of these wars, Herodotus recounts a vast context of the known world at the time, including a foray through Egypt (Book 2) from which the above quotation is taken. In his account of Egypt, Herodotus seemingly attempted to tell us everything it is possible to know about the region, from the plant life and animals and the flow of rivers to the toilet habits of the inhabitants. Indeed, it is not until well into the fifth volume that the reader encounters the events at the center of this first "history" – the wars between the Persians and Greeks in which the massively outnumbered Greeks finally defeat the Persian Empire. West defeats east. For four and a half volumes, Herodotus seems to be avoiding the subject. Or was he? What we get in these pages is a geographical and, more explicitly, geopolitical, context for the rise of the all-conquering Persian Empire. At the center of this story is how the Persian Empire outgrew its "natural" setting. The Persian Empire, he is telling us, belonged in the east, in what we now know as Asia, and by attempting to incorporate Greece it was overstretching, becoming "unnatural." His account of Egypt is just one of many digressions that tell us how human life is mapped on to natural circumstances. Nature, or environment, determines culture. The account of Persian expansion includes many stories of, not the domination of people by other people, but the domination of nature. The Persian army was said to drink rivers dry. In other places it tried to make rivers where no rivers had previously existed.

The army became so large it could not feed itself. One of its most famous defeats occurred when a huge navy, advancing on Athens, was destroyed by a massive storm. The empire had reached its natural limits and it had no right to enter "Europe." In doing so, it doomed itself. What we have in Herodotus, then, is a combination of **environmental determinism** and Greek tragedy (in which a mistake inevitably, because of fate, leads to downfall). This is a geographical theory. History depends on geography.

While Herodotus wrote detailed narrative accounts of his travels accompanied by theories concerning both the natural and human worlds, others attempted more systematic geographies. Eratosthenes (276–194 BC), the Librarian of the Library at Alexandria – calculated the circumference of the earth and developed systems of coordinates (latitude and longitude) that are the forerunners for the locational system we program into global positioning systems (GPS) today. His estimate of the earth's circumference reveals an innovative mixture of observation and theory. There are some things he knew from experience, observation, and prior knowledge. He knew, for example, that the sun would appear at its zenith (directly overhead) during the summer solstice at noon in the town of Syene (more or less on the Tropic of Cancer). In Alexandria, at the same time, he measured the angle of the elevation of the sun at 1/50 of a full circle south of the zenith. He believed that Alexandria was due north of Syene (it wasn't) and concluded that the distance between Syene and Alexandria must be 1/50 of the earth's circumference. He estimated the distance between the two cities as about 500 nautical miles and thus came up with a circumference of 46,620 km (about 16% larger than we now know it to be). This estimation was still used several hundred years later and is remarkably close (see Figure 2.1).

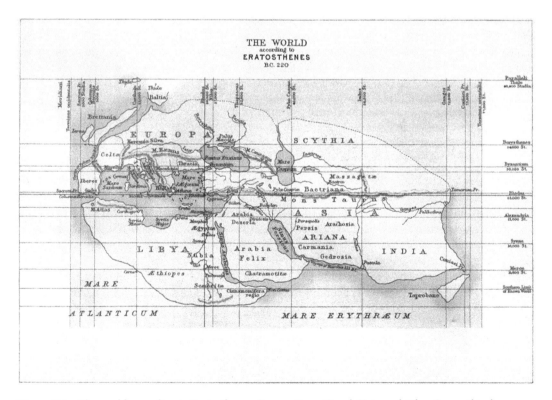

Figure 2.1   The world according to Eratosthenes. Source: From *Cram's Universal Atlas: Geographical, Astronomical and Historical* (1895).

The geographies of Herodotus and Eratosthenes are remarkable in their own right but perhaps even more remarkable in that they prefigure and illustrate two of the central ways in which geographical theory operates today. While Herodotus was busy cataloguing the areas he traveled through and heard about, Eratosthenes was busy wondering how to measure the world and provide a reliable grid of reference points for navigation. Herodotus, then, was fascinated by the particularity of different places and what made them unique (even if framed by a general theory of environment and fate). Eratosthenes, on the other hand, was interested in producing a common measure that bound the whole earth into a unified system of reference points. He was developing a universal and quite mathematical geography that was not primarily interested in the particular – in what one place was like and how it differed from the next one. What the combined work of these two scholars reveals is an emerging sense of order in the world – an interest in a whole inhabited world beyond the immediate confines of the local. The Greeks referred to this world as the *oikoumene* (later, *ecumene*) or the "inhabited earth." Just as the interest in places (Herodotus) and objective space (Eratosthenes) prefigure debate 2000 years later, so does this concept of the global *ecumene*. For it is largely the inhabited earth that interests geographers today. Humans, after all, exist in a thin layer from just below the earth's surface to just above it. This is geography's layer. Go too far beyond it and you enter the world of astronomy, go too far below and you are in the realm of geology. Geography as the study of the *ecumene* is not a bad definition of our discipline.

Kenon, chora, and *topos*

The very different geographies of Herodotus and Eratosthenes are frequently held up as examples of early geography. This is because the kinds of knowledge they produced match modern expectations about what geography is – a knowledge of places on the one hand and a science of space on the other. But in many ways it was other Greeks – particularly philosophers – who were busy asking questions that would form the basis for large parts of contemporary geographical theory. Consider for instance the concepts of *kenon,* **chora**, and **topos** developed by Plato and Aristotle.

The concept of *kenon* referred to the void in which all other things exist – a realm which is homogeneous and undifferentiated. *Kenon* is arrived at when one abstracts a thing from its surroundings. It is pure extension. This notion of eternal emptiness became the basis of scientific, abstract, notions of space. It was further developed by Descartes and Newton, among others, and forms the basis for all kinds of science that depend on abstract notions of space.

Chora comes from Plato's (428–348 BC) discussion of the process of becoming – the way in which existence takes shape out of the void of *kenon*. Becoming, in Plato's terms, is a process that involves three elements – that which becomes, that which is the model for becoming, and the place or setting for becoming (Casey 1997). This final element is *chora*, a term which implies both extent in space and the thing in that space that is in the process of becoming. It is often translated as a receptacle and differs from the void of *kenon* in that it always refers to a thing within it – it is not empty. *Topos* is often used interchangeably with *chora* in Plato but is usually more specific. While *chora* most often referred to a place in the process of becoming, *topos* would refer to an achieved place. Later, Aristotle would use *chora* to describe a country while *topos* would describe a particular region or place within it. Both *chora* and *topos* would eventually become part of geographical language through the notion of chorology (the study of regions) and topography (the shape of the land surface). Both *chora* and *topos* are different from the notion of *kenon* (the void) in that they refer to something more particular – more like place than space. While *kenon* is limitless space, *chora* and *topos* are finite and contain things (Casey 1997; Malpas 1999).

If anything, Plato's student, Aristotle (384–322 BC), had even more fundamental things to say about one of geography's most basic concepts – place. To Aristotle, place is a necessary starting point from which it is possible to understand both space (the infinite, the void) and movement and change. Place, he wrote, "takes precedence over all other things" (Aristotle in Casey 1997: 51). To understand change and motion, for instance, it was first necessary to acknowledge that the "most general and basic kind [of] change is change in respect of place, which we call locomotion" (Aristotle in Casey 1997: 51). The geographical question of "where" is absolutely fundamental to Aristotle, for everything that exists must be somewhere "because what is not is nowhere – where for instance is a goat-stag or a sphinx?" (Aristotle in Casey 1997: 51). Place comes first, to Aristotle, because every-thing that exists has to have a place – has to be located. Thus "that without which nothing else can exist, while it can exist without the others, must needs be first" (Aristotle in Casey 1997: 52). So, in Aristotle we have a very powerful philosophy of place. What could be more of a celebration of geog-raphy than the assertion that place is the most fundamental thing in existence – the starting point for all other forms of existence?

It is strange, then, that we are more likely to recognize the endless travels of Herodotus or the scientific measurements of Eratosthenes as "geography" despite the fact that human geography now is less likely to be an inventory of observations of life in Egypt and more likely to be a set of reflec-tions on what **place** means. Plato and Aristotle were just as much geographers as Herodotus and Eratosthenes. Describing Egypt, measuring the earth, and ruminating on the primacy of place are all geographies and all have elements of what we might call "theory" in them. None of them, however, referred to what they were doing as "geography."

The first geographers?

However, the man who busies himself with the description of the earth must needs speak, not only of the facts of the present, but also sometimes of the facts of the past, especially when they are notable. (Strabo, *The Geography*, Book 6.1.2)

Despite the fact that there must have been geographical theorists since time immemorial, it is gener-ally acknowledged that the first written account of geography in the western world, referred to as such, that still remains, was written by Strabo of Amasia (64 BC–AD 23). Strabo was a Greek citizen who came from what is now an area of northern Turkey. He was a wealthy and educated man who traveled to both Rome and Alexandria in order to pursue his education. By the time he wrote his "geography," he was probably a Roman citizen living in Rome (Dueck 2000; Koelsch 2004). Strabo wrote a massive 17-volume "geography" accounting for what was known about the inhabited world (*oikoumene*) at the time of the ancient Greeks and Romans. One reason we know that Strabo's was not the first to attempt something called geography is because he refers to earlier works that have since disappeared and to which we no longer have access.

At the time Strabo was writing, the Roman Empire was experiencing an extraordinary period of relative peace and prosperity under Emperor Augustus. It was in this context that Strabo sought to explain this world to the Romans. As someone from Greece, he was displaced and used this position to translate Greek ideas to the Romans. At the heart of his geography was a plea for a kind of world understanding that would accompany the peace Roman citizens were enjoying. Here, geography and an understanding of it were seen as the basis for tolerance of difference and otherness. There are many facets to Strabo's *Geography*, including some quite mathematical sections that have led some to consider Strabo as a measurer of things – a geographer of space (Livingstone 1993). Like Eratosthenes he was keen to provide measurements of the globe and distances between places. Indeed, much of the 17 volumes features endless lists of places and their relative locations. Scholarship on Strabo has often concentrated on whether his locations and measurements were accurate.

But Strabo was also, and perhaps more importantly, a geographer of place – of the lived world. Strabo argues that "the peculiar task of the geographer" is to explain "our inhabited world" (Koelsch 2004). In this sense, Strabo was a cultural geographer. Consider the following:

> The seaboard that comes next after Leucania, as far as the Sicilian Strait and for a distance of thirteen hundred and fifty stadia, is occupied by the Brettii. According to Antiochus, in his treatise *On Italy*, this territory (and this is the territory which he says he is describing) was once called Italy, although in earlier times it was called Oenotria. And he designates as its boundaries, first, on the Tyrrhenian Sea, the same boundary that I have assigned to the country of the Brettii – the River Laüs; and secondly, on the Sicilian Sea, Metapontium. But as for the country of the Tarantini, which borders on Metapontium, he names it as outside of Italy, and calls its inhabitants Iapyges. And at a time more remote, according to him, the names "Italians" and "Oenotrians" were applied only to the people who lived this side the isthmus in the country that slopes toward the Sicilian Strait. (Strabo, *The Geography*, Book 6.1.4)[1]

Here, we see Strabo providing encyclopedic details about regions and places in what is now Italy. But his cultural geography did not simply describe places and regions but attempted to synthesize these into an understanding of the interactions between these places. This was a study of empire and how the periphery related to the center (i.e. Rome). As Koelsch has put it: "we should not become bogged down in the details, book by book. Although Strabo drew on many sources (…), his genius lay in the organization of the various parts of his book and in his editorial synthesis. He incorporates each regional or chorographic segment into a systematic pattern for comparative purposes, advancing the general plan of a universal geography of the oikoumene" (Koelsch 2004: 508). Strabo was attempting to understand the relationships between specific places and a wider spatial realm at the time of the first Roman emperor. He was trying to figure out how the global interacted with the local – a task still central to geography in the twenty-first century.

Strabo's *Geography* was also about the relationships between the present and the past in place. His argument that "the man who busies himself with the description of the earth must needs speak, not only of the facts of the present, but also sometimes of the facts of the past" reflects his interest in processes over time. Later Bunbury, a nineteenth-century historian of geography, would describe these reflections on the past ages as "digressions" (Bunbury 1879), while another scholar of Strabo, Clarke, is more sympathetic, noting how the identities of present places are so often constructed on the basis of memories of the past (Clarke 1999). For some reason, Bunbury and others did not see these accounts of place histories as being as important as all the details, locations, and distances so valued by military tacticians and politicians, but it was accounts of place that were most important to Strabo. Clarke quotes a line from Book 14 of Strabo's *Geography* in which he writes: "in the case of famous places it is necessary to endure the tiresome part of such geography as this" (Clarke 1999: 202). "This" referred to the measurements and mathematics of space. This "mathematical geography" was, for Strabo, a matter of accuracy as "Every one who undertakes to give an accurate description of a place, should be particular to add its astronomical and geometrical relations, explaining carefully its extent, distance, degrees of latitude, and 'climate'" (Strabo 1912 [AD 7–18]: 13).

Many have interpreted Strabo's geography as evidence of the complicity of geography and the state (Smith 2003). Certainly, Strabo acknowledged the importance of geographical knowledge to statecraft in the Roman Empire of Augustus:

> For the sea and the earth in which we dwell furnish theatres for action; limited, for limited actions; vast, for grander deeds; but that which contains them all, and is the scene of the greatest undertakings, constitutes what we call the habitable earth; and they are the greatest generals who, subduing nations and

---

[1] See http://www.perseus.tufts.edu/cgi-bin/ptext?lookup=Strab.+6.1.4.

kingdoms under one sceptre, and one political administration, have acquired dominion over land and sea. It is clear then, that geography is essential to all the transactions of the statesman, informing us, as it does, of the position of the continents, seas, and oceans of the whole habitable earth. (Strabo 1912 [AD 7–18]: 15–16)

To Clarence Glacken, Strabo was the archetypal cultural geographer, interested as he was in the inhabited earth (Glacken 1967). Physical geography, for Strabo, was a setting for human actions. His writing on the interactions of the physical environment and the human world prefigured work done in the early twentieth century in its insistence on the ability of humankind to modify and adapt to the natural world. Glacken quotes a long passage from Book Two of Strabo's *Geography*:

Arts, forms of government, and modes of life arising from certain [internal] springs flourish under whatever climate they may be situated; climate, however, has its influence, and therefore while some peculiarities are due to the nature of the country, others are the result of institutions and education. It is not owing to the nature of the country, but rather to their education that the Athenians cultivate eloquence, while the Lacedaemonians do not; nor yet the Thebans, who are nearer still. Neither are the Babylonians and Egyptians philosophers by nature, but by reason of their institutions and education. In like manner, the excellence of horses, oxen, and other animals, results not alone from the places where they dwell, but also from their breeding. (Strabo quoted in Glacken 1967: 105)

Strabo's *Geography* (particularly the first two volumes) is also littered with theories of the physical environment. He is fascinated by the changing relations between the sea and the land. He wonders why mussels and other evidence of salt water are often found hundreds of miles inland and concludes that land and sea must have once been in different places than they were in the time of Augustus. He gives a reasonable account of wave action and erosion to partially explain this:

Accordingly, the onset of the wave has a power sufficient to expel foreign matter. They call this, in fact, a "purging" of the sea – a process by which dead bodies and bits of wreckage are cast out upon the land by the waves. But the ebb has not power sufficient to draw back into the deep sea a corpse, or a stick of wood, or even that lightest of substances, a cork (when once they have been cast by the wave upon the land) from the places on the shore that are near the sea, where they have been stranded by the waves. And so it comes about that both the silt and the water fouled by it are cast out by the waves, the weight of the silt cooperating with the wave, so that the silt is precipitated to the bottom near the land before it can be carried forward into the deep sea; in fact, even the force of the river ceases just a short distance beyond the mouth. (Strabo 1912 [AD 7–18]: 1.3.9)

Strabo's geography is often contrasted with that of Claudius Ptolemy (AD 90–168). Ptolemy was an astronomer, astrologist, mathematician, and geographer believed to have been born in Egypt before, like Strabo, becoming a Roman citizen. His eight-volume *Geographia* has little time for encyclopedic accounts of places or the action of waves. Its concerns are far more general, concerning the dimensions of the globe, calculations of longitude and latitude, and map projections. Ptolemy's *Geographia* resembles an atlas in that it is a series of maps with accompanying lists of places with their location in latitude and longitude. It is an atlas of the world known to the Romans at the time (more or less from Cape Verde in the west to the middle of China in the east). Ptolemy developed a grid of latitude and longitude which spanned the known world (see Figure 2.2). Latitude was measured from the equator and longitude from the western extremity of the known world. Ptolemy gave instructions on how to make maps scientifically using map projections and systems of location that formed the basis for cartography for years to come.

In the work of Herodotus, Eratosthenes, Plato, Aristotle, Strabo, and Ptolemy, we can see the outline of some of the key geographical questions that have informed 2000 years of geographic theory since. We can see the question of the degree to which the natural environment determined human

Figure 2.2　Ptolemy's map of the world with grid. Source: http://commons.wikimedia.org/wiki/File:Ptolemy_Cosmographia_1467_-_world_map.jpg (accessed May 31, 2012). Original located at the National library of Poland.

life; the question of the importance of the study of particular regions or places set against the search for forms of universal order and truth; the question of the role of space and place in the constitution of reality; and a whole array of more specific theories and hypotheses (such as the source of the Nile or the action of waves). But the histories of these ideas since then are far from straightforward accounts of progress. Indeed, many of these ideas were deliberately or accidentally ignored and forgotten for hundreds of years in Europe.

Medieval Geographies

The Middle Ages saw the power of the Church dispel classical learning and the rise of a worldview centered on God. This can be seen in the Mappa Mundi, a medieval map of the known world. No longer a scientific attempt to chart the inhabited world, the world is a flat disk with Jerusalem in the middle (see Figure 2.3). Arguably, the medieval period in Europe and elsewhere has been even more neglected than classical scholarship. It is common for the period in Europe to be seen as stagnant as far as geography (and many other "sciences") are concerned. Keith Lilley has argued that this neglect is a mistake and that there are important continuities between geographies (widely conceived) of the medieval period and those of the present day (Lilley 2011). He notes how non-geographers, particularly historians, have maintained a healthy interest in the world of medieval

Figure 2.3 The Mappa Mundi from around AD 1300 in Hereford Cathedral. The Hereford Mappa Mundi Trust and the Dean and Chapter of Hereford Cathedral. Source: Richard of Haldingham / Wikimedia Commans/Public Domain.

geography while geographers have largely ignored them since the early twentieth century when a series of books by geographers made the case for their importance.

> For geographers today concerned with narrating geography's history, this shifting disciplinary terrain poses some challenges: for instance, should geographers simply leave histories of medieval geographies to 'non-geographers' to write? After all, considering the interpretative challenges presented by the European vernacular and classical languages of medieval geographical texts (and images), perhaps such specialists are most qualified to do so. However, this would then surely mean that geography's medieval ancestry is yet further removed from the growing number of (modern) histories of the subject that are being written by geographers, when instead space should be being made to include them. Furthermore, in the context of differing disciplinary agendas and traditions, how do the histories of medieval geography being written by historians square up, if at all, with the recent debates among geographers on geography's past as a 'contested enterprise', and those empirical and theoretical issues that have been of concern to us? (Lilley 2011: 149–150)

Lilley argues that geographer's neglect of the medieval period risks ceding an important area of inquiry to other disciplines and, further, contributes to the sense that geography only really began in the early modern period "a hubris that surely deserves to be challenged, and in a way that forces a reconsideration of geography's history" (Lilley 2011: 150).

In a response to Lilley's paper, Veronica della Dora suggests that, in addition to having a more complete sense of geography's history, understanding medieval geographies allows us a different lens for thinking about the earth than can accompany modern ways of seeing. In this sense, her argument mirrors that of those who ask us to take seriously ways of knowing the world that exist outside of the euro-American geographical canon. Following Lilley, she writes that "as long as we continue to emphasize 'breaks', 'revolutions' and 'rediscoveries' over geographies of transmission, connections and continuities, complexities will remain overlooked at risk of producing a caricaturized history of geographical thought" (della Dora 2011: 165). "Studies of medieval spatial perceptions" she writes, "can offer useful counterpoints to the modern 'cartographic paradigm' invoked by human geographers over the last two decades" (della Dora 2011: 165). Rather than simply see medieval geographies as an important but neglected part of our discipline's history, della Dora argues, we might, instead, think of them as a "starting point for a genealogy of 'geographical enchantment', or simply an alternative model, or lens, through which to explore the poetics of earth-writing" (della Dora 2011: 165).

Veronica della Dora follows her own advice in the book, *The Mantle of the Earth*. Here, she traces the idea (or metaphor) of the "mantle" chronologically, linking sacred and profane ways of conceiving of the earth and the human place on the earth in enchanting ways. Such a way of thinking, she argues, links classical, medieval, and modern ways of thinking geographically (della Dora 2021). In della Dora's account, the Mappa Mundi is far more than an inaccurate map.

> The tension between symbol and scholastic exegesis set the precondition for the development of one of the most characteristic visual expressions of Western medieval culture: the so-called *mappae mundi*, or Christianized images of the earth … Western *mappae mundi*, which literally translates as "world cloths," were in essence "visual encyclopedias" exalting the marvel and multiplicity of creation and the place of Creator within it. Through the cartographic medium, *mappae mundi* implicitly addressed questions that were at once theological and geographical. … In other words, *mappae mundi* accomplished in drawing what contemporary theological *summae* and universal encyclopedias accomplished in words. (della Dora 2021: 58)

As a "world cloth," the *mappae mundi* took their place, in della Dora's account, in a lineage of cloaks and drapes and ideas of surfaces that have helped us humans account for our place in the world throughout history, linking the medieval map to, for instance, Native American cosmologies featuring the "new cloak of green" unfolded by the "shining one" on the new earth. Tracing these continuities enacts what della Dora calls a "poetics of earth writing."

Bearing in mind della Dora's appeal for a focus on transmission and continuities, it is possible to see both of these in both the geographical imagination of the scholars such as Albertus Magnus within Europe and the prodigious scholarship of Muslim scholars in North Africa. Even in the context of a theological cosmology, there were some who kept the learning of classical scholars alive and transmitted them into the early modern era. One such person was Albertus Magnus, a German Dominican scholar who taught, among others, Thomas Aquinas. Albertus is notable because he consistently combined theology with science and a thorough grounding in philosophy – particularly the writings of Aristotle. Albertus wrote volumes on topics across the sciences. For geographers the key text is *De Natura Locorum* (The Nature of Places). In this text, he argued that human life on earth was influenced by both astrological forces and forces that were determinedly local. He combined the universal and the particular in a way that focused on the unique combinations of cosmological and environmental influences in particular places. Because all places are marked by unique combinations of these influences, human life in these places is also unique. In his work, we can see a kind of environmental determinism. Visible differences between people, such as skin color, were attributed to environmental influences. Thus:

> Men born in stony, flat, cold, dry places are extremely strong and bony; their joints are plainly visible; they are of great stature, skilled in war and handy in waging it, and they have bony limbs. Their customs are wild and they are like men of stone. Peoples, however, of moist and cold places, have beautiful and

smooth faces, their joints are well covered over, they are fleshy and fat, not very tall and their bellies are extended. They are daring because they have such fiery hearts, but they slacken quickly at their work. They lack zeal in war. Their faces are white or yellow. People living in mountains frequently have knotty and strumous necks and throats because the water is such that too much phlegm is generated in them. (Albertus quoted in Glacken 1967: 169–170)

Albertus believed that people were intimately connected to the places they were born in. Places provide the context for human life, and human life reflects the qualities of the places that nurtured it. To move outside of the place of birth was to weaken the particular characteristics of the person. This applied to animals and plants also. Even stones, he argued, were weakened when they were moved somewhere outside of their natural place.

But Albertus was not a straightforward environmental determinist. He also believed that free will allowed people to transform their environments in order to make them more habitable. He argued, for instance, that forests were unhealthy environments because the trees stifled the circulation of air and moisture. Thus, cutting down trees to create places in clearings was a way of improving the nature of these particular places.

However, it was not in Europe that classical geographical theory prospered, but in the Arab world. The period between the tenth and fourteenth centuries was marked by extensive travels by Arab traders and scholars that were to dwarf the later and more famous travels of European explorers. This extensive travel, combined with a sound knowledge of classical scholarship, resulted in some remarkable geographies (Alavi 1965).

Around AD 982 an anonymous Persian scholar wrote the *Hudúd al-Alam* – an account of the regions of the world as known to Persian geography (Bosworth 1970). The geography covers the familiar terrain of the mountains, seas, islands, rivers, and deserts before providing an account of the inhabited world from China in the east to Spain in the west. Indeed, the area of present-day Iran and Iraq was the center of extensive geographical scholarship that was usually combined with astronomy to produce abstract mathematical accounts of the world along with long lists of places and their precise locations. In many ways this work resembled the geographies of Ptolemy. The sciences of surveying, navigation, and cartography were all developing in the ninth and tenth centuries. Muhammad B. Kathír al-Farghāní, for instance, described the world according to seven climates – work which was to inform the Renaissance in Europe many years later and particularly the work of Roger Bacon. In addition to having a deep knowledge of the works of Ptolemy and other classical scholars, the scholars of the Muslim world were also familiar with the well-developed cosmologies of India and China which they incorporated into their work.

Ibn Battutah (1304–1368) (a scholar from Tangiers) traveled almost the whole of the known world over a 28-year period and wrote an account of his travels. He traveled as far east as China, spent many years in India, and went far enough down the African coast to prove the Greeks wrong in believing that sub-Saharan Africa was too hot for human habitation (Battutah and Mackintosh-Smith 2002). Another Muslim scholar, Ibn Khaldun (1332–1406), has been noted as the father of the social sciences owing to his insights about group cohesion and the rise and fall of civilizations. Central to his grand seven-volume "History of the World" (Muqaddimah) was the notion of social conflict being at the center of the unfolding of history. Both space and time play a role in this unfolding. Geography is important because of the dichotomy of the settled and the nomadic (mapped on to the town and the desert, respectively). Social cohesion (*asabiyah*), he argued, arose within kinship networks or tribes and was supported by religious beliefs. This cohesion, however, is doomed owing to the conflicts that inevitably arise within the social and economic spheres which, in turn, lead to a new, younger, and more cohesive group becoming dominant. This pattern then repeats itself endlessly and so history happens:

… when a tribe has achieved a certain measure of superiority with the help of its group feeling, it gains control over a corresponding amount of wealth and comes to share prosperity and abundance with those who have been in possession of these things. It shares in them to the degree of its power and usefulness to the ruling dynasty. If the ruling dynasty is so strong that no-one thinks of depriving it of its power or of sharing with it, the tribe in question submits to its rule and is satisfied with whatever share in the dynasty's wealth and tax revenue it is permitted to enjoy.... Members of the tribe are merely concerned with prosperity, gain and a life of abundance. (They are satisfied) to lead an easy, restful life in the shadow of the ruling dynasty, and to adopt royal habits in building and dress, a matter they stress and in which they take more and more pride, the more luxuries and plenty they acquire, as well as all the other things that go with luxury and plenty.

As a result the toughness of desert life is lost. Group feeling and courage weaken. Members of the tribe revel in the well-being that God has given them. Their children and offspring grow up too proud to look after themselves or to attend to their own needs. They have disdain also for all the other things that are necessary in connection with group feeling. ... Their group feeling and courage decrease in the next generations. Eventually group feeling is altogether destroyed. ... It will be swallowed up by other nations. (Khaldun 1969 [1377]: 107).

Geographically this theory of history is mapped on to the idea that desert nomadic groups gradually become sedentary as they produce a great civilization. Once this has happened, decay starts to set in and the civilization is eventually conquered by barbarian (nomadic) outsiders who then develop their own "civilization" complete with arts, literacy, and learning. As they settle down, they become softer and are eventually conquered by a new set of outsiders. This is clearly a geographical theory of history revolving around the dualism of the desert (nomadic barbarian space) and the town (settled civilization). This dualism of the settled and nomadic would continue to haunt the pages of geographic theory as geographers and others attempted to come to terms with the cultural and social significance of more or less fixed and settled places and the threats and opportunities posed by various forms of movement which also have social and cultural meanings mapped on to them.

It was in the Arab world that the geography of the Greeks and Romans was kept alive and developed while Europe went through a period of theocracy that militated against learning and scholarship that was not biblical. They were to reappear, however, as central actors in the European Renaissance – a period in which classical learning was rediscovered and put back on the scholarly agenda. This was only possible because it had been kept alive in the Arab world for over 400 years. In 1410, copies of Ptolemy's *Geography* were translated into Latin as part of a general rediscovery of classical learning. Strabo's *Geography* was translated into Latin in the 1450s by order of Pope Nicholas V. Ptolemy and Strabo became central figures in artistic representations of Renaissance learning where knowledge of the world and its place in the cosmos was seen as central to the new **humanism** emerging in northern Italy. *The School at Athens*, a painting of 1509 by Raphael, for instance, shows Strabo and Ptolemy holding models of the world (a globe) and the cosmos in the School at Athens (see bottom right corner, Figure 2.4).

Columbus and other explorers were to base their journeys of exploration on Ptolemy's maps. Ptolemy shows China extending east and south much further than it actually does and this is one of the reasons for Columbus's belief that he would reach Asia by heading west. By 1569, Mercator had produced his new world map which would form the basis for two-dimensional representations of the world up until the present day.

Toward Modern Geography

The rediscovery of classical learning in the fifteenth and sixteenth centuries, alongside the continuities of the medieval period in Europe, cleared the way for the emergence of geography as a full-fledged academic discipline in the nineteenth century. It was not just the rediscovery of Ptolemy and

Figure 2.4   Raphael, *The School at Athens* 1509/1510. Source: pyty/Adobe Stock.

Strabo that inspired this, however; it was also the importance of geographical knowledge to the practices of discovery and exploration that framed knowledge of the world as imperial and colonial power.

In northern Italy and Flanders (Belgium and the Netherlands) in particular, new kinds of social relations were being formed as mercantile capitalism began to emerge as a characteristic economic form of cities that were based on trade. This trade (in luxury goods such as silk and spices) connected cities such as Amsterdam and Venice to sites around the world (particularly in South and East Asia). Traders became wealthy and influential in a new way that was not immediately connected to the power of the Church. Influence came from trade and capital rather than an established place on a social hierarchy defined by the Church and/or the aristocracy. This new class also had a new imaginative geography based on its exploitative connections with emerging world markets. To trade successfully with places on the other side of the world, it was necessary to develop ways of knowing these places, ways of navigating between places, and ways of controlling often unruly populations. Skills in map making and navigation emerged that relied on long-forgotten sciences of optics which allowed map makers to make images of places that were in some sense "objective" in that they allowed traders to get successfully from A to B. These maps were also used as decorative objects that symbolized control and ownership of land closer to home.

Consider Vermeer's painting *The Geographer* (1668–1669) (Figure 2.5). This painting shows a deeply thoughtful man standing by a window in a room surrounded by the implements of the geographer's art. There is a globe, a map on the wall, and the measuring device in his hand. The geography of the map that he is looking at is linked to the world outside by the light pouring in through the window. The painting is a celebration of new knowledge and the knowledge of the geographer in particular. Clearly, for Vermeer to be interested in painting such a figure, geography and those who practiced it had to have some level of importance that would be recognized by the viewer.

Figure 2.5   Johannes Vermeer, *The Geographer* 1668/1669. Source: © DeAgostini Picture Library/Scala, Florence.

Geographical and cartographic knowledge was at the center of this new world alongside other kinds of knowledge also represented by Vermeer elsewhere, such as astronomy and music. People who made maps were important.

One such map maker was Willem Janszoon Bleu who lived between 1571 and 1638 in and around Amsterdam. He was the son of a wealthy fish merchant who became interested in the rediscovered sciences of astronomy and mathematics. Around 1596, he qualified as a maker of globes and atlases and established his own print shop from where he manufactured and sold his maps. At this time maps and globes had become the treasured possessions of wealthy households in the area. Having a globe or map was not simply a functional necessity allowing the owner to know where places were and how to get to them, rather they signified the knowledge and power of the owner and were often displayed prominently in households. Maps and globes thus served as key objects in the new imaginative geography emerging at the time that linked science to art and commerce. Such was the success of Bleu that he was appointed the map maker of one of the world's most powerful trading and shipping companies, the Dutch East India Company. The Dutch East India Company connected Holland to its growing trading empire.

At the time, the Dutch were in direct conflict with the Portuguese over the trade in spices between what is now Indonesia and Europe. Both were attempting to develop far-flung empires based on trade. Up to 1600, the Portuguese had an effective monopoly on the trade in highly lucrative spices based on their knowledge of routes between Europe and Indonesia. In 1596, a Dutch expedition of four ships successfully (despite the loss of half the crew) brought a large cargo of spices back from Indonesia and in the following years fleets of Dutch ships became larger and increasingly profitable. In 1603, a large number of Dutch companies were bundled together as the Dutch East India Company and it was granted a monopoly on trade by the government for 21 years. In the next 100 years, the company became the largest and most successful in the world with established bases all over Asia, a standing army of 10,000 soldiers and over 30 warships. Trade and military might went hand in hand. The company was effectively the world's first major multinational corporation, with power and influence that outstripped most nations. It was also the first company to issue stocks in itself.

This was the world that Bleu entered as a map maker. He was effectively a cartographer for the world's most powerful trading entity. His maps played a key part in the combination of capitalism and imperial might that the Dutch East India Company projected. As such, he was one player in a new world of aggressive global capitalism and imperialism in which knowing the world through exploration and representation was a form of power.

One of Bleu's associates, in his later years, was a young man named Bernhard Varenius (1622–1650). Some have claimed that the geographer in Vermeer's painting is Varenius. Varenius had studied medicine in Leiden and had intended to become a doctor in Amsterdam. While in Amsterdam, however, he became acquainted with Bleu and a number of well-known Dutch explorers and navigators (including Abel Tasman, after whom Tasmania is named). It was then that he turned his attention to geography. And although he was to die at the age of 28 in 1650, the last few years of his life saw him publish a number of geographical texts including a regional geography of Japan and what was to become the first widely read geography textbook, the *Geographia Generalis* (1650). In this book, Varenius laid out an all-encompassing account of the science of geography. He divided "general geography" into three related fields: absolute, relative, and comparative geography. Absolute geography was, he argued, the mathematical facts of the world: its size, distance from sun, shape, etc. Relative geography concerned the relationships between different parts of the earth's surface resulting from the motions of the earth and its relationship to other solar bodies. Thus relative geography could explain why seasons, or the length of day, were different in different places. Comparative geography is the study of the earth's surface, the locations of different places, and how to get to them. Varenius's geography was to be the most influential geographical text in Europe for over a century and was twice revised by Sir Isaac Newton for Cambridge editions. While it might seem strange now to think of a geography textbook being translated and edited by the world's most eminent physicist, geography, in Varenius, was considered to be a branch of mathematics: "a science mixed with Mathematics, which taught [literally teaches] about the quantitative states of the Earth, and of the parts of the Earth, namely shape, place, size, motion, celestial appearances [or bodies], and other related properties" (Warntz 1989: 172). What is clear is that the *General Geography* was the first widely used geography textbook and that Newton may even have taught geography classes using it as an "assigned text." So here we have geographical ideas becoming a key part of a burgeoning university education system. Although there was no discipline of geography as such, geographical theory was right at the heart of the origins of the modern university system and was largely transmitted through Varenius's book.

The geography of Varenius reflected a broader change in the role of science and a new faith in reason and rationality. It fits within developments across the sciences and humanities that can be seen in the work of Bacon, Galileo, and Descartes, among others. In this new "**humanism**," human reason became the basis for existence. It is this reason, and particularly geographic reason, that is celebrated in *The Geographer*.

The work of Varenius reflects the by now familiar division of geographical knowledge into that which is interested in the universal and that which is focused on the particular:

> Geography, itself, falls into two parts: one general, the other special. The former considers the earth in general, explaining its various parts and general affections. The latter, that is, special geography, observing general rules, considers, in the case of individual regions, their site, divisions, boundaries and other matters worth knowing. But those who have so far written on geography have discussed at length special

geography alone … and have explained very little relating to general geography, with much that is necessary being neglected and omitted … geography itself scarcely preserves the title of science. (Varenius in Unwin 1992: 67)

Central to Varenius's arguments was the need to be more "scientific" and it is here that general geography came into its own. There is a clear sense in his work that special geography is a footnote to general geography.

General geography includes the three elements of *absolute* geography, *relative* geography, and *comparative* geography. These, Varenius argued, should be studied through particular methodologies, the techniques of reason, of measurement, mathematics, and geometry. These are the foundation of the kind of rationality evident in Vermeer's *Geographer*. Absolute geography consisted of 21 chapters covering matters such as geometry, the properties of the whole earth (such as size and motion), types of mountains, the tides, rivers, and the atmosphere. The relative geography section consisted of nine chapters concerning such things as the length of days, the seasons, sunrise, and the different climatic areas of the earth. Finally, the comparative geography section (10 chapters) focuses on the needs of sailors and ships, on navigation and longitude and latitude of places in comparison with each other. This latter part, in particular, seems directed at the new long-distance trade and colonization practices of the Dutch state.

Special geography, on the other hand, was based on observation in order to understand **chorography** (the regional) and **topography** (the local), terms derived from *chora* and *topos*. It was only when the geographer arrived at these levels that the world of humans entered the study of geography. Varenius did not live long enough to write the special geography but he did write an outline of it in *Geographia Generalis*. It is only at the very end that we see a recognizable human geography with headings such as "political government," "cities," and "virtues and vices." And, Varenius tells us, in order to understand the human world the geographer must use the evidence of the senses rather than pure reason and mathematics. Special geography, then, not only concerns a different set of subject-matter but necessitates a different methodology.

The geography of Varenius reflects already established versions of geography developed by the likes of Strabo and Ptolemy. Importantly, it also reflects the context that Varenius shared with Vermeer, Bleu, Tasman, and others. To understand how such ideas emerged, it is important to consider that Varenius lived in northwest Europe at a time when maps had become central to the new trade empire of the Dutch East India Company. This was a time when news from the other side of the world arrived in the parlors of Delft or Amsterdam. It was a time of exploration, trade, navigation, and empire. Maps were placed decoratively on walls, geographers were the subject of fine art, and geography went hand in hand with governance and commerce. There is a specific kind of geographic and cartographic imagination in the work of Varenius. It is an imagination that seeks to know the world as a set of mathematical laws. It is all-encompassing. General. It is no accident that exactly such a knowledge should emerge in a place that was at the heart of a new trade empire. Thinking of the world in this way sits snugly with a desire to include dramatically different areas of the world in an overarching calculus of control and trade.

Conclusion

This chapter has covered over 1500 years of geographic theory. Obviously much more than this was going on in that time, but this survey of early geographies sets the theoretical scene for much of what is to follow. Some of the key theoretical questions of geography were already established by 1500. First, note how the history of geographic theory so far has featured both general forms of geographical knowledge (distances, measurements, maps, latitude, longitude, etc.) and particular forms of geographical knowledge (the portraits of particular places in Strabo, for instance). This tension between the general and the particular has maintained its place at the center of geographical debate right up to the present day. It has changed form many times but can be seen, for instance, in the distinction between special geography and general geography in the work of Varenius; in the arguments about the idiographic and nomothetic (in Chapter 5); in the humanistic distinction between space and place (see Chapter 6); and in the critique of universalizing theory by postmodern geographers (see Chapter 9). Most recently it has emerged in the decolonial critique of colonial forms of seemingly all-embracing knowledge (see Chapter 14).

The second important question that emerges from these early geographies is the relationship between the human and natural worlds. There was no clear distinction between human and physical geography in the work described so far. Scholars from Herodotus to Ibn Khaldun would account for both the physical and human worlds in their geographies and often attempt to describe the interactions between them. The development of the idea of the habitable world (*ecumene*) in Greek thought set the scene for human geography centuries later which would continue to explore the notion of inhabitation – how people make the world into their home; how the natural landscape is transformed into a cultural landscape. Such questions were at the heart of cultural geography at the beginning of the twentieth century and formed a central question for humanistic geographers in the 1970s and 1980s. Indeed, the journal now known as *cultural geographies* started life in 1993 as *Ecumene*.

The third important lesson of these early geographical theories is the importance of historical and geographical context to the development of ideas. Note, for instance, how the geography of Strabo was written at a time of relative peace in the Roman Empire. An important component in the art of running an empire is understanding the world of that empire, its mathematical geography of locations and distances, its natural resources, and the customs and habits of its various peoples. Geographical theory is often constructed at the heart of empire as the center attempts to control the periphery. Geographical knowledge plays a central role in the practice of statecraft, as Strabo understood. The period between the tenth and fourteenth centuries was a time of expansion in the Islamic world and it was in North Africa and the Middle East that geography flourished in a culture that valued learning, trade, and pilgrimage. Varenius lived at the center of a rapidly globalizing trade empire that drew on geographical knowledge to navigate large distances and colonize distant lands. Travel is also a theme that links Strabo to Ibn Khaldun to Varenius. Trade and pilgrimage both involve mobility over long distances. Strabo and Ptolemy were displaced Greeks in the Roman Empire. Herodotus traveled the Mediterranean world. Ibn Battutah and Ibn Khaldun made pilgrimages to Mecca over vast distances.

By the eighteenth century, the ground had been prepared for the arrival of the modern discipline of geography as a subject taught in the major universities of Europe. As we will see in the next chapter, there is clear continuity in the kinds of theoretical ideas that the new discipline embraced and the kinds of problems that scholars had been studying for several thousand years.

References

Aitken, S. C. and Valentine, G. (2006) *Approaches to Human Geography*, Sage, London.

Alavi, S. M. Z. (1965) *Arab Geography in the Ninth and Tenth Centuries*, Dept. of Geography, Aligarh Muslim University, Aligarh.

* Battutah, I. and Mackintosh-Smith, T. (2002) *The Travels of Ibn Battutah*, Picador, London.

Bosworth, C. E. (ed.) (1970) *Hudud Al-'Alam; "the Regions of the World"; a Persian Geography, 372 A.H.–982 A.D* (trans. V. Minorsky), Luzac, London.

Bunbury, E. H. (1879) *A History of Ancient Geography Among the Greeks and Romans: From the Earliest Ages Till the Fall of the Roman Empire. By E.H. Bunbury, … With Twenty Illustrative Maps. In Two Volumes*, John Murray, London.

Casey, E. S. (1997) *The Fate of Place: A Philosophical History*, University of California Press, Berkeley, CA.

Clarke, K. (1999) *Between Geography and History: Hellenistic Constructions of the Roman World*, Oxford University Press, Oxford.

Cresswell, T. and Merriman, P. (eds) (2010) *Geographies of Mobilities: Practices, Spaces, Subjects*, Ashgate, London.

Crowe, P. R. (1938) On progress in geography. *Scottish Geographical Magazine*, 54, 1–18.

* della Dora, V. (2011) Is geography the eye of (pre-modern) history? Looking back at and looking through medieval geographies. *Dialogues in Human Geography*, 1(2), 163–188.

della Dora, V. (2021) *The Mantle of the Earth: Genealogies of a Geographical Metaphor*, University of Chicago Press, Chicago, IL.

Dueck, D. (2000) *Strabo of Amasia: A Greek Man of Letters in Augustan Rome*, Routlege, New York.

* Glacken, C. J. (1967) *Traces on the Rhodian Shore; Nature and Culture in Western Thought from Ancient Times to the End of the Eighteenth Century*, University of California Press, Berkeley, CA.

Herodotus (2007 [450 BC]) *An Account of Egypt*, Wikisource, The Free Library. http://en.wikisource.org/w/index.php?title=An_Account_of_Egypt&oldid=370023 (accessed July 12, 2007).

Hubbard, P. (2002) *Thinking Geographically: Space, Theory, and Contemporary Human Geography*, Continuum, New York.

Jones, R. (2004) What time human geography? *Progress in Human Geography*, 28, 287–304.

Khaldun, I. (1969 [1377]) *The Muqaddimah, an Introduction to History*, Princeton University Press, Princeton, NJ.

Koelsch, W. A. (2004) Squinting back at Strabo. *Geographical Review*, 94, 502–518.

* Lilley, K. (2011) Geography's medieval history: a neglected enterprise. *Dialogues in Human Geography*, 1(2), 147–162.

Livingstone, D. N. (1993) *The Geographical Tradition: Episodes in the History of a Contested Enterprise*, Blackwell, Oxford.

Malpas, J. E. (1999) *Place and Experience: A Philosophical Topography*, New Cambridge University Press, York.

Peet, R. (1998) *Modern Geographical Thought*, Blackwell, Oxford.

Smith, N. (2003) *American Empire: Roosevelt's Geographer and the Prelude to Globalization*, University of California Press, Berkeley, CA.

Strabo (1912 [AD 7–18]) *The Geography*, G. Bell and Sons, London.

Unwin, P. T. H. (1992) *The Place of Geography*, Longman Scientific Technical, Harlow.

Warntz, W. (1989) Newton, the Newtonians, and the Geographia Generalis Verenii. *Annals of the Association of American Geographers*, 79, 165–172.

Wright, J. K. (1947) Terrae incognitae: the place of the imagination in geography. *Annals of the Association of American Geographers*, 37, 1–15.

Chapter 3

The Emergence of Modern Geography

Despite the success of Varenius's geography and the beginning of university-taught geography in the seventeenth century, we have to fast forward to the late eighteenth century to see a remarkable resurgence of geography, particularly in Germany. Indeed, it is German geography that provided the bedrock for the next 100 years or so of development in the discipline. One key figure here is the German philosopher Immanuel Kant (1724–1804) who, as well as being one of the world's most important philosophers, was a teacher of physical geography. He lectured in geography for 40 years between 1756 and 1796. He never published a book on the subject but students who attended his course did, using their notes from his lectures (Elden 2009).

Kant's geography needs to be understood in relation to his wider philosophical pursuits. David Livingstone, in *The Geographical Tradition*, argues that it is not the details of Kant's teaching on physical geography that are important but his "assault on the various arguments for the existence of God" (Livingstone 1993: 115). Kant distinguished between the "nuomena" and the "phenomena." The "noumena" refers to an external reality, the world beyond the individual mind. The "phenomena," on the other hand, refers to the appearances of the world – the way we know the world through perception. While we know things as they are perceived (as they appear to us), we can never know things as they are in some pure form. A number of "phenomena," in Kant's thinking, form the basis for our perception of all other things. Foremost among these are space and time. Space and time are not things in and of themselves but ways of perceiving things. Nevertheless, they are primary ways in which we make sense of the world. They produce order out of chaos. It is worth quoting Livingstone at length to consider what this means:

> And so Kant opened up an unbridgeable chasm between mind and world, and cut knowledge adrift from the realm of external reality. Science accordingly is only valid within the "phenomenal" sphere: it deals with observation, cause and effect relations, spatio-temporal properties. But science can never break through to the shadowy world of the noumena, for science's practitioners are never free from the mental spectacles that provide the very modes in which all of us actually conceive of entities and events. The world, as it were, mirrors the mind, not the other way round. (Livingstone 1993: 116)

Geographic Thought: A Critical Introduction, Second Edition. Tim Cresswell.
© 2024 John Wiley & Sons Ltd. Published 2024 by John Wiley & Sons Ltd.

It is this belief in the impossibility of accessing the "real world" that leads Livingstone to argue that this also meant that it was impossible to argue for the existence of God, as any ideas about "cause and purpose" in the world can only be a reflection of our own perception – not something that incontrovertibly exists in the world. The third part of the argument is that if God cannot be proven to exist through recourse to the scientific study of the natural world, then science (and particularly geography) was decoupled from theology: "Mountains are formed according to natural law and it is the geographer's duty to investigate these rather than speculating on their supposed role in the divine economy" (Livingstone 1993: 117). This move to center human perception in the economy of knowledge was taken by the French theorist, Michel Foucault, as the invention of "man" as a knowing subject. As Derek Gregory has put it: "It was then [1775–1825] that 'man' was constituted as both an object of knowledge and a subject that knows, and in the process the space of European knowledge 'toppled', 'shattered', and dissolved into a radically new configuration" (Gregory 1994: 26).

Kant's belief that space and time were the fundamental categories through which we perceive the world gave geography and history particular importance in the hierarchy of knowledge. Geography covered space while history covered time. Aristotle's belief that anything that exists has to exist somewhere (and therefore the "somewhere" came first) is also reflected in Kant's **ontology**. History happens in space and therefore geography is the "substratum" of all knowledge, "the first part of knowledge of the world" (quoted in Elden 2009: 9). History is defined by geographical limits. "Without this foundation," Kant wrote, "history is scarcely distinguishable from fairy-tales" (quoted in Elden 2009: 11). Geography, in Kant, is divided into subfields among which physical geography provides the basis for all other geographies. He further went on to define a set of other subfields: mathematical, moral, political, commercial, and theological geography. This was quite extraordinary for a time and place in which there was not a formal discipline of geography. These subdivisions (except perhaps theological geography) closely resemble the sub-disciplines we know today, such as cultural, economic, and political geography. Kant also distinguished between scales of knowledge in a way that more or less follows the divisions outlined by Varenius. Geography was the description of the whole world, chorography the description of regions, and topography the description of places.

It seems strange today to think of a leading philosopher teaching a physical geography course (just as it seems strange to think of Newton, the world's leading physicist, spending his time translating and possibly teaching Varenius's geography). But just as Newton's interest in geography came from his belief that it constituted a kind of mathematics, so Kant's geography sprang from his belief that geographical ideas were an important element in "trying to enlighten his students more about the people and world around them in order that they might live (pragmatically as well as morally) better lives" (Louden 2000: 65). Kant thought it necessary for his students to know about geography (and anthropology, which he taught alongside geography) before exploring his general philosophy. Geography provided a pragmatic and moral basis for more metaphysical explorations.

As Livingstone has suggested, the specifics of Kant's teachings on mountains or rivers (for instance) is not particularly interesting, in many ways reflecting the inherited knowledge from Varenius and others. Kant, after all, is famous for hardly ever leaving home (Königsberg), so was unlikely to be involved in extensive fieldwork! There are, however, elements of his broad philosophy that have continued to haunt the discipline. Indeed, the term "neo-Kantian" has become a term of abuse. In geography this is primarily because of the notion of **absolute space**:

> "Absolute space" refers to space conceived as a given field of action; natural and social events and processes happen "in space," and their location can be measured according to some kind of coordinate system. Philosophically absolute space derived from Descartes and Kant among others; in practical terms it can be thought of as the space of private property, national territory. It is the space of Newtonian physics and of national state making alike, the space of nineteenth-century European expansion and colonization as well as Euclidean geometry. Absolute space has become, in Western societies, the space of common sense. (Smith 2003: 12)

Here, Neil Smith, in an account of the life of Isiah Bowman, a leading American political geographer of the early twentieth century, suggests that we can blame Kant for the notion of absolute space – the idea that space precedes the things that are in space and that it can be understood objectively and scientifically as an extension of geometry. This notion of space, he argues, is, in turn, responsible for a collection of things that include the **nation-state** and private property. Clearly, in Smith's view, not a good thing. But Smith is not the only one to accuse Kant of significant theoretical misdeeds. Edward Soja also has trouble with Kant and his (alleged) support for a view of space as "absolute":

> But its most powerful source of philosophical legitimacy and elaboration is Kant, whose system of categorical antinomies assigned an explicit and sustaining ontological place to geography and spatial analysis, a place which has been carefully preserved in a continuing neo-Kantian interpretation of spatiality. The Kantian legacy of transcendental spatial idealism pervades every wing of the modern hermeneutic tradition, infiltrates Marxism's historical approach to spatiality, and has been central to the modern discipline of geography since its origins in the late nineteenth century. The vision of human geography that it induces is one in which the organization of space is projected from a mental ordering of phenomena, either intuitively given, or relativized into many different "ways of thinking." (Soja 1989: 125)

Here, Soja is not so concerned with the political implications of Kant's thought but with his "transcendental spatial idealism" which seems to be found everywhere. The problem for Soja is that space is here reduced to a category of thought and has no purchase on the real world. Kant is typically associated with **idealism**, the belief that things can only really be said to exist in the world of ideas as we have no way of knowing the nature of reality. Even absolute space is not something, in Kant, that is a feature of the world but rather a feature of our thinking about it – a way of making sense of it. To Kant, the categories of thought are entirely internal to human consciousness and they are transhistorical and universal. Humans cannot help but perceive the world through the categories of space and time. This makes it impossible, for instance, to argue that an idea has to be understood in its geographical and historical specificity – to say that something is "socially produced" – which is to argue that it is not essential and universal, but rather a product of its time and place like Tic Toc or K-pop. Kant's thought thus strengthens one aspect of geographical theory – the idea that space is fundamental in some way to understanding, while weakening geographical theory on another level – the idea that place as context is fundamental in explaining things.

It may be that Kant's thoughts on space have been misunderstood and/or misrepresented. Stuart Elden reconsiders Kant's philosophy of space. "Kant, it is generally supposed," he argues, "held a view of space that was totalising, based on Cartesian geometry, absolute in the Newtonian sense. Thus we find a range of adjectival pairings common: Kantian space is Cartesian, Newtonian, and sometimes even more ahistorically Euclidean" (Elden 2009: 18) (see the quote from Smith, above, to see these pairings in action). Elden argues that, in his later years, Kant spent a great deal of time challenging Newtonian conceptions of space and was informed in this challenge by the philosopher Leibniz, who had proposed a *relative conception of space* – the view of space as the product of relations between things. In the Leibnizian view of space, space comes after, not before, the things that are in it. It is nonsensical, under an absolute (Newtonian) conception of space, to talk of the "production" of space as space must necessarily preexist everything. A relative conception of space, on the other hand, suggests that space is always a product in that it comes after the distribution of things. But remember also that Kant was arguing that space was not an object but an intuition, a way of perceiving that everyone has programmed into their brains. Absolute space is not a feature of the world but a feature of our way of making sense of the world. But Kant also argued (confusingly) in his most well-known book, *The Critique of Pure Reason*, that space comes before experience and is, indeed, necessary for us to experience anything. Space is the "ground of all intuitions . . . the condition of possibility of appearances, not as a determination dependent on them" (quoted in Elden 2009: 20). So here we have Kant, at different times, arguing that space is both a category of perception and the ground for our perceptions to be possible. To Elden there is enough doubt in Kant's various writings on space that we cannot clearly say that Kant held on to an "absolute" conception of space:

> Space and time are thus not absolute in anything like a Newtonian sense, nor does Cartesian geometry tell us about the world *as it is*. Rather they are modes of access to the way the world appears to us. Kant therefore concludes … 'space and time of course have objective reality, but not for what pertains to things outside of their relation to a faculty of cognition, but rather only in relation to it, and thus to the form of sensibility, hence solely as appearances'. (Elden 2009: 21)

Here, Kant's conception of space is less a grid-like thing out there in which everything else has to be located, but a product of the thinking and sensing body. Indeed, this kind of space has more in common with those we will encounter in the humanistic and phenomenological traditions in geography (Chapter 6) than they do with the absolute space of spatial science (Chapter 5).

Alexander von Humboldt (1769–1859) and Carl Ritter (1779–1859)

It was in Germany, or more accurately, Prussia, that geography flourished in the early nineteenth century. Alexander von Humboldt's *Kosmos* and Carl Ritter's *Erkunde* are two of the foundational texts of the discipline of geography. Alexander von Humboldt was a master of many subjects with a grounding in physical sciences, particularly geology, while Ritter has a more philosophical and human-centered approach, becoming the first professor of geography in Berlin in 1820.

Alexander von Humboldt was both an intellectual (schooled in the thinking of Kant among others) and a traveler. He worked for a while as a mine inspector across much of central Europe and in this job he learned the importance of direct observation and careful measurement – concerns he took with him on trips to South America in 1799–1804 and Siberia at the age of 60 in 1829. Humboldt is often described as a complete scholar owing to his combination of philosophy, travel, and practical know-how. Humboldt could focus on a small part of the earth's surface and attempt to link the whole world in a single account. He could look at the intricacies of a small plant or the beauty and complexity of the whole planet. Some have suggested that an aesthetic appreciation of the wholeness of the natural world lay at the heart of his geographical enterprise (Livingstone 1993: 135). Veronica della Dora has written that "[h]is goal was to uncover unity in diversity, the universal behind the particular, the invisible behind the visible" (della Dora 2021: 149).

At the age of 70, Humboldt began to write *Kosmos*, a five-volume "sketch of the physical description of the universe." For the most part the "universe" proved to be slightly too large for even Humboldt and most of the work he was able to finish concerns parts of Latin America. The work is primarily (as its subtitle suggests) a work of physical geography grounded in geology and biology. Clearly the book did not match his ambition:

> I have the crazy notion to depict in a single work the entire material universe, all that we know of the phenomena of heaven and earth, from the nebulae of stars to the geography of mosses and granite rocks – and in a vivid style that will stimulate and elicit feeling. Every great and important idea in my writing should here be registered side by side with facts. … My title is *Cosmos*. (von Humboldt quoted in Livingstone 1993: 136)

One of the things that makes Humboldt's project stand out is his concern to bring together the general and the specific – the "great and important ideas" with the relatively humble "facts." He wanted to "discern physical phenomena" in the context of a nature that was a whole, "animated and moved by inward forces" (quoted in Livingstone 1993: 137). Livingstone has argued that Humboldt left us with a number of "Humboldtian practices" including: measurement (the importance of quantifying geographical phenomena through empirical measurement with specific instruments – what physical geographers now call "fieldwork"); the importance of the region (Humboldt believed the world to be divided up into natural regions identified through their flora and fauna); and finally the use of mapping as a way of recording data that are spatially distributed.

Humboldt was, however, more than a measurer and a mapper. He was also a practitioner of poetics. Veronica della Dora has observed how Humboldt brought a poetic sensibility to his work that provided a "way of seeing" that was different from but connected to the mapping and measuring he is famous for.

> Like Goethe, Humboldt encouraged the poetic contemplation of nature. Poetic contemplation was an antidote to the cold mechanistic gaze of modern science that, Humboldt feared, was threatening the free pleasure of nature. By stretching the veil of poetry over the Andes, Humboldt aimed at making those distant sceneries (and the knowledge of the earth) accessible to a generation of bourgeois educated in the Romantic taste of the sublime. (della Dora 2021: 154)

Della Dora quotes Humboldt at length.

> I would recall the deep valleys of the Cordilleras where the tall and slender palms pierce the leafy veil around them, and waving on high their feathery and arrow-like branches, form, as it were "a forest above a forest"; or I would describe the summit of the Peak of Teneriffe, where a horizontal layer of clouds, dazzling in whiteness, has separated the cone of cinders from the plain below, and suddenly the ascending current pierces the cloudy veil, so that the eye of the traveler may range from the brink of the crater, along the vine-clad slopes of Orotava, to the orange gardens and banana groves that skirt the shore. In scenes like this [the heart is moved by] the features of the landscape, the ever-varying outline of the clouds, and their blending with the horizon of the sea, whether it lies spread before us like a smooth and shining mirror, or is dimly seen through the morning mist. … Impressions change with the varying movements of the mind, and we are led by a happy illusion to believe that we receive from the external world that with which we have ourselves invested it. (Humboldt quoted in della Dora 2021: 154–155)

So, despite his grounding in the physical sciences, Humboldt was not simply a detached observer of the natural world. Rather, he integrated the precision of measurement and close observation with a broader sense of the mystery and grandeur of the inhabited world.

Humboldt never held a formal academic position as a professor of geography. He was a wealthy man and was able to fund his expeditions himself. Unlike Ritter, he never supervised students and thus his impact is more diffuse and less easily traced through the emergent discipline. His influence was more that of a public intellectual. His work was read by, and clearly informed, Charles Darwin. Following his voyages to and through Latin America he was met by President Thomas Jefferson in the United States. Indeed, Humboldt's ideas clearly influenced the westward movement of the American "frontier" and formed the scientific underpinnings to the notion of "manifest destiny." Humboldt's love of cartographic techniques was particularly focused on lines connecting points of equal value (isopleths). An example of this is the "isotherm" (a line connecting points of equal temperature). This idea formed part of a larger theoretical edifice that predicted the location of important civilizations through history. As European colonizers moved west across North America, they claimed to be guided by "manifest destiny" – the idea that America was destined to be the ultimate civilization. In order to support this idea, William Gilpin, the first governor of Colorado, called on Humboldt's idea of the "isothermal zodiac" – a zone that stretched around the northern hemisphere approximately following the 40th degree of latitude along which the great empires of the world had formed in a progression that moved from east to west – China, India, Persia, Greece, Rome, Spain, and Britain. The conquest and colonization of the American west was, therefore, the natural culmination of the history of empire and civilization.

Humboldt was clearly a product of his time – an explorer/traveler from a colonial heartland seeking to travel across, measure, and represent distant parts of the world. His legacy has been somewhat contentious. Mary Louise Pratt's book, *Imperial Eyes: Travel Writing and Transculturation* featured Humboldt and his practices as a key figure in the project of Imperialism (Pratt 1992). Pratt argues that Humboldt, in a move characteristic of Imperial knowledge production, frames Latin America as "nature" and therefore as both radically different from Europe as the place of civilization/culture and as a site to be explored, viewed, possessed, and exploited. Humboldt clearly was implicated in colonial enterprises. His journeys were sometimes sponsored by the Spanish crown. Arguments about the value of Humboldt's ideas and how we should remember him extend into the present. The Humboldt Forum is a new museum in Berlin which is the home for non-European (often colonial) objects and artifacts. It is named after Alexander von Humboldt and his brother Wilhelm. Many have objected to the naming of the museum stating that Alexander von Humboldt was a colonialist who provided information so that the Spanish Crown could more effectively do the work of colonization. The pressure group, NoHumboldt21, put it this way.

> The exploration of the world and its populations by European "researchers" was a colonial project for many years and still affects the regimentation and exploitation of the global south to this day. One of the two people this project is named after, Alexander von Humboldt, was involved in this project to a great extent. The Spanish royalty and its overseas colonial regime, which was based on genocide and slavery, were particularly interested in the results of his expeditions in South and Middle America, and they supported him to the best of their ability. In this way, the Prussian "who really discovered America" who even stole buried corpses and shipped them to Europe, embodies colonial dominance. Humboldt is not an appropriate person to name an intercultural centre after. (NoHumboldt21 2013: npn)

Some authors have pleaded for more nuanced approaches to Humboldt's work, even seeing him as an anti-colonial thinker whose ideas were important stepping-stones to what we could now call political ecology (Sachs 2006). Historian Andrea Wulf has shown how Humboldt criticized the Spanish colonization of the lands of Indigenous people and spoke out against the exploitation of nature (Wulf 2016).

If Humboldt has sometimes been characterized as the practical arch-measurer of the physical world, then Carl Ritter has been described as his somewhat dreamy, human geography other half. Ritter shared with Humboldt the belief that the world added up to more than the sum of its parts. But Ritter paid more attention to the human/social world while always maintaining his belief that the human and physical worlds were inseparable. "The earth and its inhabitants stand in the closest reciprocal relations," he wrote, "and one cannot be truly presented in all its relationships without the other. Hence history and geography must always remain inseparable" (quoted in Unwin 1992: 17). And while Humboldt believed in the scientific unity of the natural world, Ritter was more mystical in his belief that the design of the earth, the unity in diversity, was evidence of the divine. To Ritter there was a *purpose* expressed in the order of things. This way of thinking about the world is teleological. **Teleology** refers to the explanation of something by something in the future. If we think of a normal explanation, simple laws of causality would suggest that an event has to happen before another event to cause it. A teleological explanation reverses this time line. A thing or event happens because of an event or thing that it has to become. Religious explanations often take this form. An event takes place because it has to fulfill some divine plan. But there is an element of teleology (or functional explanation) in a good deal of biology and particularly in evolutionary theory (often portrayed as the opposite of teleology in accounts of geography). Why does the bird have hollow bones in its wings? So that it can fly. The explanation comes after the event.

Like many before him, Ritter struggled with the kind of geography that is presented as a dry list of facts about places. Recall Strabo's complaint about having to endure the "tiresome part" of geography. To Ritter, geography was not just a set of lifeless facts but a living thing. Lists of facts were, to

Ritter, the old geography which he wanted to replace with a new and scientific account that focused not so much on description of individual places but on comprehending the interconnections between people and nature – teasing out the unity in diversity. And he wanted to do this in an **inductive** manner (this he shared with Humboldt). Inductive reasoning starts with particular observations and gradually builds up theories and laws (unlike a **deductive** account which uses precise observations to prove or disprove already established theories and laws). There is obviously a tension here between Ritter's scientific practice (starting with the particular and building up to the general) and his teleological beliefs which suggest his preexistent faith in a divine plan which would explain the unity he observed.

Perhaps Ritter's strongest influence on the later development of the discipline came through his emphasis on the region. Although Ritter saw the whole world as one unified entity, he also believed that it was divided into regions where it was possible to observe a smaller kind of unity between the different aspects of the region (its physical, biological, and human aspects). These regions were then also interrelated. It was this belief that proved influential with subsequent human geographers who brought regional geography and **areal differentiation** to the fore of the discipline (see Chapter 4). Indeed Ritter, as a professor of geography, had an impact on the discipline long after he had died, thanks to the students he supervised who took his ideas throughout Europe and over the Atlantic to the United States.

Like Humboldt, Ritter undertook the mammoth task of writing a multi-volume account of the world (an ambition notably missing from geographers since Ritter's time). Nineteen volumes of *Die Erdkunde* were published between 1817 and 1859. Still it was unfinished, focusing mainly on Africa and Asia and never reaching Europe or the Americas.

In Humboldt and Ritter we had perhaps the last two attempts to produce universal geographies (though see Reclus below). They had much in common. They both developed a regional perspective on the world. They both saw a kind of unity in the world (though different versions of unity) and they both believed in the necessity of studying physical and human geography in tandem. But they were also different. Humboldt was a world traveler who believed firmly in the importance of fieldwork and measurement. Ritter was happy, like many before him, to use the observations of others. While both looked at both human and physical worlds, Humboldt's training in natural science led him to emphasize the physical world while Ritter was more interested in the human world and the importance of (human) history. Humboldt is often described as the father of modern systematic geography, while Ritter is considered the father of regional geography (Unwin 1992).

Darwin, Lamarck, and Geography

Perhaps the most profound influence on the development of the modern discipline of geography was exerted by theorists of evolution. The key figures here are Charles Darwin and Jean Baptiste Lamarck (Stoddart 1966, 1986; Livingstone 1993). David Livingstone has argued that the engagement with evolutionary theory was a reflection of the need for geography to have a theoretical unity that would allow it to maintain a disciplinary identity. The writings of geographers from Strabo onward had been willfully diverse, covering just about anything in and of the world. It was for this reason that geography, taught in the tradition of Veranius or Kant for instance, could be thought of as simultaneously a branch of mathematics and philosophy. The new context of the nineteenth century saw knowledge being progressively carved up into discrete disciplines. This was a time when professors of geography were being appointed and departments of geography formed in leading universities. Geographical societies were established at the heart of national academic life. In Germany, the Berlin Geographical Society had been formed in 1828 and Chairs in Geography had been appointed in all the Prussian state universities, starting with Ritter in 1820. Maconochie had been appointed to the first Chair in Geography in Britain at University College London in 1833 following the formation of the Royal Geographical Society in 1830. In the United States, a Department of Geography, History and Ethics had been formed in the US Military Academy as early as 1818. In 1851, the American Geographical and Statistical Society was formed. Scientific specialties were becoming professionalized and institutionalized. It was in this intellectual context, Livingstone has argued, that geographers sought out intellectual foundations that would make the discipline of geography a plausible one.

The strongest conceptual foundations available in the intellectual climate of the late nineteenth century could be found in theories of evolution. Through a focus on evolution, Livingstone argues, geographers could keep "nature and culture under the one conceptual umbrella" (Livingstone 1993: 177). The traffic in ideas was not all one way. Darwin was influenced by the travels and thoughts of Humboldt in particular and was made a member of the Royal Geographical Society in 1838.

Darwin, of course, had formulated, over many years, the theory of evolution based on the process of natural selection. *The Origin of Species* was published in 1859 and represents one of the two or three most important bodies of theoretical work to have been developed in modern times (Darwin 2006 [1859]). It is one of the few sets of ideas to have reached into almost all facets of intellectual life (and well beyond that too). At the center of Darwin's theory is the notion that some creatures with particular variations from the norm may find these variations advantageous in dealing with the environment in which they find themselves. Individuals with these variations will thus be more likely to survive and pass these variations down to their offspring. This he termed "the survival of the fittest" or, more scientifically, "natural selection." Eventually, he adds, these differences will become exaggerated and new species will be formed. Thus, evolution happens.

Darwin's ideas were quickly extended by others to account for differences in human society. This is referred to as "social Darwinism." Ideas from Darwin's theory of evolution were translated, often metaphorically, to issues in social, political, and economic life. Thus, life in a capitalist society could be described as a case of "survival of the fittest." The competition between nations could also be theorized in evolutionary terms. The success or failure of industries or individual companies could similarly be described in this way. Biology was thus imported into social theory.

Darwin's great competitor in the theorization of evolution was Lamarck. Lamarck and "neo-Lamarckians" believed that characteristics could be "acquired" during a single lifetime. Rather than being the result of endless and seemingly random acts of chance (as in Darwin), evolution could happen more rapidly owing to the ability of a single organism to pass on a characteristic developed in its own lifetime. Thus, it might be possible for scientists simply to alter a species by, for instance, not allowing a type of wheat to grow too high, thus ensuring shorter wheat the following year. This idea proved to be disastrous when it was applied in the Soviet Union as an idea more easily incorporated into communist ideology than the notion of the "survival of the fittest." Neo-Lamarckians also believed that the environment could act directly on natural variation. This proved to be a seductive idea to geographers keen to find a theoretical link between humans and the environment. Two areas where these ideas were particularly influential were in the development of **geopolitics** and in the origins of *geomorphology*.

Geopolitics

Geography is clearly central to the practice of politics and warfare. Strabo, among others, noted the importance of geographical knowledge to "statecraft." Knowing the world is part of ordering and controlling it. In part this is simply a matter of cataloguing and detailing the contents of places far away. What kinds of resources are there? What are the people like? Are there physical obstacles to conquest and rule? But there is also a more conceptual link between geography and the practice of politics over large portions of the world. It was at the end of the nineteenth century that these connections were extensively theorized as what became known as "geopolitics" (Dodds and Atkinson 2000; Dodds 2007). The two figures who are centrally identified with the emergence of geopolitics are the British geographer Halford Mackinder (1861–1947) and the German geographer Friedrich Ratzel (1844–1904). Mackinder was awarded a Chair in Geography at Oxford University in 1887 and Ratzel was made Chair of Geography at Munich in 1875.

Mackinder was the first Chair of Geography at Oxford, a recognition that the discipline had finally become well and truly established (the first Chair in Geography in Britain had been awarded to Alexander Maconochie at University College London in 1833). Mackinder's geography was different from that which preceded him. He still embodied the idea of a macho geographer-as-explorer, being the first European to climb Mount Kenya in 1899. But his geography was much more than a combination of travel and description. His so-called "new geography" was informed by a familiarity with German geographic thought and a desire to think theoretically as well as empirically (Kearns 1997; Heffernan 2000). His concern, as Heffernan has put it, was to "attempt to understand how the different nations and regions of the world inter-related as elements in a holistic geopolitical structure" (2000: 32). Britain's empire was at its most expansive and Mackinder wanted to make sure that Britain remained a global power. We have seen how Strabo was writing at a time when the Greek empire was relatively stable. Varenius was working when the Dutch East India Company was establishing the largest trade empire the world had ever seen. Mackinder was similarly a geographer of empire. To Mackinder, an imperial power needed imperial citizens and they needed, first and foremost, to know their geography. What constituted this geography was an important question. Like many other commentators at the time, Mackinder believed that the world was more or less already known. The period of exploration was (almost) over and geography, in Mackinder's words to the Royal Geographical Society (founded in 1830), should now turn its attention to "the purpose of intensive survey and philosophical synthesis" (Mackinder 1904). Mackinder proceeded to provide just such a synthesis.

His argument was that the period of exploration had allowed European power to expand almost effortlessly into the furthest reaches of the globe. By colonizing large parts of Africa and Asia, he argued, the European powers had avoided colliding with each other. Now that the world had been effectively closed, divided up among European powers, these powers would turn their attention to each other in Europe and at the boundaries of empire:

> Every explosion of social forces, instead of being dissipated in a surrounding circuit of unknown space and barbaric chaos, will be sharply re-echoed from the far side of the globe, and weak elements in the political and economic organism of the world will be shattered as a consequence. (Mackinder 1904: 422)

Another issue for Mackinder was the declining importance of sea power in the new world. Empires such as that of Britain and Holland (for instance) had not been empires of lands butting up against each other but of lands connected over long distances by sea. The future, Mackinder argued, was in land empires. Resources made available by the rapid development of the railways across vast continents would make it possible to build land empires. The largest of these newly opening up land masses was Asia and it was in Asia that Mackinder believed the future of the world would be decided. He referred to it as the "pivot" of history (see Figure 3.1). By Asia, Mackinder was, in fact, referring to large parts of the Eurasian landmass (always the hardest place to conquer, but potentially the most profitable, as any player of "Risk" will tell you). The landmass that includes both Asia and Europe he referred to as the "World-Island." Typically, for a viewpoint which posits Europeans as active agents in an otherwise passive world, Mackinder did not believe that Asia (except perhaps Japan) would make its own empire. Rather he looked to the already existing imperial powers around Asia and asked which one of them could control this key area. The danger, he deduced, would come from an

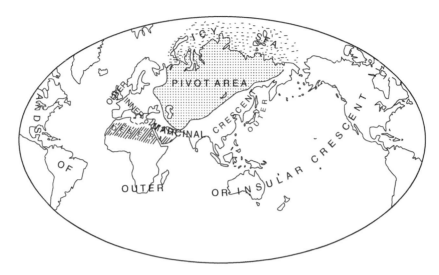

Figure 3.1 Mackinder's geopolitical map of the world in 1904. Source: From Mackinder (1904).

alliance of Germany and Russia successfully controlling Asia and heralding a new world system in which Britain would be marginalized. Clearly then, British resources should be dedicated to preventing an alliance between Germany and Russia. Geography should inform global politics.

By 1919, the reality of World War I meant that Mackinder had to revise his views slightly. Now, he said, the threat was from Russia and a divided Germany. There was, he wrote, a chasm in Europe between the east (oriented to what he called the "heartland" – a new term for the "pivot") and the west (or "coastland"). This chasm closely resembled what would become the "iron curtain" following World War II. Eastern Europe, to Mackinder, was the key:

> Who rules the East Europe commands the heartland;
> Who rules the Heartland commands the World-Island;
> Who rules the World-Island commands the World.

(Mackinder 1919: 194)

Britain and its allies, Mackinder argued, had to prevent Germany and Russia joining up by establishing "buffer states" between the two. Thus, the "World-Island" would remain free of German/Russian rule.

The other key figure in the emergence of theoretical geopolitics was the German geographer Frederick Ratzel. Ratzel is best known for the development of the concept of ***lebensraum*** – the idea that states "naturally" expand unless they are constrained by stronger neighbors. The origins of the term come from biology and particular readings of Charles Darwin as well as the social Darwinist Herbert Spencer (Keighren 2010). Originally its meaning was no more sinister than the English term "habitat." Ratzel had trained in the natural sciences and was a fan of Darwinism. Translating an idea from biology into geography was simple for Ratzel. The struggle for survival between individual creatures and species in Darwin became a struggle for space. The struggle for space between groups of humans was rooted in biology. The formation of states was, for Ratzel, rooted in nature (Heffernan 2000). The nation-state was like an organism that had its own life and behavior. Because an organism is natural its behaviors are both inevitable and normal. Normal, here, means both what usually happens and what is correct and morally right. "The state was a living geopolitical force rooted in, and moulded by, its soil. It was an organic entity, the physical embodiment of the popular will and the product of centuries-old interaction between a people and their natural environment" (Heffernan 2000: 45). The concept of *lebensraum* (living space), as developed by Ratzel and others, posited that in certain situations a nation-state was destined to expand into the space around it. This was particularly true of states with a "strong" population, society, and economy and limited space (such as Germany, according to Ratzel). States with "weak" populations, economies, and societies but large amounts of space (such as the Ottoman (Turkish) empire of the time) would, in contrast, find themselves contracting as neighbors take over space which was "naturally" theirs. As states expanded, they would reach a natural point at which they had reached their optimum extent. Reach beyond that and they too would be subject to the advances of those around them. Most of us (in Britain at least), of course, come across the idea of *lebensraum* in school history as we learn about the prelude to World War II. It was this idea that Hitler used to justify the expansion of Germany into Sudetenland (part of the modern Czech Republic) and Austria in the years immediately before 1939. While it is doubtful that Ratzel would ever have approved of such an outcome, his thought has long been associated with Nazi racial thinking (Bassin 1987). But Darwinian thinking was not only central to human geography, it also inspired and informed the birth of geomorphology as an important branch of physical geography.

Geomorphology

One of the key figures in the development of modern physical geography is William Morris Davis (1850–1934). Davis's most significant contribution to geographical theory was the idea that landforms followed a cyclical process of development (Davis 1899, 1902). They moved from youth to maturity to old age. In other words, they "evolved." These cycles of erosion became the core of teaching in the new science of geomorphology across the world (apart from in Germany) for many decades. This is how I was taught physical geography in my final two years at high school (1981–1983). Davis's theory of cycles of erosion was clearly indebted to Darwin. The idea of landforms progressing though predefined stages of evolution is a Darwinian idea, despite the lack of reference to the key Darwinian concepts of natural selection and random variability. Davis's ideas had other influences too. Haeckel, for instance, had developed the evolutionary idea that the stages of evolution could be observed in a developing embryo. A human embryo would thus pass through all the stages of evolution before becoming human at the end. The embryos of other creatures would follow a similar pattern but miss out the final stage. This process was known as "orthogenesis." These ideas were also influenced by Lemarckian views of evolution (Livingstone 1993; Inkpen 2005). While an organism could evolve along a pre-ordained pathway, this could be interrupted, slowed down, or accelerated through "acquired characteristics."

Returning to Davis, he believed that any landform could be classified according to the stage of development of its morphology as belonging to youth, maturity, or old age:

> Explanation was contained within the theory, within the evolutionary model used. Observation merely confirmed the position of a specific landscape or landform within this scheme. Stage became the explanation rather than what happened in the landscape. Significantly, however, the stage model highlighted the importance of identifying and providing an explanation for stability and change in the landscape. Both types of behaviour could be explained by reference to the same explanatory framework. Change was an essential feature of evolution and stability was illusory, based on the inability of humans to perceive alteration at a time-scale of relevance to the landscape. (Inkpen 2005: 17)

Evolution gave Davis's ideas the glow of science. The theory of evolution was perhaps the most significant scientific theory of the time and to translate this into physical geography was to make

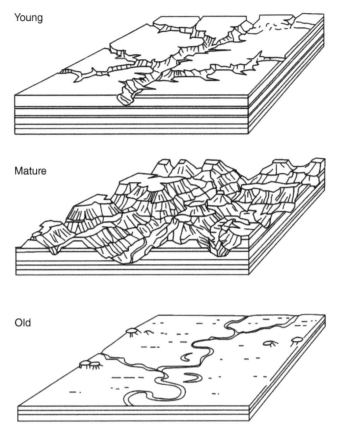

Figure 3.2   W. M. Davis's cycle of erosion. Source: From William Morris Davis Geographical Essays (1909)/ Public Domain.

physical geography, and geomorphology in particular, look scientific too. This was despite the fact that Davis's science was, in fact, highly qualitative, more likely to be accompanied by quite beautiful hand-drawn diagrams than any calculations or experiments (Figure 3.2).

The evolutionary thought that underlies the cycle of erosion can also be seen in Davis's wider ideas about the relations between the physical world (physiographic) and the world of living things (ontographic). Davis believed in controlling physical forces and human responses. This much he made clear in his presidential address to the Second Annual Meeting of the Association of American Geographers in 1909:

> I am disposed to say that any statement is of geographical quality if it contains a reasonable relation between some inorganic element of the earth on which we live, acting as a control, and some element of the existence or growth or behavior or distribution of the earth's surface organic inhabitants, serving as a response; more briefly, some relation between an element of inorganic control and one of organic response. (Davis 1909; http://www.colorado.edu/geography/giw/davis-wm/1909_ge/1909_ge_ch01_ body.html (accessed December 3, 2008))

Davis references the work of Ratzel and Reclus (see below) approvingly in his defense of geography as the study of causal relations between the inorganic and organic elements of the earth. Although Davis is remembered chiefly as a physical geographer, he lamented the fact that he had not done geography "properly" as he suggested in a letter to his student, Isiah Bowman (who became one of the most preeminent political geographers in the United States): "The chief thing I wish to emphasize is that you should develop geography proper, physiography and ontography properly combined, and not simply physiography as I have done too much" (Beckinsale 1976: 448; Livingstone 1994). Davisian geomorphology would later come to be seen more as a cultural pursuit of geographers than as serious science. The geomorphologist Dorothy Sack has shown how a paradigm change occurred as younger scientific geographers such as Arthur Strahler began to insist on a more mathematical systems-based approach to physical geography around 1950 (see Chapter 5) (Sack 1992). The new approach came to be associated with the geologist and geomorphologist, G.K. Gilbert. Gilbert did not base his approach to understanding physical landscapes on evolution, but rather on discovering the relations between observable phenomena. His work was informed by an understanding of the science of mechanics and emphasized continuous mechanical processes in the formation of landscapes. What Davis and Gilbert shared was a reliance on versions of "science" and what was considered scientific practice to bolster their theories of the physical landscape. By the 1960s, Davisian versions of geomorphology were seen as outmoded. Davis's ideas were enormously influential in fields well beyond geomorphology partly because he was a very good teacher who attracted a variety of acolytes including Ellsworth Huntington who went on to become one of the fiercest advocates of environmental determinism.

Environmental Determinism

Perhaps the key problem that lay at the heart of geography in the nineteenth and early twentieth centuries was the relationship between the natural and human worlds. This issue had clearly concerned the proto-geographers of the classical and medieval worlds, who had developed schemes of elaborate climatic zones in which different peoples were said to display different attributes depending on the climate they had to contend with. The issue of "people–environment" relationships continues to inhabit the heart of the discipline up to the present day. This label is often used to describe a significant strand of learning in North American departments of geography. Scholars describe themselves as "people–environment" geographers. Exactly how people relate to the natural world is currently a contested question, with answers ranging from an environment that determines human life to one in which nature itself is "socially constructed."

But if we go back 100 years, there was one strand of thinking that was dominant – **environmental determinism**. When an explanation is deterministic, it is one in which one set of things directly causes, in a linear and one-directional way, another set of things. Environmental determinism is the belief that the natural environment causes and explains the human/cultural world.

David Livingstone has argued that a "moral economy of climate" formed a key part of an imperial and racist imagination right up until the middle of the twentieth century. Over and over again, readers of geography journals, as well as thousands of school children and university students reading textbooks, were subjected to the idea that climatic regions were reflected in a moral topography of race. Consider the address to the British Association made by the Scottish geologist, Joseph Thompson, in 1886:

> It is a fact worthy of our attention that, as the traveller passes up the river [Niger] and finds a continually improving climate … he coincidently observes a higher type of humanity – better-ordered communities, more comfort, with more industry. That these pleasanter conditions are due to the improved environment cannot be doubted. To the student with Darwinian instincts most instructive lessons might be derived from a study of the relations between man and nature in these regions. (Thompson quoted in Livingstone 1994: 139)

Here, Thompson recruits Charles Darwin into a determinist argument about the effect of climate on such things as "order" and "industry." Over 60 years later a similar sentiment can be found in a climatology textbook by Austin Miller:

> The enervating monotonous climates of much of the tropical zone, together with the abundant and easily obtained food supply, produce a lazy and indolent people, indisposed to labour for hire and therefore in the past subjected to coercion culminating in slavery. (Miller quoted in Livingstone 1994: 141)

Here, climate (and the enslaved themselves) are to blame for slavery. Livingstone describes how such beliefs were used to make the claim that White people are incapable of performing manual labor in the tropics and therefore need to have non-White people to perform these tasks (a claim apparently still being made in geography textbooks in the 1950s).

Perhaps the key figures in the theory of environmental determinism were the American geographers Ellen Semple (1983–1932) and Ellsworth Huntington (1876–1947), and the Australian geographer Griffith Taylor (1880–1963). Behind all of these figures, and many more besides, was the influence of Charles Darwin. Darwin, you will recall, had already influenced the very different geographies of Ratzel and Kropotkin (see below). Ratzel's *Anthropo-Geographie* (1889) had argued that human activity was, for the most part, determined by the physical environment, and particularly by climate. Kant had made similar arguments about the "bedrock" of physical geography. One of Ratzel's admiring students was Ellen Churchill Semple (Keighren 2010).

Semple attempted to translate Ratzel's ideas in *Anthropo-Geographie* into a North American sensibility. Her book, *Influences of Geographic Environment*, had the full title *Influences of Geographic Environment on the Basis of Ratzel's System of Anthropo-Geography* (Semple 1911). In it she separates the world into key environmental types (mountains, rivers, coasts, etc.) and describes the kinds of people who live in and with them. In the Preface, she makes it clear that her aim is to bring the thought of Ratzel ("the great master who was my teacher and friend during his life, and after his death my inspiration" (viii)) into conversation with the English-speaking world. Although she is clearly inspired by Ratzel's thought, she wanted to leave aside the organic theory of society as the theory had, by 1911, "been widely abandoned by sociologists" and had, therefore, "to be eliminated from any restatement of Ratzel's system" (vii). The very first paragraph makes her intentions clear:

> Man is the product of the earth's surface. This means not merely that he is a child of the earth, dust of her dust; but that the earth has mothered him, fed him, set him tasks, directed his thoughts, confronted him with difficulties that have strengthened his body and sharpened his wits, given him his problems of navigation or irrigation, and at the same time whispered hints for their solution. She has entered into his bone and tissue, into his mind and soul. On the mountains she has given him leg muscles of iron to climb the slope; along the coast she has left these weak and flabby, but given him instead vigorous development of chest and arm to handle his paddle or oar. In the river valley she attaches him to the fertile soil, circumscribes his ideas and ambitions by a dull round of calm, exacting duties, narrows his outlook to the cramped horizon of the farm. (Semple 1911: 1)

Although Semple claims that she is not making an argument based on "geographical determinism," the remainder of the book proceeds to paint a picture of human life conditioned by the physical environment. Part of the basis for such claims is derived from Kant's argument that geography comes before history. "'Which was the first, geography or history?' asks Kant. And then comes his answer: 'Geography lies at the basis of history'. The two are inseparable" (Semple 1911: 10). Semple outlines four ways in which the physical environment impacts on human life:

1. Direct physical effects of the environment. Physical factors such as climate, distance from sea, and altitude, she argues, directly influence human characteristics such as stature, skin color, and color of hair (which, she argues, gets lighter at higher altitudes).

2. Psychical effect of environment. The environment, she argues, can explain cultural aspects of human life such as religion, literature, and modes of thought. "The Eskimo's hell is a place of darkness, storm and intense cold; the Jew's is a place of eternal fire. Buddha, born in the steaming Himalayan piedmont, fighting the lassitude induced by heat and humidity, pictured his heaven as Nirvana, the cessation of all activity and individual life" (Semple 1911: 40–41).
3. Environmental influence on economic and social development. Here, Semple charts how the amount of basic resources, securing the necessities of life, allows for different levels of human development, from nomadic pastoralists to urban civilizations.
4. Environmental influences on the movements of people. Semple argues that the natural environment produces barriers (such as mountain ranges) and routes (such as rivers) that enable or constrain the movements of people.

The range of environmental influences in Semple's book is enormous. It is not simply a matter of climate producing racial characteristics, as had been argued by thinkers since the classical period. Environment, for instance, is given as an explanatory factor in the development of attitudes to slavery in pre-civil-war America. Opposition to slavery in New England is explained by the "soil and boulder-strewn fields of New England," while the pro-slavery attitudes of the South are rooted, she argues, in the "rich plantations of tidewater Virginia and teeming fertility of the Mississippi bottom lands" (Semple 1911: 11). Small, isolated, places such as Greece, she insists, encourage the development of rich civilizations that then find themselves cramped and unable to progress. In contrast, "Open and wind-swept Russia, lacking these small, warm nurseries where Nature could cuddle her children, has bred upon its boundless plains a massive, untutored, homogeneous folk, fed upon the crumbs of culture that have fallen from the richer tables of Europe" (Semple 1911: 12). Mountain environments, meanwhile, militate against cultural development while rich river valleys produce philosophy and poetry. In Switzerland, the overpowering presence of majestic nature "paralyzes the mind" and results in little or no artistic or poetic development. "French men of letters, by the distribution of their birthplaces, are essentially products of fluvial valleys and plains, rarely of upland and mountain" (Semple 1911: 19) while, "[m]ountain regions discourage the budding of genius because they are areas of isolation, confinement, remote from the great currents of men and ideas that move along the river valleys" (Semple 1911: 20).

Semple is clearly sensitive about the difficulties involved in thinking and writing about "race." Indeed, one of her points appears to be to challenge theories of race based on heredity though a focus, instead, on the influences of environment. By comparing different "ethnic stocks" in similar environments, she argued, it was possible to assign any similarities (of culture, industry, or social relations) "to environment and not to race" (vii). Nevertheless, the effect of the whole book is to conclude that people who live in broadly temperate environments are industrious and thrifty, while those who live in more tropical environments are more likely to be lazy and wasteful. Racial thinking reenters through the proxy of "nature."

This is reflected in the works of Ellsworth Huntington, who wrote *The Character of Races* in 1927 (Huntington 1927). In this book, he conducted a survey of well-known academics to gather their thoughts on the characteristics of European people. The result was a set of characteristics such as "genius," "health," and "civilization" mapped onto climatic energy in the physical environment (Livingstone 1994). While Semple traced the effects on human life of a multitude of environmental factors (climate, barriers, routes, etc.), Huntington was much more focused on the effects of climate. In the tropics, he argued, "evolution had stagnated" and the occupants of those lands, particularly Black Africans, had been left in a primitive state:

"Their characteristics," he went on, "are those which unspecialized man first showed when he separated from the apes and came down from the trees. It is not to be expected that such people should ever rise very high on the scale of civilization." The tropical world was thus relegated to the moral margins of history and its people cast out from the mainstream chronicle of evolutionary advance. (Livingstone 1994: 231)

Such thinking is replicated in the work of the Australian geographer, Griffith Taylor. Taylor, who was also known as an explorer who had accompanied Scott on his ill-fated second expedition to Antarctica, constructed elaborate maps or "climographs" in which human life was mapped on to variations in temperature and humidity.

Before accompanying Scott, Taylor had accompanied William Morris Davis on field trips to the Alps and was clearly influenced by Davis's enthusiasm for Darwinian approaches. Taylor was made Australia's first Chair of Geography at the University of Sydney in 1920. He originally conducted work on glaciation in Antarctica but is best known for his study of the relationship between climate and settlement in Australia. At a time when the Australian government was attempting to draw people to the country with predictions of populations in excess of 100 million, Taylor argued that there was a limit of 30 million owing to climatic conditions. This made him increasingly unpopular and he moved to the University of Chicago in 1928. He would later become Canada's first full professor of geography when he moved to the University of Toronto. His key work was *Environment and Race: A Study of Evolution, Migration, Settlement and Status of the Races of Man* (Taylor 1927). In this book, he developed his idea of zones and strata. First, a site where evolution takes place is established, he argued:

> … then, after a reasonable lapse of time, the various differentiated classes will be found to be arranged in zones … so that the most primitive is at the margins, and the most advanced at the centre of the series of zones. Thus the earliest class will have covered the greatest area in its migrations; but fossil evidence will be found buried in the deepest stratum, under the later strata at the centre of evolution. (Taylor 1951: 447)

Thus, successive groups of people, each more "evolved" than the one preceding it, migrate out from a central core to marginal spaces, leaving the "center" to the most evolved group or race. The least evolved could be found furthest from the core in horizontal space. The furthest margin, according to Taylor, was Africa, where the most "primitive" people could be found. Black people, he suggested, were pretty close to being Neanderthal. Taylor was particularly unapologetic about his belief in determinism:

> The writer is a determinist. He believes that the best economic programme for a country to follow has in large part been determined by Nature, and it is the geographer's duty to interpret this programme. Man is able to accelerate, slow or stop the progress of a country's development. But he should not, if he is wise, depart from the directions as indicated by the natural environment. – What the Possibilist fails to recognise is that Nature has laid down a 'Master Plan' for the World – This pattern will never be greatly altered; though man may modify one or two per cent of the desert area, and extend the margins of settlement. It is the duty of geographers to study Nature's plan, and to see how best their national area may be developed in accord with temperature, rainfall, soil, etc., whose bounds are quite beyond our control in any general sense. (Taylor 1951: 160–162)

It was not just Darwinian approaches that linked Taylor to Davis. Like Davis, Taylor was also adept at field sketches and produced vivid visual material to make his case. Heather Winlow has shown how Taylor used elaborate diagrams and maps to give his racist ideas the veneer of respectable science. She considers Taylor's "Migration-zone classification of the races of man" (Figure 3.3) noting that "Taylor's 'zones and strata' theory was elaborated through idiosyncratic cartography as evidenced by a map entitled 'Migration-zone classification of the races of man' (. . .). A triangular feature in the bottom left corner of the map indicates his racial taxonomy, shown as a number of 'strata' ranging from the Negrito to the Late Alpine races" (Winlow 2009: 398). Winlow considers the use of "scientific" symbols on Taylor's maps, such as the use of isopleths (lines connection points of equal value usually associated with weather maps), showing how they were used to construct a language of legitimacy which contributed to erroneous racist ideas. In other words, the maps did not record

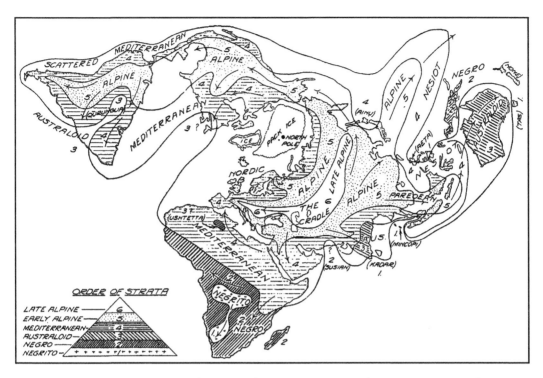

Figure 3.3 Migration-zone sources of the races of man. Source: From Taylor (1937)/University of Toronto Press.

"race," rather they constructed race and ideas of racial difference. This was further supported by language derived from geology such as "strata" and "shatter belts."

As James Blaut has argued, many geographers and geographical textbooks assume that environmental determinism has been safely relegated to a rather unsavory past (Blaut 1999). In fact, revised versions of such beliefs are very much alive. Perhaps the biggest selling geographer of the present day, Jared Diamond, a professor of geography at the University of California, Los Angeles, has made a version of a determinist argument in his book *Guns, Germs and Steel*. In it he argues that broadly European societies were destined to succeed at the expense of others owing to a combination of environmental factors (Diamond 1997). Part of his point in making this argument is to dismiss arguments that this has been due to any notion of racial superiority. But like others such as Griffith Taylor before him, he simply replaces one argument rooted in nature (race) with another (environment) – an argument that overlooks or diminishes the influences of colonialism or capitalism for instance. Forms of environmental determinism also inform some kinds of environmental theory, such as "bioregionalism" (see Chapter 4) which argues that the world is divided up into regions based on natural features such as watersheds and, further, that human communities would do well to stick to these natural regions. Environmental determinists such as Griffith Taylor remain heroes for many involved in environmental and green politics. Indeed, there are many in the modern environmental movement who would wholeheartedly agree with Ellen Semple's observation that "Man has been so noisy about the way he has 'conquered Nature', and Nature has been so silent in her persistent influence over man, that the geographic factor in the equation of human development has been overlooked" (Semple 1911: 2). More recently, we have seen the rise of "climate determinism" as a new and specific form of considerably more sophisticated environmental determinism. Climate change has rightfully been viewed as perhaps the most important issue of the age. Delineating the current and potential effects of climate change on human populations has been central to the arguments used to make a sometimes skeptical public sit up and take note. Often such arguments point to Africa and other parts of the Global South to make their case. As Sarah Radcliffe *et al.* have argued, this emphasis can lead us to "overlook the political interpretations that have long held sway in the scholarly understandings of famine, conflict and land shortages" (Radcliffe *et al.* 2010: 100). Locating apparently "natural" problems in Africa repeats the colonialist error of seeing "nature" elsewhere and erasing the complicated histories that the Global North is centrally implicated in. As Radcliffe *et al.* point out, forms of environmental or climate determinism are appealing because they seem simple and universal – they provide one big powerful explanation for a host of local circumstances.

> To the extent that the discipline of geography articulates an account of global climate change and its various local and regional ramifications (from melting polar ice to agrarian transformations in the Sahel), it too is entangled in the power of such engaging simplifications. In other words, geography's profile – as a field of research endeavour, as a claimant on public funds, and as a 'relevant' subject in instrumentally orientated times – rests in part on its credentials as a contributor to the global climate change debate. The challenge for geography lies in creating a discipline that reflects the complexity of – rather than simplifies – accounts of the world, without losing sight of a synthetic account. (Radcliffe *et al.* 2010: 100)

Focusing on climate as a singular and deterministic cause is powerful as it can serve to refute forms of climate change denial. Geography, though, can insist on nuance and context (not sexy terms in public debate) and link seemingly natural processes to clearly social, economic, and cultural factors in sophisticated and locally specific ways.

Anarchist Alternatives

Geographical theory has clearly served elites well. Imperial domination, warfare, the spread of capitalism, and the colonization of large parts of the earth's surface have gone hand in hand with the emerging discipline of geography. Almost all of the geographical ideas discussed up to now have been nurtured in places at the center of power – in ancient Greece and Rome, in the expanding Islamic world of the Middle Ages, in the imperial powers of Britain, France, and Germany. In some important sense, the ideas reflect this fact. Empires need detailed knowledge of their colonies. Both the specific knowledge of what makes parts of the earth's surface unique and the more general kind of knowledge that reduces all of the world to a set of mathematical principles can serve militaries as much as they can serve traders who seek to profit from the new opportunities that domination provides.

Geography, then, was very much a tool of the state, as Strabo, among others, has reminded us. The nineteenth century, however, saw the emergence of a full-fledged alternative geography: a set of geographic ideas that arose from outside of the state and posed significant challenges to the status quo. As well as being a time of expanding empires centered on Europe, the nineteenth and early twentieth centuries were times of extraordinary conflict and upheaval. Workers were forming collectives to challenge the power of both their employers and the state. A series of revolutionary movements (particularly in France and Russia) produced periods of great unrest and social transformation. One of the movements that arose was **anarchism**. Geography never played a more central role in the emergence of a fully formulated social and political theory than it did with the emergence of anarchism. Indeed, anarchism may be the only major political movement that can claim to have geography and the ideas of geographers right at its center. This can be seen in the work of two profoundly important thinkers: Elisée Reclus (1830–1905) and Peter Kropotkin (1842–1921).

The philosophy and theory of anarchism is much misunderstood, often stereotyped as a simple love of chaos and celebration of violence. In fact, anarchism (in its many guises) is an opposition to institutionalized authority (in the church, in the state, in the workplace, etc.). Such authority, anarchists argue, gets in the way of more fundamental forms of human cooperation and creativity. Anarchists seek to replace large-scale authority with collective decision making and small-scale communal living and working.

Elisée Reclus was born into a religious family in the south of France. His father was a minister. He attended lectures by Ritter, witnessed the Irish famine, and became involved in the 73-day Paris Commune (1857) following which he was exiled to Switzerland where he met Peter Kropotkin. He wrote both *The New Universal Human Geography* (19 volumes) and *Man and Earth* (6 volumes) (Reclus 1876a, b). He wanted to show how the world's resources could be redistributed for the benefit of all and how cooperation across borders could erode the power of the imperialist state. To Yves Lacoste, the work of Reclus represents the "epistemological turning point in the development of geographic thought." Geography up to Reclus, he suggests, "was linked essentially to the state apparatus, not only as a tool of power, but also as an ideological and propagandistic representation. Reclus turned this tool against the state apparatus, the oppressors and the dominant classes" (quoted in Reclus *et al.* 2004: 61).

Central to Reclus's thought was the relationship between the natural and human worlds. "Man," he argued, "is nature becoming self-conscious." Throughout his work he focused on the natural world that precedes human existence (the natural conditions of life) and the human social world ("the artificial sphere of existence"). While the former is more or less necessary, the second is open to infinite adaptation by humans themselves. Reclus made a distinction between "the facts of nature, which are impossible to avoid, and those which belong to an artificial world, and which one can flee or perhaps even completely ignore. The soil, the climate, the type of labour and diet, relations of kinship and marriage, the mode of grouping together, these are the primordial facts that play a part in the history of each man, as well as of each animal. However, wages, ownership, commerce, and the limits of the state are secondary facts" (Reclus *et al.* 2004: 24). One of the ways in which the natural conditions of life and the artificial sphere of existence interact is in the concept of the **region** (see Chapter 4). The natural make up of a region (which is itself defined by nature) influences human life and society in that region. This in turn produces distinctive regional cultures which are visible in the landscape. For this reason Reclus's thought has had a profound influence on modern **political ecology** and particularly bioregionalist thought. In a somewhat contradictory manner Reclus also believed that global humanity was increasingly becoming "one." He saw the effect of developments in transportation and communication that we might now refer to as globalization. But rather than seeing this as the development of a global economy (for instance) he preferred to think of it as the development of a global consciousness – the recognition of a common humanity – a planetary citizenship:

> So bounded are not the confines of the planet, that it everywhere benefits by the same industrial appliances; that, thanks to a continuous network of postal and telegraphic services, it has been enriched by a nervous system for the interchange of thought; that it demands a common meridian and a common hour, while on all sides appear the inventors of a universal language. Despite the rancors fostered by war, despite hereditary hatreds, all mankind is becoming one. Whether our origin be one of manifold, this unity grows apace, daily assumes more of a quickening reality. (Reclus *et al.* 2004: 55)

Peter Kropotkin was a Russian aristocrat (hence his nickname – the Anarchist Prince) who was trained in physical geography and conducted extensive surveys of the landforms of Siberia and glacial deposits in Finland. While undertaking such research he became fascinated by some of the people he lived and worked with, particularly the Cossacks in Siberia. He admired the way that they lived nomadically and without state regulation. When offered a Chair in Geography at St Petersburg, Kropotkin refused it, preferring instead to develop his own line of independent thought and become involved in activism that led to his arrest in Russia and then France. Such was his reputation among geographers that the British Royal Geographical Society wrote to have him released. Once free, Kropotkin moved to Britain; 100,000 attended his funeral.

Kropotkin didn't just want to describe geography but to change it. He believed in a geography based on small-scale communal life where decision making and power were localized and each member of the commune would know the others. In addition, he argued, such an organization of society would pay more attention to the effect they were having on their natural environment. The thought of Kropotkin (and Reclus) remains an inspiration for regional ecological thinking to this day (see Chapter 4). Typical of Kropotkin's thought was his belief that large-scale economic organization under capitalism was an alienating experience which worked against naturally cooperative forms of social life. The factory divided people from each other and alienated them from the production process. He described this in *Fields, Factories and Workshops* (1899):

> Political economy has hitherto insisted chiefly upon division. We proclaim integration; and we maintain that the ideal of society – that is the state towards which society is already marching – is a society of integrated, combined labour. A society where each individual is a producer of both manual and intellectual work; where each able-bodied human being is a worker, and where each worker works both in the field and in the industrial workshop; where each aggregation of individuals, large enough to dispose of a certain variety of natural resources – it may be a nation or rather a region – produces and itself consumes most of its agricultural and manufactured produce. (Kroptkin quoted in Blunt and Wills 2000: 11)

To make a "scientific" case for his theory of human cooperation, Kropotkin developed the theory of **mutual aid** in which he argued for a society that was naturally cooperative. As with Davis, Semple, and others, he was inspired by Darwin, but in a very different way. The theory of mutual aid involved a reinterpretation of Darwin's work which had been taken by "social Darwinists" to suggest that humans, like other species, were involved in a struggle for survival in which only the fittest would survive. Kropotkin was impressed with Darwin's thesis but, in *Mutual Aid*, argued for the importance of cooperation in nature between members of the same species:

> It is not love, and not even sympathy (understood in its proper sense) which induces a herd of ruminants or of horses to form a ring in order to resist an attack of wolves; nor love which induces wolves to form a pack for hunting; not love which induces kittens or lambs to play, or a dozen of species of young birds to spend their days together in the autumn; and it is neither love nor personal sympathy which induces many thousand fallow-deer scattered over a territory as large as France to form into a score of separate herds, all marching towards a given spot, in order to cross there a river. It is feeling infinitely wider than love or personal sympathy – an instinct that has slowly been developed among animals and men in the course of an extremely long evolution, and which has taught animals and men alike the force they can borrow from the practice of mutual aid and support, and joys they can find in social life. (Kropotkin 1972: 21)

Kropoktin had been inspired by his time spent with the Cossacks in Siberia. In them, he saw a level of self-organization and collective struggle that led him to conclude that it was cooperation and not competition between individuals that lay at the heart of human life. Once he started to look, he saw this cooperation between members of the same species throughout nature, from the colonies of ants to the herds of horses and packs of wolves. This mutual aid, he believed, was the key fact of evolution and could form the basis of a theory of society:

> In the practice of mutual aid, which we can retrace to the earliest beginnings of evolution, we thus find the positive and undoubted origin of our ethical conceptions; and we can affirm that in the ethical progress of man, mutual support – not mutual struggle – has had the leading part. In its wide extension, even at the present time, we also see the best guarantee of a still loftier evolution of our race. (Kropotkin 1972: 251)

Kropotkin's master work, *Mutual Aid*, is organized around an evolutionary account of the development of cooperation. It starts with the world of animals, moves through "savages," "barbarians," "the medieval city," and ends with "mutual aid amongst ourselves." Although this follows a fairly predictable pattern for the time, moving from the supposedly least to most developed life forms, the hierarchy is problematized by the recognition of the often superior quality of group life in supposedly savage primitive life. But mutual aid can also be seen in the formation of labor unions, in the family, and in the grinding life of modern urban slums. This sense of cooperation only weakens, he argues, when the state gets involved. His conclusion, though, is uplifting:

> In short, neither the crushing powers of the centralized state nor the teachings of mutual hatred and pitiless struggle which come, adorned with the attributes of science, from obliging philosophers and sociologists, could weed out the feeling of human solidarity, deeply lodged in men's understanding and heart because it has been nurtured by all our preceding evolution. (Kropotkin 1972: 245)

Just as W. M. Davis recruited versions of evolutionary theory to make geomorphology appear scientific, so Kropotkin used evolution to produce a "scientific" basis for the social theory of anarchism. If a study of biology could reveal natural cooperation, then forces (such as institutionalized authority) that stood in the way of such cooperation could be said to be "unnatural."

Anarchist geography (like environmental determinism or geopolitics) is not confined to some "early modern" period in geography's history. While anarchism has not been a major approach in geography since the late nineteenth century, it has re-emerged in recent years as anarchist influenced political movements such as the Zapatista movement in Mexico and the Occupy Movement that emerged in 2011 have become both more visible and influential. This has also been informed by leading anarchist scholars and writers outside of geography such as Colin Ward, Hakim Bey, and James C. Scott. Simon Springer has made the case for the importance of anarchism in the practice of an anti-colonial, anti-state, and anti-capitalist geography, arguing that anarchism is "a theory and practice that seeks to produce a society wherein individuals may freely co-operate together as equals in every respect, not before a law or sovereign guarantee—which enter new forms of authority, imposed criteria of belonging, and rigid territorial bindings—but before themselves in solidarity and mutual respect" (Springer 2012: 1606). Springer rejects, however, the reliance on science (Darwinian or otherwise) that marked the early anarchism of Kropotkin and Reclus and argues instead for a postanarchism that does not rely on the illusory foundations of the natural world.

In a discussion of the potentials of anarchist thought and politics, Jenny Pickerill makes the case for "prefigurative politics" – ways of being together that exist outside of the empires of capitalism and/or the state.

> There is a huge range of post/non/alter-capitalist spaces to be employed here, including eco-communities, squats, online spaces, pop-up shops, secular halls and social centres, but informal spaces can also be used, such as people's homes, or local community spaces such as village halls, allotments and meeting spaces above shops or in charity offices. (Pickerill in Araujo *et al.* 2017: 633)

These ways of being often seem small and invisible but, Pickerill argues, they present important anarchist alternatives to dominant ways of being.

> …there is much in the world that exists and flourishes on the edges of capitalist encroachment. It is in these 'unseen' spaces that alternative imaginaries are built and experimental ideas tested, not just as radical spatial interventions but also in our everyday lives in our homes and workplaces. Creative new ways of being and acting are practiced. (Pickerill in Araujo *et al.* 2017: 633)

This reflects Springer's call for an embrace of the "here and now" in thought and action acknowledging the possibility that "we can instantaneously refuse participating in the consumerist patterns, nationalist practices, and hierarchical positionings that confer legitimacy on the existing order and instead engage a 'do it yourself' culture centered on direct action, non-commodification, and mutual aid" (Springer 2012: 1616). The rediscovery of anarchist geographers and geographies has contributed to a general move to engage with what Federico Ferretti (2019) has called "Other Geographical Traditions" (OTGs) that have the capacity, in Innes Keighren's terms, to "inspire our admiration and to signal to future possibilities" (Keighren 2018: 15).

Conclusions

While environmental determinism, as it was espoused in the late nineteenth and early twentieth centuries, has largely been rejected for its simplistic science and racist overtones, it provides evidence for a deep-lying way of thinking that has structured all manner of geographical (and other) theory. This is the importation of the idea of "nature" and the "natural" into discussions of the human world. Biology has been a particularly strong breeding ground for such metaphorical forays. To Ratzel, the nation-state was an "organism." To Kropotkin, the essentially cooperative nature of human society was rooted in the behavior of birds and bees. Such thinking carried over into the "ecological" thinking of the Chicago School of Sociologists which formed in the early decades of the twentieth century and went on to influence geographers right up to the present day.

What is at stake in such thinking? Why compare humans, or cities, or states to organisms? One answer can be found in the slippery meanings of the word "nature" (Castree 2005). If something can be said to be natural then it is simply the "way things are." Nature has been constructed as an idea (in the western world at least) that refers to the nonhuman, the world that is beyond our control. Nature does not need our intervention to happen. It just does. On top of this meaning there is another meaning. The natural is often taken to mean the normal and the normal is also the morally correct. Nature thus forms the basis of a moral theory. What is natural is "the way things are" and, simultaneously, "the way things should be." Thus, to say that a strong state expands into a weak one (i.e. invades it) is made to seem both inevitable and morally justified. We could not stop such a thing happening even if we wanted to. If we want to argue, as Kant, Semple, and others have done, that some races are "inferior," then to say they are "naturally inferior" because of the environment is simultaneously to argue that all manner of ill-treatments are justified morally. To argue otherwise is to argue against nature.

Another possible reason for the justification of social theories through recourse to natural science is the belief that human science, to be taken seriously, has to be every bit as scientific as natural science. Science, we are told, is neutral, objective, and value free. Thus, the more the study of human life can be made to resemble, or draw its theories and methodologies from, natural science, the more seriously it will be taken. This concern with the scientific will be one that reemerges throughout this book and particularly with the development of spatial science in the late 1960s (see Chapter 5).

References

Araujo, E., Ferretti, F., Ince, A., Mason, K., Mullenite, J., Pickerill, J., Rollo, T., and White, R. J. (2017) Beyond electoralism: reflections on anarchy, populism, and the crisis of electoral politics. *ACME: An International E-Journal for Critical Geographies*, 16(4), 607–642.

Bassin, M. (1987) Race contra space: the conflict between German geopolitik and National Socialism. *Political Geography Quarterly*, 6, 115–134.

Beckinsale, R. P. (1976) The international influence of William Morris Davis. *Geographical Review*, 66, 448–466.

Blaut, J. (1999) Environmentalism and eurocentrism. *Geographical Review*, 89, 391–408.

* Blunt, A. and Wills, J. (2000) *Dissident Geographies: An Introduction to Radical Ideas and Practice*, Longman, Harlow.

Castree, N. (2005) *Nature*, Routledge, London.

Darwin, C. (2006 [1859]) *On the Origin of Species by Means of Natural Selection, or, the Preservation of Favoured Races in the Struggle for Life*, Dover Publications, Mineola, NY.

Davis, W. M. (1899) *The Peneplain. American Geographer*, 23, 207-39.

Davis, W. M. (1902) *Elementary Physical Geography*, Ginn, Boston, MA.

Davis, W. M. (1909) An inductive study of the content of geography, in *Geographical Essays* (ed. D. W. Johnson), Ginn, Boston, MA, pp. 3–22.

della Dora, V. (2021) *The Mantle of the Earth: Genealogies of a Geographical Metaphor*, University of Chicago Press, Chicago, IL.

Diamond, J. M. (1997) *Guns, Germs, and Steel: The Fates of Human Societies*, W.W. Norton & Co., New York.

Dodds, K. (2007) *Geopolitics: A Very Short Introduction*, Oxford University Press, Oxford.

Dodds, K. and Atkinson, D. (eds) (2000) *Geopolitical Traditions: A Century of Geopolitical Thought*, Routledge, London.

Elden, S. (2009) Reassessing Kant's geography. *Journal of Historical Geography*, 35, 3–25.

Ferretti, F. (2019) Rediscovering other geographical traditions. *Geography Compass,* 13(3), e12421.

Gregory, D. (1994) *Geographical Imaginations*, Blackwell, Oxford.

Heffernan, M. (2000) Fin de siecle, fin de monde? On the origins of European geopolitics, 1890–1920, in *Geopolitical Traditions: A Century of Geographic Thought* (eds K. Dodds and D. Atkinson), Routledge, London, pp. 27–51.

Huntington, E. (1927) *The Character of Races as Influenced by Physical Environment*, Scribner's Sons, New York.

Inkpen, R. (2005) *Science, Philosophy and Physical Geography*, Routledge, London.

Kearns, G. (1997) The imperial subject: geography and travel in the work of Mary Kingsley and Halford Mackinder. *Transactions of the Institute of British Geographers*, 22, 450–472.

* Keighren, I. M. (2010) *Bringing Geography to Book: Ellen Semple and the Reception of Geographical Knowledge*, I. B. Tauris, London.

Keighren, I. M. (2018). History and philosophy of geography II: the excluded, the evil, and the anarchic. *Progress in Human Geography*, 42(5), 770–778.

Kropotkin, P. A. (1972) *Mutual Aid: A Factor of Evolution*, Allen Lane, London.

* Livingstone, D. (1994) Climate's moral economy: science, race and place in post-Darwinian British and American geography, in *Geography and Empire* (eds A. Godlewska and N. Smith), Blackwell, Oxford, pp. 132–154.

* Livingstone, D. N. (1993) *The Geographical Tradition: Episodes in the History of a Contested Enterprise*, Blackwell, Oxford.

Louden, R. B. (2000) *Kant's Impure Ethics: From Rational Beings to Human Beings*, Oxford University Press, Oxford.

* Mackinder, H. (1904) The geographical pivot of history. *Geographical Journal*, 23, 421–442.

Mackinder, H. J. S. (1919) *Democratic Ideals and Reality: A Study in the Politics of Reconstruction*, Henry Holt and Company, London.

NoHumboldt21 (2013) *Stop the Planned Construction of the Humboldt Forum in Berlin Palace!* https://www.no-humboldt21.de/resolution/english/ (accessed June 28, 2022).

Pratt, M. L. (1992) *Imperial Eyes: Travel Writing and Transculturation,* Routledge, New York.

Radcliffe, S. A., Watson. E. E., Simmons, I., Fernández-Armesto, F., and Sluyter, A. (2010) Environmentalist thinking and/in geography. *Progress in Human Geography*, 34(1), 98–116.

Reclus, E. (1876a) *The Earth and Its Inhabitants: The Universal Geography*, J.S. Virtue, London.

Reclus, J. J. Â. (1876b) *Nouvelle Gâeographie Universelle*, Hachette, Paris.

Reclus, E., Clark, J. P., and Martin, C. (2004) *Anarchy, Geography, Modernity: The Radical Social Thought of Elisée Reclus*, Lexington Books, Lanham, MD.

Sachs, A. (2006) *The Humboldt Current: Nineteenth-Century Exploration and the Roots of American Environmentalism,* Viking, New York.

Sack, D. (1992) New wine in old bottles: the historiography of a paradigm change. *Geomorphology*, 5, 251–263.

Semple, E. C. (1911) *Influences of Geographic Environment on the Basis of Ratzel's System of Anthropo-Geography*, Holt, New York.

Smith, N. (2003) *American Empire: Roosevelt's Geographer and the Prelude to Globalization*, University of California Press, Berkeley, CA.

Soja, E. W. (1989) *Postmodern Geographies: The Reassertion of Space in Critical Social Theory*, Verso, New York.

*Springer, S. (2012) Anarchism! What geography still ought to be. *Antipode*, 44(5), 1605–1624.

Stoddart, D. (1966) Darwin's impact on geography. *Annals of the Association of American Geographers*, 56, 638–698.

Stoddart, D. (1986) *On Geography and Its History*, Blackwell, Oxford.

Taylor, T. G. (1927) *Environment and Race: A Study of the Evolution, Migration, Settlement and Status of the Races of Man*, Oxford University Press, Oxford.

Taylor, T. G. (1937) *Environment, Race, and Migration*, University of Chicago Press, Chicago, IL.

Taylor, T. G. (1951) *Geography in the Twentieth Century: A Study of Growth, Fields, Techniques, Aims and Trends*, Methuen, London.

*Unwin, P. T. H. (1992) *The Place of Geography*, Longman Scientific & Technical, Harlow.

*Winlow, H. (2009) Mapping the contours of race: Griffith Taylor's zones and strata theory. *Geographical Research*, 47(4), 390–407.

Wulf, A. (2016) *The Invention of Nature: The Adventures of Alexander von Humboldt, the Lost Hero of Science*, John Murray, London.

Chapter 4

Thinking About Regions

The regional concept constitutes the core of geography. This concept holds that the face of the earth can be marked off into areas of distinctive character; and that the complex patterns and associations of phenomena in particular places possess a legible meaning as an ensemble which, added to the meanings derived from a study of all the parts and processes separately, provides additional perspective and additional depth of understanding. This focus of attention on particular places for the purpose of seeking a more complete understanding of the face of the earth has been the continuous, unbroken theme of geographic study through the ages. (James 1929: 195)

Two questions have motivated geographical thinking over the past 2,000 years more than any others. These are "what is the connection between the human and physical worlds?" and "how can we account for spatial difference?" The second of these questions lay at the heart of what Varenius called "special geography" with its focus on particular areas of the earth's surface and the qualities that make these areas different and unique from the areas around them. It is this line of enquiry that lies at the heart of human geography over the first half of the twentieth century. The concept that exemplifies this pursuit is **region**. At first glance the idea of a region seems a fairly vague one. It might, for instance, be thought of as a synonym for "area." One question that immediately arises is at what scale a region exists. As a word that is used in everyday speech, it most often suggests a subdivision of something bigger. Regional art, for instance, signifies artistic practices, styles, and products that are associated with a particular bit of a nation – say the Midwest of the United States. But we would be unlikely to label a national art form regional even if a nation is part of a larger continental or global whole. We can even think of a house as having particular regions if we happen to be a sociologist trained in the work of Erving Goffman who might describe the living room, drawing room, or lounge of a house as the "front region" while the kitchen or bedrooms are part of the "back region" – not for public consumption (Goffman 1956). As well as being part of something larger, regions also seem to include smaller units within them – sub-regions, locations, places. There is something about the idea of a region, then, that suggests in-betweenness. It is vague. While Britain may have its own regions such as the "southeast" or "East Anglia," it is also part of a region called Northwest Europe or even

---

Geographic Thought: A Critical Introduction, Second Edition. Tim Cresswell.
© 2024 John Wiley & Sons Ltd. Published 2024 by John Wiley & Sons Ltd.

"the west" or "the Global North." As a concept then, region denotes an area of an indeterminate size that is itself part of a larger whole but includes smaller units within it. What we choose to refer to as a region depends very much on what it is we are talking about.

If I was to ask where you live, or where you were born, you would probably give me the name of a place straight away. I was born in Cambridge and live in Edinburgh. You would be unlikely to name a region unless the place you were born or live in is unknown to me. In that case you might locate a place in a region I have heard of – say, the Pacific Northwest of the United States. But if I ask you instead what region you live in, you might be slightly puzzled. The problem with such a question is that there is no fixed answer. I live in Edinburgh which is in the Central Belt of Scotland. It is also in Northwest Europe, in Europe, in "the west," and in the "north" depending on your view of world development geography. Edinburgh is also divided into regions. I live in Edinburgh South (the constituency of my Member of Parliament), in the neighborhood of Morningside. These are very different from other parts of Edinburgh. One way to think about regions, popular in political geography, is simply to define them as those areas that have been designated as such through forms of government and regulation. In this sense Morningside does not really exist. My local government is the government of the City of Edinburgh. It used to be in the county of Midlothian that was also known as Edinburghshire until 1921. I am represented by both a Member of the Scottish Parliament (as a devolved region/nation) and a Member of Parliament (for the UK). There is no one correct answer to what region I am in. Regions are slippery in this way.

Theoretically there are a number of problems we encounter when we think, write, and talk about regions. Perhaps the core problem is the question of the value of thinking about the particular and the specific. Theory tends to imply an ability to make generalizations across instances of things. An interest in regions that is simply content to say that each region is unique and distinct and is happy to delineate and describe that uniqueness seems strangely atheoretical. This problem highlights a central tension not only in human geography but in most of human thought. This is the tension between the universal and the specific. On the one hand there are those who argue that it is only in the universal that true scientific thought can be located. When we encounter a unique thing there is no basis in that for the development of laws which may allow us to understand things in general. When we say that water freezes at 0 degrees centigrade, we are not saying that one particular puddle will freeze, but that all water everywhere will freeze. When we posit a theory of gravity, it is not a theory about why one apple fell from a tree but why all things on earth will fall to the ground (given no other force being applied, of course). On the other hand there are those who argue (like Richard Hartshorne below) that the particular uniqueness of regions is in some way fundamental and that geography is in a privileged position because of the centrality of the region in the discipline. This tension is, of course, an extension of the tradition of general and special geography we have been tracing through the previous chapters.

We will trace this tension through this chapter and it will reappear throughout the book. Another tension that marks the engagement with the idea of region is the question of whether or not a region is actually a thing in the world which is waiting to be recognized or is simply a more or less arbitrary construction of "society" or, indeed, the discipline of geography. Is a region an object or, instead, an idea? To some, the principal problem with the idea of the region is that it does not relate to some clear "thing" that is recognizable in reality. As long ago as 1954, Derwent Whittlesey wrote that "It is an intellectual concept, an entity for the purposes of thought, created by the selection of certain features which are relevant to an areal interest, or a problem, and by the disregard of all features which are considered to be irrelevant" (Whittlesey 1954: 30). Again, this question of **essentialism** (the belief that things have objective realities with particular sets of characteristics that make them what they are) and **social constructionism** (the belief that things are invented, produced, or constructed by particular people in society) is also one that runs through the rest of this book. A third question concerns the very idea that regions are unique, particular, and internally homogeneous.

While critiques of regional geography from the late 1960s onward suggested that it was a hopeless interest in the particular at the expense of the general, more recent theorists of regions have noted how they are both internally differentiated and, importantly, connected to, and in some sense, created by, various kinds of connection to things that lie outside of the region. We will explore this *relational* approach to regions toward the end of the chapter. At yet another level it is frequently argued that while regions may have once been unique and more or less homogeneous internally, forces within modernity and capitalism have effectively leveled out these differences, not in theory, but in actuality. Regions simply matter less than they did. Finally there is a question of politics. Some have suggested a **critical regionalism** that challenges the universalizing tendencies of modern globalized capitalism by confronting it with the particular. Here, the region is configured as a geography of resistance. Others, however, have argued that regionalism represents a retreat into the particular which is necessarily reactionary and exclusive. Here, regionalism is associated with forms of nationalism, and even fascism. First let us consider some of the traditions in regional thinking that have marked the twentieth century.

Approaching the Region

It is often the case that accounts of the history of geography posit regional geography as a particular and familiar kind of geography that dominated the discipline for the first half of the twentieth century and then, under challenges from more "scientific" forms of geography, was effectively killed off. What I argue here is that there have been many versions of interest in something called the region and that this interest is ongoing. The tradition of regional geography that was dominant in the first half of the twentieth century was largely descriptive. A typical account of a region would include the physical features, the predominant climatic conditions, forms of architecture, types of economic activity, and cultural and religious beliefs and customs. They might include the kinds of cuisine associated with a region, modes of dress, leisure, and ritual practices. The main aim of such accounts was to show how regions were unique, how they were different from one another. "By taking a defined and perhaps relatively small area," Michael Chisholm wrote, "the geographer could attempt to understand the totality of phenomena and their interrelationships" (Chisholm 1975: 33). This differed from an alternative process that started with a particular object or process of interest (say levels of immigration, or amounts of foreign investment in an economy) and then described how these were geographically, or regionally, variable. In the second approach, regions still play a role but only as instances of whatever is being examined. All the other facets of the region (climate, cuisine, rituals, etc.) become, at best, only marginally relevant.

The former kind of analysis formed the center of a variety of geographical enterprises in the early twentieth century. It did so, however, in quite different ways. Let us consider four important traditions in a broadly conceived engagement with the region. These are not meant to be exhaustive, but suggestive of the richness of the possibilities that regional thinking has to offer. While they are linked by the centrality of the region concept, they are also marked by their differences in approaches to the region.

Vidal de la Blache and French regional geography

… that which geography, in exchange for the help it has received from other sciences, can bring to the common treasury, is the capacity not to break apart what nature has assembled, to understand the correspondence and correlation of things, whether in the setting of the whole surface of the earth, or in the regional setting where things are localized. (Martin *et al.* 1993: 193)

Paul Vidal de la Blache (1845–1918) was a prominent French geographer who, along with his colleagues and students, developed a distinctive way of thinking about the geography of late-nineteenth- and early-twentieth-century France. He sought to show how France was divided up into distinctive, unique regions where everything from climate to cuisine marked one region off from the ones next to it. Vidal was the central figure in French geography. He was heavily influenced by the German geographical traditions of Ritter, Ratzel, and von Humboldt. Vidal developed some of the work of Ratzel (see Chapter 3) in the concept of "*milieux de vie*," a term which referred to broadly global values. **Genres de vie** (ways of life) described unified and identifiable patterns of living (the subject of historical and cultural geography). The *genres de vie* included all the ideas, behaviors, and things we might now associate with "culture" as a whole way of life. These include distinctive traditions, cuisines, buildings, agricultural practices, institutions, and less tangible habits of a people. These ways of life were associated with another concept – that of "*pays*" or distinct physical regions where civilization became visible in the *paysage* (landscape).

Vidal is best known for his account of the regional geography of France, Le *Tableau de la géographie de la France* (Vidal de la Blache 1908). This book set the standard for what became known as regional monographs – a particular kind of text that set out to describe the complex layering of both natural and cultural phenomena in a particular place. It was a model that traveled across the world and influenced, in particular, the work of Carl Sauer in Berkeley. Vidal's ideas about the practice of regional geography were further developed in his posthumous book, *Principles of Human Geography*, edited and completed by his student, Emmanuel de Martonne (Vidal de la Blache *et al.* 1926). In this book, Vidal combines his interest in the specificity of regions with his theorization of the relations between people and the land. The unity of people and land that is apparent in the *pays*, he argues, is not one in which the ways of life are determined by the hard facts of the natural environment – climate, soil, topography, etc. – but one in which people make choices about how to best utilize the natural attributes that define a region. In this way, Vidal developed a critique of environmental determinism that came to be known as **possibilism**. People, he argued, do what they do, not because of the raw demands of nature, but because of their way of life, their *genre de vie*. Thus, Vidal placed culture firmly at the center of geography. In this sense, regions, and the differences between them, provided a theoretical tool to question a universal theory (human life is determined by climate) through a focus on the specificity of particular sections of the earth.

The concept of "*genre de vie*" was central to Vidalian geography. As Anne Buttimer has put it, *genre de vie* could "encompass spatial and social identity, a label which could designate those groupings whose economic, social, spiritual, and psychological identity had imprinted itself upon the landscape" (Buttimer 1971: 53). To Buttimer, Vidal's key concept found a way to bypass the dead-ends in which other disciplinary concepts had found themselves:

> In other words, the actual conditions of a society could not be explained solely in terms of cultural evolution, as was suggested in anthropological literature; nor could the evolution of economic organization provide the entire picture. Certainly an overemphasis on the place factor, as the disciples of Ratzel had done, could not explain the empirical forms of society. What was needed, then, was a notion which echoed the integration of place, livelihood, and social organization in a group's daily life. *Genre de vie* seemingly encompassed all these characteristics. (Buttimer 1971: 53)

And *genre de vie* also necessitated a rethink of how culture (or civilization) was related to nature. In Vidal's work, nature is central. By questioning the one-way explanations of environmental determinism, Vidal was not refuting the importance of natural forces in the processes of human geography. Rather he was arguing for the study of unique syntheses of people and the world. *Genre de vie* was both produced by and productive of a natural habitat. Repeatedly Vidal would show how places with very similar natural environments were transformed into very different worlds by the varying ways of life that had developed on them. Thus, as Buttimer points out, "early British Columbia, Tasmania and Chile had similar milieux but entirely different genres de vie" (Buttimer 1971: 55). Vidal, while happy to talk about people through the lens of "race" in ways that do not sit comfortably with the twenty-first-century reader, was keen to show how race was not a product of environment:

> … with regard to social groups, the types which become dominant in the march of progress and continue to develop are those which originally resulted from the collaboration of nature and man, and gradually became more and more emancipated from the direct influence of environment. Man has devised certain modes of living. With the help of materials and substances which nature supplied, he has succeeded, little by little, handing down methods and inventions from generation to generation, in building up a systematic regime for stabilizing his existence, one which moulds the environment to his liking. (Vidal de la Blache *et al.* 1926: 185)

Vidal's work came to dominate French geography and his students went on to produce a rich set of regional studies in what became known as a Vidalian tradition. This continues to the present day and can be seen, for instance, in the account of regional geography by contemporary French geographer Paul Claval, which leans heavily on Vidal's insights (Claval 1998). His influence is also notable in the advent of **humanistic geography** in the mid-1970s (see Chapter 6), where his emphasis on ways of life held particular attractions for Anne Buttimer, who became one of the foremost interpreters of the Vidalian tradition.

Hartshorne and the chorological point of view

In the United States, the major statement on the theoretical underpinnings of geography in the first half of the twentieth century was Richard Hartshorne's *Nature of Geography* (1939). As with Vidal, Hartshorne's major influences were German, particularly the work of the German geographer, Alfred Hettner. Hartshorne sought to define the discipline of geography through an analysis of its history as well as through a rigorous process of positive theory construction. Hartshorne's answer to the question of what geography was (a question that recurs endlessly and somewhat tiresomely throughout its history) was that its substantive and theoretical focus was on "areal differentiation." Geography, in Hartshorne's words, "interprets the realities of areal differentiation of the world as they are found, not only in terms of the differences of thing from place to place, but also in terms of the total combination of phenomena in each place, different from those at every other place" (Hartshorne 1939: 462). It was geography's task, therefore, to provide a synthesis of relevant information that describes and accounts for the differences between one place (or region) and another. At the heart of this exercise in synthesis lay regional geography, or, borrowing from the Greek, **chorology**. While Vidal's geography was based on the detailed exploration of particular "*pays*" in France (mostly, it has to be said, by his students), Hartshorne's plea for chorology rested mainly on a reading of geography through history (particularly German geography) and his own reasoning. Hartshorne's arguments for regional geography are thus more philosophical and less grounded than Vidal's.

Hartshorne used the idea of the region to make a case for geography's exceptionalism. While most disciplines have a formal object of study, he argued, geography's purpose was synthetic. It involved the totality of things found in a particular region or place. This is a view he takes from his reading of Hettner:

> The goal of the chorological point of view is to know the character of regions and places through comprehension of the existence together and interrelations among the different realms of reality and their varied manifestations, and to comprehend the earth surface as a whole in its actual arrangement in continents, larger and smaller regions, and places. (Hettner quoted in Hartshorne 1939: 13)

While history was the study of specific times, geography was the study of particular regions. As well as drawing on Hettner, Hartshorne was developing the philosophical notions of Immanuel Kant here. Kant had argued that the two fundamental categories within which everything else must happen were space and time. Hartshorne went on to suggest that this makes geography and history particularly important, primal endeavors that form the basis for all other sciences. Hartshorne is often interpreted as making a case for geography as an **idiographic** science – a science of the particular – as opposed to a **nomothetic** one. This is slightly unfair. Hartshorne did try to suggest that geography needed to balance the specific and the general (Agnew 1989; Unwin 1992). "A geography which was content with studying only the individual characteristics of its phenomena and their relationships and did not utilize every opportunity to develop generic concepts and universal principles," he wrote, "would be failing in one of the main standards of science" (Hartshorne 1939: 383). The findings of regional geography, he argued, would inform the construction of systematic understandings and vice versa:

> The ultimate purpose of geography, the study of areal differentiation of the world, is most clearly expressed in regional geography; only by constantly maintaining its relation to regional geography can systematic geography hold to the purpose of geography and not disappear into other sciences. On the other hand, regional geography in itself is sterile; without the continuous fertilization of generic concepts and principles from systematic geography, it could not advance to higher degrees of accuracy and certainty in interpretation of its findings. (Hartshorne 1939: 468)

Here, Hartshorne recognizes the value of systematic geography but insists that it is in the practice of regional geography that geography remains distinctive. It is clear here and elsewhere that despite the notional balancing of the specific and the general, it was in the region that geography's identity lay. In the quotation above, it is in the region that Hartshorne locates the "purpose of geography" and without it the discipline would simply disappear. Hartshorne makes his case for geography as the study of the specific again and again:

> That geography is a field of knowledge which is concerned to know and understand individual cases follows directly from its function as the study of places. The concept of place, like that of person or event, is in its essence a concept of the specific. (Hartshorne 1959: 157)

Hartshorne believed that geography should concern itself with whatever occurred in a particular place or region. It was the fact that such phenomena (climate, soil, industry, etc.) occurred together that was of interest to geography. And it was this togetherness that lay at the heart of geographical research:

> Geography does not claim any particular phenomena as distinctly its own, but rather studies all phenomena that are significantly integrated in the areas which it studies, regardless of the fact that those phenomena may be of concern to other students from a different point of view. (Hartshorne 1939: 372)

Geographers, Hartshorne continued, should not confine themselves to individual phenomena (just the climate, just the industry, just the soil) but should consider how they coexist, combine, and integrate to produce something unique, never the same anywhere else.

Hartshorne's development of a chorological approach in human geography formed the most substantial philosophical account of what the pursuit of geography was and should be in the first half of the twentieth century in the United States, and the idea of the region was at the heart of it. His pursuit of the region, though, is clearly different from the deeply descriptive account of French regions in the work of Vidal and his students. There are no such descriptions of regions in Hartshorne's work. His is a philosophical and methodological enterprise rather than an immediately practical one.

Soviet regional geography

At the same time as Hartshorne was writing, regions were playing a quite different role in the Soviet Union. Regional geography was particularly strong in the Soviet Union. Following the 1917 revolution, the discipline of geography was set to work for the purposes of planned economic production. Lenin strongly repudiated any notion that the natural environment could impose limits on the development of human economies. This was seen as neo-Darwinism and bourgeois. He wanted geographers to discover the rational regions upon which the new Soviet economy could be planned. One geographer in particular, Nikolai Baranskiy, was central to Soviet geography. As a personal friend of Lenin, he was in a good position to direct the future of Soviet geography in the decades following the revolution. Baranskiy was a strong believer in the role of regions in economic planning and he was able to influence the State Planning Commission (GOSPLAN) in their planning of the national economy. In 1922, they described the importance of the region to economic planning:

> A region should be a distinctive territory that would be economically as integrated as possible, and, thanks to a combination of natural characteristics, cultural accumulations of the past, and a population trained for productive activity, would represent one of the links in the entire chain of the national economy. … Regionalization will thus help establish a close link between natural resources, working skills of the population, and assets accumulated by previous cultures and new technology, and yield an optimal productive combination by insuring a division of labor among regions and, at the same time, organizing each region as a major economic system, thus evidently insuring optimal results. (GOSPLAN quoted in Martin *et al.* 1993: 243)

GOSPLAN divided the USSR into 21 rational regions which were then studied in depth by young geographers, economists, and others. While there was some argument over how best to study geography in university, it was the arguments of Baranskiy that won. It was only a regional approach, he argued, that could effectively contribute to the development of the Soviet economy. In 1934, the Presidium of the Committee on Higher Technical Education of the Central Executive Committee of the USSR directed the university sector to teach geography regionally, demanding: "Most of the content of economic geography should be taught on a regional basis. In the course on the economic geography of the U.S.S.R. in particular, at least seventy percent of the time should be devoted to economic regions" (quoted in Saushkin 1966: 30–31). Such an approach was later supported and developed by V. A. Anuchin who argued for an approach which focused on the regional interplay of physical geography, settlement patterns, and the economy (see Martin *et al.* 1993). Baranskiy promoted the regional approach to the economy in the Soviet Union as an alternative to the "spontaneous process" of capitalism.

> The inter-regional economic ties, just as all other aspects of socialist economy, are now organized in a planned way, each region fulfilling definite, strictly fixed parts of the country's general plan. The geographical division of labour, which under capitalism, develops as a spontaneous process, has turned in the U.S.S.R. into a method of planned territorial organization of labour based on its regional specialization for the purpose of attaining maximum effectiveness. (Baranskiy 1959: 411)

Soviet economic geography rested on the belief that capitalist economies distorted the real distribution of productive forces through processes of imperialism and market forces. This, it was argued (in common with many Marxist commentators in the west), meant that some regions had to remain peripheral to imperial centers which reaped the rewards of raw materials located hundreds or thousands of miles away. This, in the eyes of Soviet geographers, was how Tsarist Russia had developed:

> The distribution of industry did not correspond to the distribution of raw materials. The economic map did not tally with the map of natural resources. There was no coal strata or oil cupolas or veins of ore at

the base of the industry concentrated in the centre. Under Leningrad there was a swamp; under Moscow there was clay. ... The fuel which the centre burned was not its own. It was obtained in the outlying districts and was brought from afar: the coal from Ukraine, the oil from the Transcaucasus, the timber from the north. (Mikhaylov 1937: 43–44)

In a socialist planned economy, they argued, economic planning should be based on the actual, objective, distribution of **productive forces** (a term used to describe raw materials, arrangements of space, labor power, knowledge, technology, and machinery combined). Factories should be located close to raw materials and transport costs should be kept to a minimum. As one key text put it:

The rate and direction of the economic development of the principal economic regions of the country has tremendous significance. In order to raise the most important branches of the national economy to the level of the economically more advanced capitalist countries, it is necessary to raise the level of all the economic regions of the country. It is necessary to utilize more fully their natural and labor resources. (Balźak and Harris 1949: 140)

The idea of region remained central to Soviet geography long after it had fallen out of favor in the world of Anglo-American geography. One reason for this was a strong belief in the ontological existence of regions that precede our description and definition of them. Regions, to Baranskiy and others, really, objectively, existed. They could be located and demarked by recognizing the correspondence of natural resources and pools of labor power and expertise. Capitalism, they argued, paid no heed to the existence of these regions and thus they inevitably produced horribly distorted and imperialist geographies. Recognizing the reality of regions, they insisted, would produce a geography that was both just and efficient.

Bioregionalism

Soviet geographers were not the only people who have insisted on the objective existence of regions as things in reality, which precede our definitions, descriptions, and analyses of them. Another group of people who see regions as having a pre-discursive existence are bioregionalists. Although they have had very little influence in the discipline of geography (Parsons 1985; McTaggart 1993), their ideas are intensely geographical.

The term **bioregionalism** was coined by Allen Van Newkirk in Canada in the mid-1970s. Bioregionalism, in his eyes, referred to "regional models for new and relatively nonarbitrary scales of human activity in relation to the biological realities of the natural landscape" (Van Newkirk quoted in Aberley 1999: 22). The concept was further developed by Berg and Dassman who wrote that the term bioregion referred to the "geographical terrain and terrain of consciousness – to a place and the ideas that have developed about how to live in that place" (Berg and Dasmann 1977: 399) and who held that the way to delineate a bioregion was to first study "climatology, physiography, animal and plant geography, natural history and other descriptive natural sciences" but that the fine-tuning of a bioregion's boundaries are "best described by the people who have lived within it, though human recognition of the realities of living-in-place" (Berg and Dasmann 1977: 399).

Bioregions, then, are based on a very specific coming together of biologically defined realities (most often watersheds) and the people who live and work within them. Importantly, these regions cannot be defined from outside, by others, but organically, from inside, where there is a unique combination of nature and culture. At the root of the concept, though, is the fact of an objective region rooted in nature that is more real than the artificial political boundaries of counties, states, or nations. Kirkpatrick Sale, in the best-known and most widely read account of bioregionalism, published by the influential American conservationist organization, the Sierra Club, argued that the world could be divided into a series of spatially nested "natural regions" ranging from the scale of the earth itself to relatively small river watersheds (Sale 1985). But bioregionalism is not some dry scientific concept of natural scientists. Connected to this notion of objective regions are a variety of, more or less anarchist, political visions of self-regulating populations who are, like the natural world around them, determinedly local. What's more, the regional cultures that have and might develop in these regions would value the other inhabitants, both fauna and flora, of that region in a way that resembles how some believe Indigenous communities have valued such co-inhabitants. In other words, bioregionalism starts with the actually existing division of the world into regions based on ecology but maps on to that both new forms of human sociality and governance and, finally and importantly, a spirituality that is in touch with all the forms of nature that surround them. At the core of bioregionalism is the idea that regions are not mere constructs of the human imagination but things that, in themselves, are agents in the world:

> Bioregionalism is the world at work on itself, getting something done which the world knows to be in need of doing. It gets the work done through ideas, through words written and spoken, through organization, discipline, practice and politics. But from first to last, it is the world's work, and the world either knows or will figure out how to get it done. (McGinnis 1999: xv)

Advocates of bioregionalism frequently note how this is not a new idea but, importantly, an idea that is rooted in pre-modern or Indigenous relationships with the natural world:

> Bioregionalism is not a new idea but can be traced to the aboriginal, primal and native inhabitants of the landscape. Long before bioregionalism entered the mainstream lexicon, indigenous people practiced many of its tenets. Increasingly, however, population growth and new technologies, arbitrary nations/state boundaries, global economic patterns, cultural dilution and declining resources are constraining the ability of indigenous (and nonindigenous) communities to maintain traditions consistent with their past. (McGinnis 1999: 2)

There is undoubtedly a good deal of nostalgic romanticization of Indigenous culture in bioregional theory. Modern things are "arbitrary" and morally suspect while traditional ways are to be admired as being in touch with the pulsing identity of the region. Bioregionalists consistently advocate a "return" to place, region, and vernacular culture.

One of the principal figures in the development of bioregional thought is the poet and ecological activist Gary Snyder. Snyder questions the kinds of regions that are the products of cartographers, politicians, and landowners:

> [W]e are accustomed to accepting the political boundaries of counties and states, and then national boundaries, as being some sort of regional definition; and although, in some cases, there is some validity to those lines, I think in many cases, and especially in the Far West, the lines are quite often arbitrary and serve only to confuse people's sense of natural associations and relationships. So, for the state of California ... what was most useful originally for us was to look at the maps in the *Handbook of California Indians*, which showed the distribution of the original Indian culture groups and tribes (culture areas), and then to correlate that with other maps, some of which are in Kroeber's *Cultural and Natural Areas of Native North America* ... and just correlate the overlap between ranges of certain types of flora, between certain types of biomes, and climatological areas, and cultural areas, and get a sense of that region, and then look at more or less physical maps and study the drainages, and get a clearer sense of what drainage terms are and correlate those also. All of these are exercises toward breaking our minds out of the molds of political boundaries or any kind of habituated or received notions of regional distinctions. ... People have to learn a sense of region, and what is possible within a region, rather than indefinitely assuming that a kind of promiscuous distribution of goods and long-range transportation is always going to be possible. (Snyder quoted in Aberley 1999: 17)

Possibly the central natural fact in defining bioregions is the existence of watersheds. In his essay "Coming into the Watershed," Gary Snyder relates a drive he took with his son, Gen, across California:

> So we had gone in that one afternoon's drive from the Mediterranean-type Sacramento Valley and its many plant alliances with the Mexican south, over the interior range with its dry pineforest hills, into a uniquely Californian set of redwood forests, and on into the maritime Pacific Northwest: the edges of four major areas. These boundaries are not hard and clear, though. They are porous, permeable, arguable. They are boundaries of climates, plant communities, soil types, styles of life. They change over the millennia, moving a few hundred miles this way or that. A thin line drawn on a map would not do them justice. Yet these are the markers of the natural nations of our planet, and they establish real territories with real differences to which our economies and our clothing must adapt. (Snyder 1995: 220)

It is this sense of "real" regions that permeates Snyder's essay. "California," he notes, with the hard-headed conviction of a good social constructionist, is "a recent invention with hasty straight-line boundaries that were drawn with a ruler on a map and rushed off to an office in D.C." (Snyder 1995: 221–222). And then, abandoning the skeptical view of the social constructionist, he outlines how California consists of six regions each "of respectable size and native beauty, each with its own makeup, its own mix of birdcalls and plant smells." But it does not stop there, as each of these "natural regions" is linked in his eyes to "a slightly different lifestyle to the human beings who live there (and) different sorts of rural economies, for the regional differences translate into things like raisin grapes, wet rice, timber, cattle pasture, and so forth" (Snyder 1995: 222).

> A watershed is a marvelous thing to consider: this process of rain falling, streams flowing, and oceans evaporating causes every molecule of water on earth to make the complete trip once every two million years. The surface is carved into watersheds – a kind of familial branching, a chart of relationship, and a definition of place. The watershed is the first and last nation whose boundaries, though subtly shifting, are unarguable. (Snyder 1995: 229)

The bioregion, based on a watershed, is clearly defined here as a real, as opposed to arbitrary, thing. Based on these real things, bioregionalists build a moral philosophy suggesting that we should live locally and "return to place." They also build a political philosophy of local self-governance. But at the root is a "natural" system on which all is based and in this sense it reinvents elements of environmental determinism (though usually without the racist implications) and the kind of regional negotiation with nature found in the *pays* of Vidal.

Critiquing the Region

The fixation on the idea of the region in early-twentieth-century human geography led to a number of ongoing critiques. These came to a head in the 1950s and 1960s as young geographers in Britain and America began to argue for a more systematic, more scientific geography. To them, regional geography seemed to be quaintly archaic, overly descriptive, and lacking in ambition. The following catalog of regional geography's failures is fairly typical:

> … too much emphasis on the unique and similar; too little generalization and therefore no applicability of new insights to other regions; not enough use of accepted research methodology and techniques from other social sciences; an overabundance of eclecticism and unfounded choices of 'problems', 'themes' and 'interdependencies'; and, especially in Germany and France, too much research aimed at the study of landscapes and not enough at the study of relevant social problems and phenomena. (Hoekveld 1990: 11–12)

In addition, Hoekveld and others argued, the world was simply different now (i.e. the 1980s). While a regional emphasis may have made sense in Vidal's France, modernity and urbanization had intervened, making it very difficult to separate out the particular attributes of regions from the global flows and currents that passed in and out of regions. The Brazilian geographer, Milton Santos made this point arguing that "[i]n the contemporary conditions of the world economy, the *region* is no longer a living reality characterized by internal coherence. It is … principally defined from the outside. Diverse criteria alter its limits. Under these conditions, the region in itself has ceased to exist" (Santos 2021: 18). It was difficult for those critical of the regional tradition to see how it was possible to relate those things that were internal to a region to those which were external. Regional geography in its traditional guise was rooted in a deep symbiotic relationship between humans and nature in a particular area. And yet, as Ray Hudson and others argued, regional uniqueness was often itself a reflection of broader social forces surrounding class, gender, and other social divisions that combined to bring regions into being (Hudson 1990).

To Ron Johnston, regional geography had been evidence of a strong tradition of empiricism in a human geography that had been simply content to list things as they came together in a particular area of the earth's surface. Regional geographies were, at their best, readable, personal accounts of the essence of a region – not much different, perhaps, from contemporary travel writing or travel guides even. However enjoyable such accounts may be, Johnston found much of what passed as regional geography "almost always ultimately disappointing because its goal is insufficient: it can inform and entertain but can it advance knowledge?" (Johnston 1990: 123). To Hudson, the goal of "describing the unique characteristics of regions" was "a perfectly respectable though intellectually limited objective" (Hudson 1990: 67). There was, in other words, an apparent lack of intellectual ambition in regional geography. It might be "enjoyable," "perfectly reasonable," or even "entertaining" but was ultimately "intellectually limited," "disappointing," and "passé."

Traditional regional geography certainly maintained its advocates. In his presidential address to the Association of American Geographers in 1981, John Fraser Hart referred to regional geography as "the highest form of the geographer's art" (Hart 1981). "The highest form of the geographer's art," he wrote, "is producing good regional geography – evocative descriptions that facilitate an understanding and an appreciation of places, areas, and regions" (Hart 1981: 2). The address is an attack on the new kind of "scientific" geography that had been powerful for at least a decade leading up to it (see Chapter 5). But it also questioned the belief that regions are things that exist independently. "This brand of 'scientific geography'," he wrote, "was predicated on two assumptions: first, that regions are real things, 'objective realities' that are simply waiting to be discovered … and second, that the world is covered with a network of identifiable areas, or 'true geographic regions' that are homogeneous with respect to all significant geographic variables." The true purpose of good regional geography, he insisted, was to understand and then describe regions, not to draw lines around them. And good regional geography also had to be well written. "Good geographical writing," he wrote, "is an art." (Hart 1981: 27). Predictably, perhaps, Hart's address was met with dismay by some of the "scientific" geographers that he was criticizing. They argued that Hart's address was asking for a return to the "descriptive morass from which we have recently emerged" (Golledge 1982: 558).

So ideas of the region as it had been approached in the early part of the twentieth century were critiqued from a number of angles. It was not generalizable and thus not scientific, it was descriptive and not explanatory, it was often atheoretical in the sense that no particular theory of the region was in evidence in this descriptive enterprise, it was concerned with the rural past in a world that was urbanized and no longer static, it was, at best, entertaining, but lacking in ambition. Despite all this, however, an interest in regions has continued to mark geography and its sister disciplines, albeit in radically different ways.

New Regional Geographies

Since the decline of what we might refer to as "traditional" regional geography in the 1950s, there have been a number of calls for "new regional geographies" led by political geographers. The first of these emerged in the 1980s and was informed by social theory (particularly **structuration theory**) and time-geography (see Chapter 10). Indeed, many of those who had been chastising regional geography for its lack of ambition were simultaneously arguing for a new regional geography that did not rest on the "mere" description of all that could be cataloged in a given area. Instead, they sought to couple the idea of the region with the insights of theoretical developments from elsewhere.

Central to new regional geographies was the belief that regions are social constructs. Rather than presupposing the existence of regions, political geographers wanted to show how regions emerged both from social processes in the world and from the kinds of analysis that social scientists, themselves, conducted. Regions came into being at the end of these processes, not at the outset. Hudson referred to regions, in a very untraditional way, as "discrete and distinctive socially produced time-space envelopes" (Hudson 1990: 72). Regions, he argued, are indeed unique and particular but this uniqueness and particularity is not innate but socially produced. Furthermore, regions are not bounded and inward looking but linked by social relationships into unequal systems that transcend individual regions (relations of class, gender, etc.). While regions are socially produced, they are, nonetheless, implicated in the production and reproduction of these wider systems of inequality. So, to put it somewhat crudely, the division of England into "the south" and "the north" is at the same time a product of economic forces that necessitate the existence of particular classes and a tool in the continuing operation of such economic processes. The so-called north/south divide in England, seen in this way, is both a product and a producer of economic, social, and cultural differentiation. The same could be said of the north/south divide in Italy or the United States. Regional differentiation, therefore, was seen as part of the wider social processes inherent in a global capitalist economy, not just as a result but as part of the process itself. In this sense the traditional interest in the region was meshed with more recent engagements with social theory in such a way that geography was seen as having something special to offer the wider world of grand theorizing through its interest in the region. It was exactly this coming together of formally disparate worlds that led Roger Lee to write that "A regional geography which is capable of demonstrating the effectivity of place in social process intensifies the significance of geography in social explanation and assists in the development of social theory that is sensitive to the inherent significance of space in social affairs" (Lee 1990: 104). The region, then, is no longer seen as existing, fully formed, in the world waiting to be discovered, mapped, and quantified, rather it is something in process that is active in the construction of all manner of social, cultural, and economic differences. Region here is not the end product of other processes that come first (natural, economic, cultural, etc.) but an active force in the constitution of these processes. This, of course, transforms geography from a subject that simply describes end products to a discipline that is at the heart of explaining how such abstract notions as the "economic," "cultural," "social," and even "natural" come into existence.

Since the 1980s, political geographers have continued to subject the idea of the region to scrutiny. In part this has been prompted by the kind of observations about changes in the world that political geographers might be expected to make. While much of the twentieth century witnessed the gradual erosion of regional difference in the face of wider national, and then global forces, the last few decades of the twentieth century and the beginning of the twenty-first century have seen the apparent resurgence of the region as a political and cultural force. While regional identities and allegiances had long been seen as regressive, reactionary, and nostalgic, the very same identities reappeared as progressive and liberatory. One of the most important of the theses that emerged in relation to this reappearance of the region was the idea of the "hollowing out of the state" (Jessop 2002). Sociologists, geographers, and political scientists noted how the state (in many ways the key geographical basis for authority in the nineteenth and twentieth centuries) was losing its key functions to both supranational bodies (the EU, NAFTA, UN, etc.) and regional political assemblies. This "new regionalism" has not been seen as nostalgic but as a new way of developing economies in the developed world. The development of devolved government in Scotland and Wales is one example of this. Across Europe all manner of regional political movements are seeking to take functions away from the state. The term "Europe of the Regions" is used within EU policy circles to refer to a tier of policy-making bodies that lie between the local municipality and the nation-state.

This observation of new forms of political regionalization in the world has been coupled with a revival of the region *in theory*. So what kind of theory does new regionalism demand? Answers can be found in the work of political geographers. Anssi Paasi, for instance, has argued that:

> …we need *critical views* on regions that recognize and theorize them as social and geohistorical processes. Regions and their borders are contested political, economic and cultural constructs, perhaps even sort of *social* contracts, that are performed in various social practices and discourses, particularly in governance and politics/policies. (Paasi 2022: 12)

These new political geographers clearly do not subscribe to the idea that regions exist as pre-given identities, already in the world with attached distinct identities. Rather they are interested in the coming into being of regions at a number of scales. Nevertheless, once brought into being, regions are seen as real things that have effects and consequences for those living in and outside of them.

Anne Gilbert argued that the fact that regions are socially constructed means that regions are kinds of projects for certain dominant groups in society who mobilize in order to produce the established definition of a region. At the same time, groups seeking to counter forms of domination can resist through counter definitions of a region.

> If a group within this structure is strong enough to impose standardization in a certain area at a certain time, the regional entity emerges, and its differentiation from other areas is sharpened. If the groups within a given regional structure are instead too weak to generate some sort of unity, they are integrated by groups dominant at other scales and the regional differentiation associated with the former disappears. The regional whole comes from the power of certain groups to impose their values and norms upon the majority and the cultural solidarity necessary to the specification of an area. (Gilbert 1987: 2017)

Inspired by social theoretical work in the 1980s that sought to understand the interplay of large social structures and individual agency, Gilbert saw the region as an outcome of such entanglements as well as a tool in the production of dominant structures in society. Gilbert's work was notably used by Clyde Woods in his account of the regionalization process in the Mississippi Delta (Woods 1998a). Woods refers to the groups involved in the production and contestation of regions as "regional blocs." The blocs do not have singular identities but are rather composed of alliances "between disparate ethnic, gender, class, and other elements" (Woods 1998a: 26). Their goal is to assert control over both the material resources of the region but also the cultural, social, and economic institutions that defined it. The idea of a "bloc" is borrowed from the Italian Marxist activist and theorist, Antonio Gramsci who insisted that the projects of blocs were never complete or finished, but always subject to resistance from other blocs. In the case of the Mississippi Delta, this meant the resistance of working-class African Americans.

> Attempts by working-class African Americans to establish social democracy within a plantation-dominated economy provided a material basis for an ethic of survival, subsistence, resistance, and affirmation from the antebellum period to the present. The kin, work, and community networks that arose from these efforts served as the foundations of thousands of conscious mobilizations designed to transform society. (Woods 1998a: 27)

Woods refers to the dominant bloc in the Mississippi Delta region as the "plantation bloc," gesturing towards the long history of plantation economies and slavery in the deep south. The dominance of the plantation bloc, Woods argues, is structured around "an economic monopoly over agriculture, manufacturing, banking, land, and water; a fiscal, administrative, and regulatory monopoly over local and county activities; and an authoritative monopoly over the conditions and regulation of ethnic groups and labor" (Woods 1998b: 79). Woods traces waves of regional bloc activity from origins in a plantation economy based on the labor of enslaved African people, though Reconstruction and Jim Crow right up to the establishment of the Lower Mississippi Delta Development Commission (LMDDC) in 1988. In each case he shows how the region was being constructed economically, socially, and culturally in ways that reproduced the hegemony of the plantation bloc but were continually contested by an alternative regional bloc defined my what Woods calls **Blues epistemology** – a set of relations, lived experiences, and cultural meanings arising out of working-class African American life (see Chapter 14). In this sense, the region of the Mississippi Delta is seen as an ongoing project.

> The development path of the Mississippi Delta plantation bloc must be abandoned. At its best it reproduces permanent social crisis that daily plows under communities, families, and aspirations. At its worst, it permanently holds open the gates to new forms of segregation and slavery. … [A] new development path does not have to be invented. For over a century and a half, working-class African American communities and their allies have continuously experimented with creating sustainable, equitable, and just social, economic, political, and cultural structures. Just as it was inevitable that a new plantation bloc mobilization would occur, it was also inevitable that the long-suppressed alternative would be resurrected. (Woods 1998a: 288)

Woods' account of the Mississippi Delta reveals how regions do not just exist but are the products of effort and struggle by institutions (such as the LMDDC), social movements, and communities.

When we argue that regions are produced through social processes that is not the same as saying that they are made out of nothing or simply imagined into existence. Woods describes a series of waves of regional definition where each draws on what came before. Similar processes can be seen in the southwest of England and the North of Italy. Martin Jones and Gordon MacLeod ask us to pay attention to the way practical politics is often based on the identification of discrete bounded spaces within which regions are given a territorial expression:

> there *may* be certain circumstances in which, as an object of analysis, the region should be taken as a practical and 'prescientific' bounded territorial space that has been institutionalized through particular struggles and become 'identified' as such a discrete territory in the spheres of economics, politics and culture. (Jones and MacLeod 2004: 437)

They consider insurgent regional political movements such as the Cornish independence movement as well as more formal attempts to establish regional governance in England. Regional movements, they argue, utilize popular, regional mythologies (such as the ancient but nonexistent region of "Wessex") to argue for the necessary existence of political regional identities. They show how activists in the southwest of England have argued for the creation of a Wessex region. The very act of naming is here implicated in the possible production of a region, as "Wessex" is more specific sounding (and carries a sense of tradition, however artificial) than the term "South West" (Jones and MacLeod 2004).

A similar, if less politically comfortable process is described by Benito Giordano in his account of the construction of a mythical region of Padania in the north of Italy by the right-wing Italian political party the Lega Nord (Northern League) (Giordano 2000). He reveals how the Lega Nord have propagated the idea that the Italian state does not represent the people of the north of Italy and is in the hands of the people from the south. The proposed new region of Padania is constructed in opposition to the region of the south, which is described as corrupt, lazy, and "Mediterranean" in distinction to the more "European" values and practices of the north. As is often the case, one region (coded as positive) is constructed in relation to another (coded as negative). To bring Padania into existence, the Lega Nord suggests that it have its own flag, anthem, and government. Like "Wessex," the very act of producing a name, Padania, is implicated in the production of a region.

In both of these cases we can see how regions are being constructed (with limited success in both cases) through the definition of boundaries, the designation of particular values and cultural norms to the people who live inside those boundaries, the attempted development of new tiers of government, and the simultaneous encoding of other regions in negative ways (as "not us"). All the while, however, it is clear that "'Padania' is constituted by its (networked and highly globalised) economic geographies – not any bounded and unchanging identity" (Amin 2004: 35).

The Finnish political geographer Anssi Paasi has also been a long-time advocate of the importance of regions to the practice of politics and his work helps us to understand these forms of regionalization. To Paasi there is no pre-given model for regions. There is no driving force (global capitalism, nature, cultural identity, etc.) that brings regions into being. Rather "(a) region is comprehended as a concrete dynamic manifestation of social (natural, cultural, economic, political, etc.) processes that affect and are affected by changes in spatial structures over time" (Paasi 1986: 110). Regions emerge for a number of contingent reasons and gradually become institutionalized. As this process occurs, the region develops distinctive territorial and symbolic form as well as systems of regional governance and a sense of identity or regional consciousness. And the region is never finished but is always changing, always becoming. Regional geographers, then, would study this process of the emergence of regions.

Within this renewed attention to the idea of the region is a tension between *territorial* and *relational* understandings. A territorial understanding focuses our attention on the ways in which regions are produced as bounded entities within which particular things occur or are allowed. This comes close to a traditional idea of regions as clearly differentiated entities in the world that are qualitatively and quantitatively distinctive. A relational understanding insists that regions need to be understood through their relations with other regions, elsewhere. Rather than focusing on boundaries, a relational approach insists on the importance of the flows that constitute a region as particular and distinctive. Paradoxically, it is argued, a region is produced through and by things that lie beyond it.

Consider the region referred to as the "south east" of England. This is a region with London at its heart. It is the region closest to mainland Europe, and is commonly thought of as the richest part of the country. In some senses it has a commonsense existence. Although it does not have a formal governmental structure (unlike Wales or Scotland), it does have an existence in the minds of most people in the United Kingdom. In the late 1990s, a group of geographers asked the simple question of where the south east was.

In answering it, they revealed how the south east is internally extraordinarily diverse. While it does include the extreme wealth of the city of London and the booming property markets of the capital and home counties, it also includes some of the poorest and most deprived areas in London's East End (resting right up against those areas of extreme wealth). While it includes a world city, it also includes rural areas. Medium-sized towns like Oxford and Cambridge provide jobs in science and high technology, while cars are still being made in Oxford. This supposed region, in other words, was not internally homogeneous and the wider processes of neoliberal capitalism were producing new inequalities all the time:

> Some areas were by-passed, structurally excluded by the operation of this particular form of uneven development. Others – our hot spots – saw growth piled upon growth. Yet even within them another geography of inequality was secreted, a geography distinguishing between parts of towns and villages,

a micro-geography of daily lives within buildings. These geographies were the spatial expression of the inequalities produced at the heart of the neo-liberal, "free market," project itself. (Allen *et al.* 1998: 89)

In addition to being internally diverse, the region, as the end of the quotation above suggests, was being produced by flows that passed in and out of any notion of a tightly bounded region. It was, in large part, produced and defined by its outside. The south east is very much on the move. People, both migrants and tourists, enter and leave in enormous numbers. This is one of the most important areas in the world for the global flows of capital. The recent collapse of large parts of the global banking and finance industries has seen large numbers of people lose their jobs overnight, leading to declines in the value of property in an area that has seen massive rises in property values over the past decade. And this is not primarily because of anything that has happened in the south east of England but due to decisions made thousands of miles away in the United States and elsewhere. And because the finance sector in the south east is one of the largest in the world, any downturn in its fortunes will, in turn, affect places and regions all over the world. The south east, in other words, is thoroughly connected.

This observation of the south east's internal differentiation and external connectedness led these geographers to develop what they called a "relational" approach to the region. This approach, they argue, "has the effect of thinking the social and the spatial together from the outset" and further, it recognizes that "space and spatiality is socially constructed" and that this process of construction is "constantly evolving":

> It is the recognition of the continuousness of this process of construction that has led us to think in terms of space-*time*. Any settlement of social relations into a spatial form is likely to be temporary. Some settlements will be longer lasting than others. The general lines of some elements of the dominance of London have broadly endured for centuries; the creation of Milton Keynes provoked a relatively sudden and sustained reorganization of spatialities over at least a half of England; the negotiation of the work:home boundary between the high-tech scientist and his partner may evolve from week to week. In principle, therefore, it is important at least to be aware of the essential temporality of all spatial figurations. It is in this sense that the-south-east-in-the-1980s was a distinct space-time. (Allen *et al.* 1998: 138)

Such an approach is not unique to the south east of England but is one which, they argue, should provide a model for the study of regions in general:

> ... an adequate understanding of the region and its futures can only come through a conception of places as open, discontinuous, relational and internally diverse. In short, regions are a construction of space-time: a product of a particular combination and articulation of social relationships stretched over space. To see them as anything less, is to settle for an inadequate understanding of contemporary regional geographies. (Allen *et al.* 1998: 143)

Some geographers have continued to argue for a relational approach to regions over the past decade. Ash Amin, for instance, has been puzzled by what he sees as the continued territorial, bounded sense of region both in academic writing and in the politics of the new regionalism where it is frequently argued that "region-building and regional protection is the answer for local economic prosperity, democracy and cultural expression" (Amin 2004: 35). Instead of attempting to assert control over a given territory, Amin argues, local actors need to assert control over networks and nodes that increasingly define life in the twenty-first century. It is not the space of the region that needs to be controlled, but the spaces of flow. Attempts to wrest power away from decision makers at a wider scale through the production of regional territories, he argues, are illusory as it is impossible to "govern a 'manageable' geographical space. There is no definable regional territory to rule over" (Amin 2004: 36). Instead, Amin suggests a politics of connectivity that recognizes that we may have more in common with someone the other side of the world than we have with our immediate neighbors. Territoriality defined economies, identities, or politics simply no longer make sense to Amin.

Critical Regionalism

The region has even found itself at the heart of postmodern theory. The architectural critic, Kenneth Frampton, wrote an influential paper about postmodern architecture and urban form. At the heart of his paper was the notion of "critical regionalism." The problem with the postmodern cityscape, he argued, was that it was being dominated by "universal" values and technologies such as land speculation, the dominance of the automobile, and the prioritization of the sense of vision over all other senses. Architecture, he argued, tended to be a mask for marketing and social control based on the need for a "façade" to cover up "the harsh realities of this universal system" (Frampton 1985: 17). Frampton argued that architecture needs to adopt a critical regionalism. He compares the universalist impulse to produce a space devoid of regional qualities with an architecture that, while using modern and non-region-specific techniques, is nonetheless able to utilize those things that make a region different, unique:

> Here again, one touches in concrete terms this fundamental opposition between universal civilization and autochthonous culture. The bulldozing of an irregular topography into a flat site is clearly a technocratic gesture which aspires to a condition of absolute **placelessness**, whereas the terracing of the same site to receive the stepped form of a building is an engagement in the act of "cultivating" the site. (Frampton 1985: 26)

Critical regionalism is regionalist because it draws on the natural resources of a region (light, topography, color, etc.) but it is critical because it draws away from an unquestioning belief in regional uniqueness which can end up in reactionary nostalgia marked by the endless repetition of vernacular architectural forms. Critical regionalism does not simply deny the universal but uses the universal in the construction of a regional architecture. "The fundamental strategy of Critical Regionalism," Frampton writes, "is to mediate the impact of universal civilization with elements derived *indirectly* from the peculiarities of a particular place." Key to this mediation is the inspiration provided by the region in which architecture is being practiced. This inspiration may come from "such things as the range and quality of the local light, or in a *tectonic* derived from a peculiar structural mode, or in the topography of a given site" (Frampton 1985: 21). Consider the following section of Frampton's essay in which he equates the power of the visual and the development of perspective as part of a western universalizing tendency and suggests a heightened sense of varied human perception as a way of getting in touch with the regional aesthetic:

> … Critical Regionalism seeks to complement our normative visual experience by readdressing the tactile range of human perceptions. In so doing, it endeavors to balance the priority accorded to the image and counter the Western tendency to interpret the environment in exclusively perspectival terms. According to its etymology, perspective means rationalized sight or clear seeing, and as such it presupposes a conscious suppression of the senses of smell, hearing and taste, and a consequent distancing from a more direct experience of the environment. This self-imposed limitation relates to that which Heidegger has called a "loss of nearness." In attempting to counter this loss, the tactile opposes itself to the scenographic and the drawing of veils over the surface of reality. Its capacity to arouse the impulse to touch returns the architect to the poetics of construction and to the erection of works in which the tectonic value of each component depends upon the density of objecthood. The tactile and the tectonic jointly have the capacity to transcend the mere appearance of the technical in much the same way as the place-form has the potential to withstand the relentless onslaught of global modernization. (Frampton 1985: 29)

Frampton is well aware of the dangers of the region, its potential to become a little twee at best, and reactionary at worst. He insists that critical regionalism, unlike simple regionalism, is not advocacy for a seemingly ahistorical vernacular style. Critical regionalism embraces modernity and the technology that comes with it in constructing an architecture that self-consciously works with what the region has to provide.

Conclusion

What we see in Frampton's essay is the tension between the universal and the particular that lies at the heart of the history of geography and, specifically, regional geography. How do you advocate the region without simultaneously withdrawing into romantic nostalgia? Is it possible to challenge the production of global space by producing the regional and how do you do this without simply withdrawing from the world into a clearly bounded and exclusive space? This tension between global space (often considered to be abstract and placeless) and the lively significance of the regional is a tension worked through over and over again. The question of region and how to think about this is clearly not simply a preoccupation of "regional geographers" in the early part of the twentieth century. It is at the same time much older than that (going back to Plato's and Aristotle's discussion of *chora* and *topos*) and still ongoing in a series of lively reflections in geography and beyond. In one such reflection, David Matless takes us back to the early days of regional geography (Matless 2015). In *The Regional Book,* Matless provides a series of descriptions of points in the region of the Norfolk Broads in the east of England. Each short piece of prose poetically accounts for elements of the region without explanation or exegesis. It is, like the early days of regional monographs, mostly descriptive, but it adds a layer of creativity to this exercise. It does not attempt to present a comprehensive view of a region but provides forty-four fragments of snapshots of recurring elements of the region's distinctive landscape in a way which evokes a sense of place through a focus on details. It suggests the re-emergence of "the highest form of the geographer's art" in a spare and affecting way. Elsewhere, Matless gives an account of his purpose in writing *The Regional Book.* In part, he wanted to get away from the idea that regional description is a dry and conservative practice of the past that does not stand up to more recent energetic forms of theorization:

> The aesthetic characterisation of older work as simple and limited might also be questioned. There was more to old regional geography than plain description. Even if 'plain description' is an outcome desired and achieved, that denotes an aesthetic accomplishment worthy of appreciative understanding, if not necessarily emulation. And the old hat might even be worth trying on. (Matless 2017:12)

He wants to practice regional description in order to enact "respectful looking."

> Geographical description concentrates attention, gathers experience, observes and describes. The accounts in *The Regional Book* are mono-vocal, deliberately singular, responding to an assumed request to describe a landscape without reflex resort [sic] to the commentary of others. … Descriptions cut against, while also recognizing, and occasionally inhabiting, forms of attention, and styles of seeing, claiming conventional authority. (Matless 2017: 22)

Matless's work points to a re-invigorated practice of regional writing that revels in the possibilities of description and the forms of attention in entails. Just as versions of the local and the region can be used for political purposes, both conservative and progressive, so creative accounts of regions play their part in reenchanting worlds that have been overlooked as geographers look to bigger processes in a world of globalization.

References

Aberley, D. (1999) Interpreting bioregionalism: a story from many voices, in *Bioregionalism* (ed. M. McGinnis), Routledge, London, pp. 13–42.

Agnew, J. A. (1989) Sameness and difference: Hartshorne's *the nature of geography* and geography as areal variation, in *Reflections on Hartshorne's The Nature of Geography* (eds J. N. Entrikin and S. Brunn), Association of American Geographers, Washington, DC, pp. 121–139.

* Allen, J., Massey, D. B., and Cochrane, A. (1998) *Rethinking the Region: Spaces of Neo-Liberalism*, Routledge, London.

* Amin, A. (2004) Regions unbound: towards a new politics of place. *Geografiska Annaler. Series B, Human Geography*, 86, 33–44.

Balźak, S. S. and Harris, C. D. (1949) *Economic Geography of the USSR*, Macmillan, New York.

Baranskiy, N. N. (1959) *The Economic Geography of the USSR*, Foreign Languages Publishing House, Moscow.

Berg, P. and Dasmann, R. (1977) Reinhabiting California. *Ecology*, 7, 399–401.

* Buttimer, A. (1971) *Society and Milieu in the French Geographic Tradition*, Published for the Association of American Geographers by Rand McNally, Chicago.

Chisholm, M. (1975) *Human Geography: Evolution or Revolution*, Penguin, Harmondsworth.

Claval, P. (1998) *An Introduction to Regional Geography*, Blackwell, Oxford.

Frampton, K. (1985) Towards a critical regionalism: six points for an architecture of resistance, in *Postmodern Culture* (ed. H. Foster), Pluto, London, pp. 16–30.

Gilbert, A. (1987) The new regional geography in English and French speaking countries. *Progress in Human Geography*, 12(2), 209–228.

Giordano, B. (2000) Italian regionalism or "Padanian" nationalism—the political project of the Lega Nord in Italian politics. *Political Geography*, 19, 445–471.

Goffman, E. (1956) *The Presentation of Self in Everyday Life*, University of Edinburgh Social Sciences Research Centre, Edinburgh.

Golledge, R. (1982) Commentary on "The highest form of the geographer's art". *Annals for the Association of American Geographers*, 72, 557–558.

* Hart, J. F. (1981) The highest form of the geographer's art. *Annals for the Association of American Geographers*, 72, 1–29.

* Hartshorne, R. (1939) *The Nature of Geography: A Critical Survey of Current Thought in the Light of the Past*, The Association of American Geographers, Lancaster, PA.

Hoekveld, G. (1990) Regional geography must adjust to new realities, in *Regional Geography: Current Developments and Future Prospects* (eds R. J. Johnston, J. Hauer, and G. A. Hoekveld), Routledge, London, pp. 11–31.

Hudson, R. (1990) Re-thinking regions: some preliminary considerations on regions and social change, in *Regional Geography: Current Developments and Future Prospects* (eds R. J. Johnston, J. Hauer, and G. A. Hoekveld), Routledge, London, pp. 67–84.

James, P. (1929) Towards a further understanding of the regional concept. *Annals of the Association of American Geographers*, 19, 67–109.

Jessop, B. (2002) *The Future of the Capitalist State*, Polity, Oxford.

Johnston, R. J. (1990) The challenge for regional geography: some proposals for research frontiers, in *Regional Geography: Current Developments and Future Prospects* (eds R. J. Johnston, J. Hauer, and G. A. Hoekveld), Routledge, London, pp. 122–139.

* Jones, M. and Macleod, G. (2004) Regional spaces, spaces of regionalism: territory, insurgent politics and the English question. *Transactions of the Institute of British Geographers*, 29, 433–452.

Lee, R. (1990) Regional geography: between scientific theory, ideology, and practice (or, what use is regional geography), in *Regional Geography: Current Developments and Future Prospects* (eds R. J. Johnston, J. Hauer, and G. A. Hoekveld), Routledge, London, pp. 103–121.

Martin, G. J., James, P. E., James, E. W., and James, P. E. (1993) *All Possible Worlds: A History of Geographical Ideas*, John Wiley & Sons, Inc., New York.

Matless, D. (2015) *The Regional Book*, Uniformbooks, Axminster.

Matless, D. (2017) Writing regional cultural landscape: cultural geography on the Norfolk Broads, in *Reanimating Regions: Culture, Politics, and Performance* (eds J. Riding and M. Jones), Routledge, London, pp. 9–25.

McGinnis, M. V. (1999) *Bioregionalism*, Routledge, London.

McTaggart, W. R. (1993) Bioregionalism and regional geography: place, people and networks. *Canadian Geographer*, 37, 307–319.

Mikhaylov, N. (1937) *Soviet Geography*, Methuen, London.

Paasi, A. (1986) The institutionalization of regions: a theoretical framework for understanding the emergence of regions and the constitution of regional identity. *Fennia*, 164, 104–146.

* Paasi, A. (2022) Examining the persistence of bounded spaces: remarks on regions, territories, and the practices of bordering, *Geografiska Annaler: Series B, Human Geography*, 104(1), 9–26.

Parsons, J. (1985) On "Bioregionalism" and "Watershed Consciousness". *The Professional Geographer*, 37, 1–6.

Sale, K. (1985) *Dwellers in the Land: The Bioregional Vision*, Sierra Club Books, San Francisco.

Santos, M. (2021) *For a New Geography*, University of Minnesota Press, Minneapolis.

Saushkin, Y. G. (1966) A history of Soviet economic geography. *Soviet Geography*, 7, 3–104.

Snyder, G. (1995) *A Place in Space: Ethics, Aesthetics, and Watersheds: New and Selected Prose*, Counterpoint, Washington, DC.

Unwin, P. T. H. (1992) *The Place of Geography*, Longman Scientific & Technical, Harlow.

Vidal de la Blache, P. (1908) *Tableux De La Géographie De La France*, Librairie Hachette, Paris.

Vidal de la Blache, P., Martonne, E. D., and Bingham, M. T. (1926) *Principles of Human Geography*, Holt, New York.

Whittlesey, D. (1954) The regional concept and the regional method, in *American Geography: Inventory and Prospect* (eds P. James and C. F. Jones), Syracuse University Press, Syracuse, NY, pp. 19–68.

Woods, C. (1998a) *Development Arrested: The Blues and Plantation Power in the Mississippi Delta*, Verso, London.

* Woods, C. (1998b) Regional blocs, regional planning, and the blues epistemology in the Lower Mississippi Delta, in *Making the Invisible Visible: A Multicultural Planning History* (ed. L. Sandercock), University of California Press, Berkeley, pp. 78–99.

Chapter 5

Spatial Science and the Quantitative Revolution

In 1948, the geography department at Harvard was closed. A year earlier, Edward Ackerman had been granted an associate professorship in geography. This had upset a geologist, Marland Billings, who had complained that Ackerman's appointment had taken much-needed resources away from geology (it was actually a department of geology and geography). Billings protested to the Provost, Paul Buck, claiming that geography was not really a proper discipline anyway. It was not intellectually rigorous, he claimed – not really science. Harvard's geography department was unusual in that it was made up of predominantly human geographers. It included the advocate of a regional approach, Derwent Whittlesey, and some scholars who were starting to develop more quantitative aspects of the discipline, Edward Ullman, and Ackerman himself. A committee was formed to explore the nature of geography and its place at Harvard. As a result, geography was closed and Ackerman and Ullman were fired. Geography was labeled intellectually inadequate (Smith 1987: 14). In addition to this account of a department closing due to intellectual arguments about the adequacy or otherwise of "soft" approaches to geographical knowledge, there was the added factor that Whittlesey, and some colleagues, were gay and that their ways of conducting regional geography was associated with their sexuality in a deeply homophobic academic environment. Proper men, it seemed, should do proper science (Mountz and Williams 2023). As is always the case, seemingly pure academic arguments are bound up with persistent systematic asymmetries of power within the discipline which inevitably intersect with the development of geographic thought.

The closure of geography at Harvard formed part of the context for developments in theory over the following two decades. By the middle of the twentieth century, geography had a reasonably comfortable existence in universities across the world. It was a discipline with a clear heritage whose purpose (in human geography anyway), for the most part, was to describe regions. Sometimes it was also concerned with explaining why one region differed from the ones around it. There were, of course, other traditions in the discipline. Physical geography did not always follow a regional model.

Geographic Thought: A Critical Introduction, Second Edition. Tim Cresswell.
© 2024 John Wiley & Sons Ltd. Published 2024 by John Wiley & Sons Ltd.

Other geographers, especially in North America, were more focused on the relationship between humans and nature. But regional geography was dominant.

This was behind the declaration of geography as lacking in scientific rigor. Imagine a biologist who spent her career describing a wolf. Just one wolf. She would be able to tell you many things about that wolf, from what it ate to how far it could see, from the color of its coat to how many cubs it might have. But she would have little interest in other wolves and still less in foxes or dogs and none at all in any other living thing. She would spend her career telling people about this one wolf and why it was different from every other living thing. It was a special wolf. Or imagine a chemist who knew all about one element on the periodic table – say strontium – but nothing about any other element.

In science, such a fixation on the singular, the unique, the particular, is absurd. These fixations might make interesting accounts in the humanities but they would not constitute science. Why not? Because science involves being able to make general statements. When we say that water freezes at zero degrees Celsius, we mean that all water, everywhere, and throughout history, has frozen and will freeze at zero degrees Celsius. This science is general and certain. It explains all such events in the past and predicts all such events in the future. If, suddenly, some water, somewhere, did not freeze at zero degrees, we would have to come up with a reasonable explanation (perhaps it is contaminated with salt) or change our mind.

In 1948, human geography could not lay claim to this kind of general statement. For the most part, geographers (even many physical ones who were happy to describe landforms) practiced regional geography. They accumulated facts about the Upper Midwest of the United States or the Balkans (for instance) and presented them as a portrait of a particular place. The last great attempt to produce general "laws" in geography had been made by environmental determinists such as Huntington, Semple, and Taylor (see Chapter 3) and these had been discredited. Geographers did not want to repeat that mistake. What was to happen over the next 20 years (with Ackerman and Ullman very much at the center of things) was a determined attempt to prove the Provost of Harvard wrong. Geography could be a science and could make law-like statements about the geographical world.

There had always been voices questioning the focus on the particular. The Scottish geographer P. R. Crowe in 1938 for instance had written that:

> The idea that Man's social reactions are measurable finds every inch of the way contested. This is perhaps the reason why modern geography shows a tendency to retreat upon things, to become a morphological analysis of all sorts of objects that nobody had previously thought worthy of study. (Crowe 1938: 2)

Crowe was reflecting on regional geography and its apparent obsession with describing things as though they were fixed "things." Crowe insisted on the dynamism of human life and asked geographers to pay attention to the "circulation complex in the broadest possible manner," suggesting that geography then might become a concern with "the sum total of human space-relationships with a vast area" (Crowe 1938: 13). So we can see two themes in Crowe's call for "progressive geography" 50 years after Mackinder had established geography in Britain as a university discipline. These themes are the necessity of measuring things and a refocus away from fixed "things" and toward more dynamic "processes." Critiques such as Crowe's were to be repeated throughout the next few decades. Kimble, for instance, was to make a similar point about processes and circulation when he wrote:

> From the air it is the links in the landscapes, the rivers, roads, railways, canals, pipe-lines, electric cables, rather than the breaks that impress the aviator … regional geographers may be trying to put boundaries that do not exist around areas that do not matter. (Kimble [1951] quoted in Johnston and Sidaway 2004: 62)

The importance of measurement and what we might broadly called "quantification" has deep roots in geography's history. We have seen how the early history of geographical theory was balanced between the intimate description of individual places or regions on the one hand, and the measurement of the earth and the relations between these places on the other. Ptolemy and Varenius are two examples of geographers who sought, through measurement and other means, to escape the confines of the particular and embrace a "general geography." The rise of spatial science and the quantitative revolution, then, was a new version of an age-old argument for a "general geography."

The main targets for the new spatial scientists were the region and regional geography. The region was limiting and passé, they argued. It was getting in the way of real science and giving geography a bad name in the halls of prestigious universities like Harvard. Besides, the real world of the mid twentieth century could not be easily divided up into regions anyway. The world was urbanized, while most regional geography looked backwards to largely rural and agricultural lifestyles. So many people, things, and ideas moved in and out of any declared region that it made it very difficult to even make claims to uniqueness. National and international economies, the nation-state, global cities, communication systems, and financial flows all militated against easy regional differentiation and could not be described, understood, or explained by regional analysis. Describing the unique characteristics of regions, Hudson suggested, was "a perfectly reasonable though intellectually limited objective" (Hudson 1990: 67).

The dramatic alternative to regional geography that developed in the 1950s and 1960s has become known as "quantitative geography," "theoretical geography," or "spatial science." At its core is a move away from a fixation on the particular and specific characteristics of regions and toward the making of claims that could be generally applied. Accompanying this was a move away from broadly descriptive and often qualitative methodologies and toward the use of mathematics and models.

To spatial scientists, theory was necessarily mathematical, a point made by William Bunge: "The creation of theory is difficult because the scientist must successfully identify the purely logical symbols of mathematics with a set of observable facts" (Bunge 1962: 2). Mathematics, Bunge tells us, is a pure language that is logical and free from contradiction. A theory that is expressed in mathematical terms is thus both aesthetically pleasing and suitably clear. "Thus, science, striving for clarity, ultimately is forced to use mathematical forms" (Bunge 1962: 2). David Harvey makes this point too, but adds a caution about the difficulty of linking the clarity of mathematics with the messiness and confusion of the human world:

> The geographer's fear of explicit theory has not been entirely irrational. The practical problems of extending the scientific method into the social and historical sciences are considerable. Similar problems arise in geography. The complicated multivariate analysis which geographers were trying to analyse (without the advantage of experimental method) is difficult to handle. Theory ultimately requires the use of mathematical languages, for only by using such languages can the complexities of interaction be handled consistently. Data-analysis requires the high-speed computer and adequate statistical methods, and hypothesis testing also requires such methods. (Harvey 1969: 76)

Here we have a kind of theory of theory. For the first time a significant number of geographers were talking about theory and about the possibility of a "theoretical geography." What they meant by theory, however, was very different from many of the definitions I have outlined in the Introduction to this book and probably unrecognizable to the majority of human geographers practicing today. To understand why Harvey and Bunge (among many others) believed that "theory ultimately requires the use of mathematical languages," we need to understand the philosophy of **positivism**.

Positivism

The philosophical approach known as positivism, and its stricter version, logical positivism, is often claimed as the philosophical bedrock for the quantitative revolution and spatial science. In fact, very few geographers seem to have read any work by positivist philosophers or even claimed to have been informed by the philosophy of positivism. In some senses positivism was given as a philosophical justification for quantification after the fact. Despite this, human geography is often described in terms of positivist geography and post-positivist geography. So what is positivism?

Positivism in general is a belief that only things which can be experienced through the senses can count as true knowledge. Its most important proponent was the sociologist and philosopher Auguste Comte (1798–1857). Comte believed that the highest and purest form of the search for truth was "positive" science. This, he argued, would replace earlier searches for truth in religion and philosophical metaphysics. The kinds of beliefs and knowledge produced in these ways, Comte claimed, could not be proven or disproved, unlike the knowledge produced by positive science.

Positivism would be based on five principles (Gregory 1978). First, scientific knowledge should be based on the observable reality (*le réel*). When two observable phenomena could be linked through regular and repeated association, it would then be possible to explain and predict their occurrence by relating one to the other. One *causes* the other. No unobservable, supernatural, metaphysical, or even structural forces could be said to cause anything. To Comte, the observable world and kinds of causation that linked elements of the observable world were the highest order of reality. If anything existed beyond this world, then they were beyond the reach of scientific knowledge and there was little point in asking questions about them at all. Such questions, he suggested, could not be answered. The second principle was that scientific knowledge has to be supported by some sense of shared reality (*la certitude*). If one scientist was to make a claim about what exists and about causation, it needed to be accessible to other scientists who could then confirm these observations. Third came the necessity for scientific knowledge to be based on the construction of testable theories (*le précis*). Value judgments, which were incapable of verification or falsification, were to be separated from observable facts and testable theories. All scientific knowledge should be in some sense useful and potentially applied (*L'utile*). The final principle of Comte's was that scientific knowledge was in progress, unfinished, and in a process of becoming whose endpoint was a utopian unification of theories that might produce a perfect society (*le relative*). In this sense, scientific progress was social progress and as scientific knowledge became more perfect, all the reasons for warfare and human misery (reasons which were based on unscientific knowledge) would disappear. Given this utopian endpoint it should come as no surprise that Comte attempted (ironically) to form a "positive religion."

Logical positivism is a philosophy born in Vienna in the years leading up to World War I. Logical positivists believe in the primacy of the empirical world – the world of observable facts, as well as our ability to be rational about what we can sense. They allowed for some kinds of explanation that were not based on observable facts by distinguishing between *analytic* and *synthetic* statements. Analytic statements could be validated without pointing to observable reality. Three plus three equals six is an analytical statement that is simply true by definition. Indeed, mathematics consists almost entirely of analytical statements. Synthetic statements are those that require observable verification. A key belief of the logical positivists was the notion of verifiability. Logical positivists believed that a statement is only "cognitively meaningful" if there are procedures available to decide if it is true or false. Statements that are religious or metaphysical in character cannot be verified and are thus not cognitively meaningful. While it is hard to conclusively verify the truth of a statement, it is more straightforward to prove that it is false (as a universal law is proved false by just one case that does not fit). Thus the philosopher Karl Popper replaced the doctrine of verifiability with a notion of falsifiability. Science, he suggested, should proceed through attempts to falsify knowledge. Once proved false, new kinds of knowledge would have to emerge to take the place of those which had been falsified. This is made clear by Wilson in a commentary on "Theoretical Geography": "Theories are never proved to be generally true. The ones in which we believe represent the best approximation to truth at any one time. To achieve this status, they must be tested and, to date, they must not have been contradicted. We expect, then, that theories will be subject to constant development and refinement" (Wilson 1972: 32).

While, as with any philosophical approach, there are arguments over what positivism is exactly, there are some core assertions that can be outlined. Positivism is a set of rules about what counts as proper scientific knowledge. The best kind of science, it suggests, is one that can be formulated mathematically and can be proven to be logically coherent. Elements of any truly scientific claim should be testable and potentially falsifiable. The statement "God exists," for instance, is neither provable nor falsifiable and therefore does not count as a scientific statement. Positivists also tend to believe that knowledge is a cumulative enterprise through which we progressively learn more about the world and therefore begin to understand it better. Science, they would argue, is free from subjective or cultural beliefs and is therefore transcultural and transhistorical. It is not explained by its context but by its logical coherence and testability. Importantly, positivism insists on the difference between facts and values and it is only facts (which are observable, provable, and measurable) that are important in the construction of knowledge. Because positivism insists on the centrality of things that we can sense or observe, it is clearly *empiricist*. **Empiricism** is suspicious of abstract ideas and theories that are discounted as "unscientific." Positivism, then, is a philosophy of science. It has its roots in a suspicion of both religious and metaphysical belief and insists on the centrality of observable "positive" data that can usually be measured and made into either law-like statements or mathematical models or both. Facts exist in the world and they can be sensed, described, measured, and modeled. Both laws and models can be tested empirically and if they are not proven wrong then they can be used to predict future events.

If we take positivism in its purest form, then it is the case that very little quantitative geography is truly positivist. As Derek Gregory has outlined, geographers have rarely directly taken from positivist philosophy but nonetheless managed to loosely follow some of its basic precepts (Gregory 1978). The notion of law (that is by necessity universal in its application) does not translate well from hard physical science to even physical geography, let alone human geography. This point was made by David Harvey when he suggested that "if we employ very rigid criteria for distinguishing scientific laws, then we can scarcely expect geographical statements to achieve such a status" (Harvey 1969: 107). Instead, Harvey suggests, geographers have been happy to embrace looser or softer notions of law and a willingness to act *as though* geographical phenomena were governed by universal laws even when exceptions could always be found.

One of the Vienna Circle who instigated logical positivism was Gustav Bergmann, a philosopher and psychologist who moved to the United States to avoid Nazi persecution and took up a professorship at the University of Iowa. It was here that he met the geographer Frederick Schaefer. Bergmann completed a paper Schaefer was to become known for posthumously – "Exceptionalism in geography" (Schaefer 1953). This paper took on the claims for geography as the study of areal differentiation, made by Hartshorne in particular, and suggested a new **nomothetic** geography in its place. It was this paper that was eagerly embraced by both Bunge in his book *Theoretical Geography* (1962) and Haggett in *Locational Analysis in Human Geography* (1965), two of the key texts in the "quantitative revolution."

The General and the Specific

Occasionally there are published debates in geography that crystallize differences in theoretical approach and become key readings for all students in future years. Perhaps the most well known of these is the debate between Richard Hartshorne (as the leading exponent of a regional method) and

Fred Schaefer (a vigorous critic of the regional method and proponent of a "nomothetic" approach). Schaefer's paper was published posthumously in 1953. In it he claimed that Hartshorne's arguments for regional geography were *exceptionalist*. By this he meant that Hartshorne was arguing that geography (and history) were not like other sciences because they involved the integration or synthesis of all manner of data in a particular unit of space (or time if you are an historian). In this sense, Hartshorne was building on Kant's insight that space and time are the pre-given foundations for everything else and thus have a special (or exceptional) role in the development of knowledge. Schaefer argued both that geography was not exceptional in this sense (as other sciences did deal with uniqueness in a variety of ways) and, more importantly, that geography should be more concerned with the generalizable. Geography, he argued, needed to be a proper science, and a proper science was one that was able to generate laws in the same manner as the physical sciences. Human geography, he suggested, had been **idiographic** (concerned with the specific and unique) in nature while it should and could be **nomothetic** (concerned with the general and universal). "Geography," he insisted "has to be conceived as the science concerned with the formulation of laws governing the spatial distribution of certain features on the surface of the earth" (Schaefer 1953: 227). This plea for the development of spatial "laws" reflects a particular kind of ambition. The interests of regional geographers could seem parochial. Even the best regional geography was only ever likely to say anything about the region in question. Meanwhile economists, following Adam Smith, were busy making general laws of economic behavior and natural scientists were asking questions about the very conditions of life itself. Describing the soils and folkways of the Upper Midwest just seemed small-minded. Schaefer clearly wanted to be a scientist. Scientists get respect.

Science has always been prestigious. The "harder" the science, the more prestigious it is. Physics is often seen as the pinnacle of this hierarchy with its constant search for laws which might explain large portions of the known and unknown universe. Natural sciences such as biology, chemistry, and geology share some of this glory. But the closer we get to human life the lower down the hierarchy we fall. Social science has always had this problem. Anthropology, sociology, and human geography are all seen as, in the words of David Mercer "'soft', 'emotional' and 'undisciplined' subjects" (Mercer 1984: 157). Academic science expressed ideas more or less mathematically and with a degree of generalizability. Human geography up until the 1950s was on the very "soft" end of a hierarchy of sciences and, to many (including the Provost of Harvard), was not a science at all.

This sense of seeking the respect that would come from being a bona fide science is clearly expressed by Richard Morrill:

> A strong (pre)conditioning element was an as yet vague discontent with the low status of geography – in our universities, in science, in the eyes of our peers in other fields; with the lack of funding; with our dependence on unglamorous teaching careers; and in the provision of descriptive service courses for others. We were deeply disturbed by the fundamental question of whether geography deserved to be in the university at all, but we became determined to show that it did. (Morrill 1984: 59)

So a need for spatial science emerged from both theoretical and practical limitations of thinking of the world in regional terms (when the world was so dynamic and so many processes cut through and across regions) and from a perceived need for intellectual ambition. Regional geography appeared to be lacking in both scope and ambition.

Central to the origins of spatial science was a rejection of the idea that geographical enquiry was essentially idiographic – concerned with the particular. This, you will recall, was the argument made forcefully by Hartshorne in *The Nature of Geography* in which he wrote that "no universals need be evolved, other than the general law of geography that all its areas are unique" (Hartshorne 1939: 468). This view is well summarized by David Harvey in *Explanation in Geography* (Harvey 1969). Hartshorne was leaning on Kant when he suggested that "Geography and history fill up the entire circumference of our perceptions: geography that of space, history that of time" (Hartshorne 1939: 135). Harvey describes the inferences that can be drawn from Hartshorne's statement as follows (paraphrased):

i.   As everything occurs in space, then there is no limit to what geographers can study.
ii. If there is no limit to what geographers can study, then geography is not defined by its subject-matter.
iii. If geographers are concerned with potentially everything, as it occurs in a spatial location or region, then we are primarily concerned with the uniqueness of these occurrences and locations.
iv. If locations are unique, then the description and interpretation of these locations and what occurs within them cannot be accomplished through reference to general laws. Geography requires an idiographic method.

Harvey argues that this form of argument was popular in the 1930s because it provided a philosophical case for the rejection of environmental determinism – the last, and ill-fated, significant attempt to generate law-like models in human geography. Simply put, it seemed better to make small-scale claims about small-scale places than to fall into the trap of making spurious generalizations that could be easily ridiculed.

Many of the key texts in spatial science take aim at Hartshorne's notion that the only law of geography was that "its areas are unique." The transport geographer Edward Ullman (one of those fired from Harvard), for instance, wrote in 1953 that:

> Up to now I have not referred to geography as the study of the significance of areal differentiation, the current catholic definition of geography after Hartshorne, Hettner, James and others. This concept is implied in the spatial or 'science of distribution' point of view. I cannot accept areal differentiation as a short definition for outsiders because it implies that we are not seeking principles or generalizations of similarities, the goal of all science. (Ullman 1953: 60)

This argument is repeated by Peter Haggett in *Locational Analysis in Human Geography* when he argues that while the work of Vidal, the Berkeley geographer Carl Sauer, and others on particular areas or regions is an important part of the discipline's heritage, "their very success led geographers to overlook the equal need for comparative studies. Areal differentiation dominated geography at the expense of areal integration" (Haggett 1965: 3). In response, Haggett, following the work of Bunge (1962), makes the following claim:

> Let us consider the trivial case of two pieces of white chalk and imagine them lying on the desk in front of us. If we pick them up and examine them closely, we shall see that they are not identical in every detail. Then to describe them both as 'white chalk' is surely an error. To be accurate each piece should be given a special and unique identifying term; but in practice we assign the two objects to the same class of 'white chalk'. To do other than this is to abandon all our descriptive terms and to be reduced to saying, in Bunge's phrase, '. . . things are thus'. (Haggett 1965: 3)

One of the ways in which geographers sought to make general claims about the world was through models. One of the key texts of the quantitative revolution in geography was Chorley and Haggett's *Models in Geography* in which they note how it is the humanities that are concerned with the specific while the sciences are concerned with the general. Geography, on the whole, they suggested, should be a science:

> In short, every individual is, by definition, different, but the most significant statement which can be made about modern scholarship in general is that it has been found to be intellectually more profitable,

satisfying and productive to view the phenomena of the real world in terms of their 'set characteristics', rather than to concentrate upon their individual deviations from one another. (Chorley and Haggett 1967: 21)

Models, they suggest, are "a simplified restructuring of reality which presents supposedly significant features or relationships in a generalizable form" (Chorley and Haggett 1967: 22).

Perhaps the most insistent opponent of regional geography was William Bunge in his book *Theoretical Geography* (Bunge 1962). Bunge correlates the need to be scientific (and thus concerned with generalization) with the need to be theoretical. "Since science places great emphasis on theory," he wrote in the foreword to his book, "it follows that a scientific geography would in turn be greatly concerned with theory. … this book is theoretical because there are many books on geographic facts and none on theory and it is for the purpose of redressing this lack of scientific balance that the book is so heavily theoretical" (Bunge 1962: x). Bunge is passionate in his distaste for the old geography of lists of facts. He wanted geography to be a science based on space and the kinds of order that get produced in space:

Geography today has not lost its old interests but it has greatly deepened them. Geography is still interested in the earth's surface as the home of man, still concerned with the *where* of life, but nowadays a simple memory of locational facts, a gazetteer, will no longer do. The student must have a deeper understanding of *why* objects are located where they are. The earth is not randomly arranged. Locations of cities, rivers, mountains, political units are not scattered helter-skelter in whimsical disarray. There exists a great deal of spatial order, of sense, on our maps and globes. (Bunge 1962: xv)

This idea of geography as a list of facts about particular places – as a "gazetteer" – clearly troubled Bunge. He was after something far more profound and powerful – something which could be used to explain and predict. "Are geographic phenomena unique or general?" he asked. "If they are unique, they are not predictable and theory cannot be constructed" (Bunge 1962: 7). So here we have a logic that links generalizability to theory. In these terms, theory is a scientific and positivist enterprise. Anything concerned with the unique is, by necessity, atheoretical (or even anti-theoretical).

This concern with making geographical knowledge generalizable also saw a rise in concern with the *spatial*. If traditional geography had been about regions, or, sometimes, places, then the new geography was going to be about space. Regions and places prompted thoughts about their uniqueness and particularity, thoughts that seem to have their proper home in the humanities. Writing about space (a key term for philosophers and physicists alike) appeared to allow a greater profundity and, importantly, generalizability. Bunge aligned space and the spatial to the nomothetic, while place remained idiographic:

The *space* versus *place* dispute is a direct consequence of their positions on general versus unique. Hartshorne is pessimistic as to our ability to produce geographic laws, especially regarding human behavior. Schaefer has done us a great service in sweeping away our excuses and thereby freeing us from self-defeat. (Bunge 1962: 12)

In the end, Bunge was unequivocal: "Only by the complete rejection of uniqueness can geography resolve its contradictions" (Bunge 1962: 13). He was dismissive of the arguments for uniqueness. Yes, regions or places could be said to be unique, but so could anything, and this does not stop us from habitually treating them as if they were members of classes which shared a common definition:

The colors red and orange are not unique. They share the common differentiating characteristic of being colors, and with a broadening of either of their class intervals, they can be made to include each other. Similarly, England and Scotland are both classes in a category called location. They are both regions on the earth's surface and on a smaller scale the two regions can be placed in the common regional class of the British Isles. (Bunge 1962: 18)

At the heart of the turn toward spatial science, then, was a critique of uniqueness and a quest for the general. Geography was to become a proper science and that involved finding spatial laws. From the late 1950s onward, geography underwent what has become known as the "quantitative revolution" as geographers donned white lab coats and grappled with complicated equations in an effort to be taken seriously. This did not happen evenly and was associated with particular places such as the University of Washington where Edward Ullman went after being fired from Harvard. It was largely Anglo-American in its geography but its impact was felt everywhere that geography was taught. It still loomed large on the syllabus when I undertook my undergraduate degree in the mid-1980s. Spatial science was informed by, and produced, a large number of theories and hypotheses. In the rest of this chapter, I focus on the kinds of models developed on the interface of economics and geography that informed and inspired the new spatial scientists – particularly central place theory – and the attempts to understand and theorize movement that lay at the heart of spatial science.

Central Place Theory

The quantitative geographers of the 1950s and 1960s found some inspiration in the general models of earlier economists and geographers. An early example was the model created by amateur economist (and farmer) J. H. Von Thünen (1783–1850) to explain and predict agricultural land use. The model is appealing to geographers as it locates the workings of an (admittedly rudimentary) economy in space. As with all models, it begins with a set of assumptions. In this case, he assumed an "isolated state" that has no contact with anywhere else and is surrounded by wilderness. The land is assumed to be flat and featureless (an isotropic plain) and transport opportunities are dispersed evenly over the surface (there are no roads or canals or rivers). Farmers are presumed to act in a way that will maximize profits. If all these things were true, Von Thünen argued, then a pattern of rings would emerge around a city, with each ring having a particular land use (see Figure 5.1). Close up to the city would be land uses that were intensive and involved the production of things that needed to be close to the point of consumption – such as milk. At the opposite extreme would be ranching, as animals needed more land and could be walked to the city before slaughter. His model was a neat

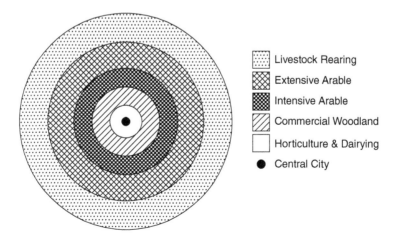

Figure 5.1   Johan Heinrich von Thünen's isolated state.

way of illustrating the importance of both land value and transportation costs and their impact on the geography of the areas surrounding cities.

Another model was the industrial location model produced by the geographer and sociologist Alfred Weber which attempted to explain and predict the location of industry based on the minimization of costs. The most direct inspiration for many of the central players in the quantitative revolution, however, was the central place theory of Walter Christaller and August Lösch. Bunge, for instance, was quite clear about this: "If it were not for the existence of central place theory, it would not be possible to be so emphatic about the existence of a theoretical geography independent of any set of mother sciences" (Bunge 1962: 133). He dedicated *Theoretical Geography* to Christaller.

Central place theory was developed by Christaller in the period leading up to and during World War II. Christaller based the development of central place theory on observations of settlement patterns in southern Germany. The theory was published in 1933 (Christaller 1966 [1933]). Christaller was unhappy with the explanations often given for the location of towns. These tended to emphasize the uniqueness of the particular location of a place and the natural affordances it offered. In other words, explanations were often unique to the settlement in question. Such explanations he, suggested, often began:

> … as a rule with topographical and geographical conditions and then explains simply that here a town "had to originate," and, if the location is favorable, that here especially a town "had to develop favorably." But there are numerable locations, where no town is found, that are equally, or even more favorable. In fact, towns may be found in very unfavorable spots, and those towns, circumstances permitting, may even be fairly large. Neither the number, nor the distribution, nor the sizes of towns can be explained by their location in respect to the geographical conditions of nature. (Christaller 1966 [1933]: 2)

Christaller was looking for just such an explanation – a general one that would allow geographers to predict and explain the sizes, numbers, and distributions of towns. "Up to now," Christaller wrote in 1933, "no one has obtained clear, generally valid laws" (Christaller 1966 [1933]: 2). The search for laws was central to Christaller's quest. Economists, he argued, had produced laws of the economy. "If there are now laws of economic theory, then there must be also laws of the geography of settlements," he insisted, "economic laws of a special character, which we shall call *special economic-geographical laws*" (Christaller 1966 [1933]: 3). These laws would be different from the acknowledged laws of the natural sciences but "no less valid." He suggested that perhaps the term "tendencies" might be used instead of laws. Christaller carefully notes the relationship between theory and evidence in his work. He explains that he will begin with theory that is derived **deductively**; "it is, therefore, unnecessary to begin with a description of reality. Hence, the theory has a validity completely independent of what reality looks like, but only by virtue of its logic and 'the sense of adequacy'" (Christaller 1966 [1933]: 4). As with much of what was to become spatial science, Christaller was insistent that while his theory concerned geography and economics, it was based on principles that were equally valid in the physical sciences:

> The crysallization of mass around a nucleus is, in inorganic as well as organic nature, an elementary form of order of things which belong together – a centralistic order. This order is not only a human model of thinking, existing in the human world of imagination and developed because people demand order; it in fact exists out of the inherent pattern of matter. (Christaller 1966 [1933]: 14)

In addition, and also in common with much supposedly hard, objective, and disinterested science, Christaller saw a certain beauty in the order and simplicity in the central facets of his model. The hexagonal patterns he came up with do have a stark, modern beauty to them. It was the notion of "centrality," however, that caught Christaller's attention:

The stronger and more purely the location, form and size express the centralistic character of such community buildings, the greater is our aesthetic pleasure, because we acknowledge that the congruence of purpose and sense with the outer form is logically correct and therefore can be recognized as clear. (Christaller 1966 [1933]: 14)

Logic, here, was not just rational and correct, but beautiful and productive of "aesthetic pleasure." Hardly the stuff of hard-nosed objectivity.

The model that Christaller came up with describes and attempts to predict where settlements are located, their size and number. It assumes a uniform distribution of a basic population of small settlements on a flat and uninterrupted surface (an isotropic plane). Accompanying this are a number of other assumptions:

- An evenly distributed population (people do not live in clustered centers, this is where services are provided).
- Evenly distributed resources.
- All consumers have equal purchasing powers and they will always buy things from the nearest supplier.
- Transportation costs are equal in all directions and proportional to distance.
- No excess profits (perfect competition).

Given these presumptions, it then predicts the location of different size settlements (say cities, towns, and villages) based on the least average distance of movement for consumers of particular goods. The pattern that emerges from such assumptions is a hexagonal market area. A hierarchy of settlements is produced, with a set number of villages for each town and a set number of towns for each city. If a village has 7 consumers the town has 49 and city has 353. Particular types of market activities need a minimum number of consumers to exist, so cities have more activities than towns or villages and larger settlements will include the activities available in smaller ones. These settlements, which provide services to an evenly distributed population, are known as "central places."

Key to the theory are two basic concepts. The concept of "threshold" asserts that a certain minimum population is necessary for the provision of particular goods or services. Piano tuning, as a service, for instance, may need a large number of people with pianos within a given distance for the service to be viable. A piano tuner is unlikely to be located in a large rural area where only one person plays piano. In the same area, however, there are likely to be a number of post-offices or grocery stores as everybody needs to post letters and eat food. The second concept is the "range" of goods or services. This refers to the average maximum distance people will travel to purchase goods and services. They may travel quite a long way for a piano tuner but are more likely to stay close to home for bread and milk. Christaller's theory was developed, with more complicated mathematics about demand, by August Lösch in 1954 (Lösch 1954).

Central place theory, for Christaller, was significantly more than descriptive. Rather than simply being a model to predict the location of settlements, Christaller used it to make the world conform to the model. Areas where settlements did not conform to the expectations of the theory were seen to be in some sense "imperfect" or "irrational." During World War II, Christaller worked for Himmler's SS as a planner in the Reich Working Group for Territorial Investigation. Here he was asked to develop his theories for the planning of the new territories of eastern Europe (see Figure 5.2). He was involved in producing (rather than explaining or predicting) an ideal spatial order – a normative pattern for settlement distribution. The normal positivist and scientific separation of facts and values becomes strained. The pattern was no longer how things *are* but how they *should be*. The new parts of eastern Europe under Nazi control (predominantly Poland) became a blank space upon which hexagons could be imposed. The population could be removed (4.5 million people were

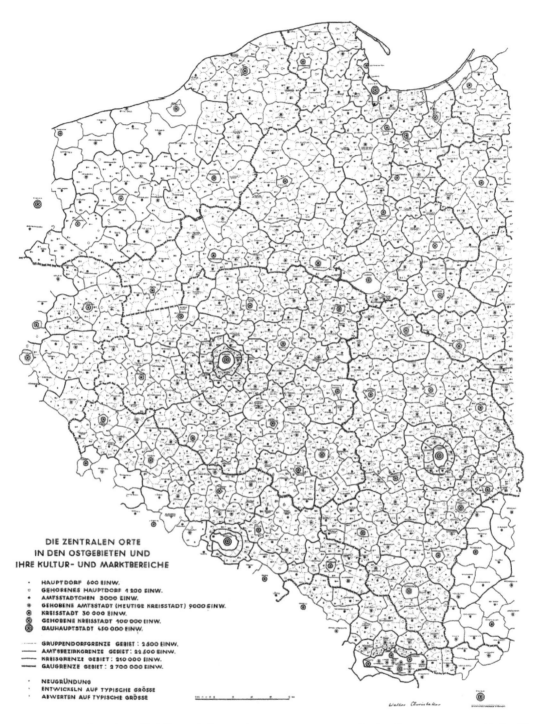

Figure 5.2 Walter Christaller's "central place in the eastern regions map for the SS reorganization of Poland."
Source: From Christaller (1941).

scheduled for removal), some to labor camps, some to concentration camps, and some to other parts of the Reich. No planning laws were in the way. Villages and towns could be erased or built out of nothing. The model could be realized through force (Ward 1992). In addition to the planning of new model towns and cities across the east, Christaller's model for the rational location of settlements was also used to plan the location of the death camps and train lines that would be used for the execution of the Holocaust in which over six million Jews, gypsies, gay people, and others were murdered (Rossler 1989).

We can see, in central place theory, aspects of what was to become full-fledged spatial science some 20 years later. It attempts to provide a general model in place of explanations that depended on particular contexts for the location of settlements. It uses rudimentary mathematics to make its case (made less rudimentary in Lösch's version). It rests on a number of assumptions about the world that reduce the effects of particularity (the isotropic plain, equal access to transport, etc.). It is all-encompassing in its ambition, linking the human world to observable phenomena in the natural world, and it displays a certain kind of aesthetic beauty in its simplicity.

Another realm in which spatial science was to have an impact was in the theorization of movement and transport.

Spatial Science and Movement

Movement was never more central to geography's concerns than during the heyday of spatial science in the late 1960s and early 1970s. Crowe's earlier concern that the measurability of "man's social reactions" would always be contested would have been eased. Central to the concerns of spatial science was not "things," as Crowe might put it, but processes and movements. Bunge made the case that it was movement and pattern that lay at the heart of a new theoretical geography:

> Notice that the examination of a location involves the notion of movement. Even such static features as mountains and sea coasts are explained by movements over long geologic periods. In many ways patterns and movements are interrelated as are the chicken and the egg with one causing the other. Does the location of the river valley cause the movement of the river or is it the other way? Obviously one operated on the other. Thus, theoretical geography, the geography of explanation, is interested in both movements and patterns. (Bunge 1962: xvi)

While ideas about circulation (for instance) were present in the regional geography of Vidal, movement was never a central issue for regional geographers. If you are busy dividing the world up and placing boundaries around regions, movement seems a little awkward. It tends to make regionalization messy. In spatial science, however, movement played a central role. At the heart of these approaches was transport geography. Although there are many realms in which spatial science was active, a focus on transport geography illustrates some of the key theoretical and methodological approaches inherent in spatial scientific research.

To Bunge, in his manifesto for "Theoretical Geography," movement was the key geographical fact to be explored, modeled, theorized, and explained. It was in the study of movement (as well as central place theory), he argued, that theoretical geography had made its most significant advances. Chapter 5 of his book is called "Toward a general theory of movement." It is rooted in the question posed by Edward Ullman: "What makes objects move over the earth's surface?"

> It can be argued that Ullman's question encompasses all geographic theory, since in explaining how an object acquires its location it is difficult to avoid the notion of movement. Even such 'static' features as mountains and seacoasts are explained in terms of movements taking place over long periods. (Bunge 1962: 112)

One of the advantages of thinking about movement, to Bunge, was that it linked human and physical geography together through apparently universal theories and laws. The movement of people could be equated to the movement of electricity or the movement of fluids. To a positivist and a spatial scientist, it is important that generalizations are as general as possible. So laws that could be applied to people and to water were particularly valued kinds of laws.

Edward Ullman, in his paper, "The role of transportation and the basis for interaction," had argued that things would move from one place to another due to the three principles of *complementarity*, *intervening opportunity*, and *transferability*. These three factors built on the earlier observation that things moved because of "areal differentiation" – the principle at the heart of Hartshorne's regional approach:

> It has been asserted that circulation or interaction is a result of areal differentiation. To a degree this is true, but mere differentiation does not produce interchange. Numerous different areas in the world have no connection with each other.
>
> In order for two areas to interact, there must be a demand in one and a supply in the other. Thus an automobile industry in one area would use the tires produced in another but not the buggy whips produced in still another. Specific complementarity is required before interchange takes place. (Ullman 1956: 867)

The principle of complementarity was, Bunge pointed out, one which could be applied to spatial movement theories in general (not just ones concerned with economics and transport such as that proposed by Ullman). There was, in general, a tendency to move from places of abundance to places of scarcity. This is what water tries to do when a river breaks its banks. This is what electricity does.

In addition to this, Ullman continued, there has to be no other source of supply (of tires in this instance) that is closer to the place of demand. This is what he meant by "intervening opportunity." This notion has roots in the larger principle of **least effort** described by Zipf (1949). Between any two points there is only one route that involves the least effort (or is fastest, or cheapest). This is what water tries to do as it finds its way downhill to the sea. Similarly, spatial scientists would argue, this is what companies do when they transport goods, or what people do when they make their way from home to work. Finally it had to be possible, given the costs and time involved in getting from one place to the other. This is what he meant by transferability.

These kinds of generalizations, and the laws they allow scientists to develop, were at the heart of spatial science and, in particular, their discussions of movement. Consider the following:

> For the most part the demand for transport is *derived*. With exceptions such as motorists who simply drive into the country, passengers on cruise liners and "railfans," transport is used as a means to an end: the movement of people and goods from where they are to another place where, for the time being at any rate, their satisfaction of value will be enhanced. Transport creates *utilities of place*. (White and Senior 1983: 1)

The same logic is repeated at the end of the book:

> Transport exists for the purpose of bridging spatial gaps, though these gaps can be expressed not only in terms of distance but also of time and cost. It is the means by which people and goods can be moved from the place where they are at the moment to another place where they will be at a greater advantage; goods can be sold at a higher price, people can get a better job, or live in the sort of house they prefer, or go for a holiday at the seaside. In short, people and goods are transported from one place where their utility is lower to another where it is higher. Transport as a fundamental human activity may thus be effectively studied in spatial terms: geographical methods are basic to such study and are of practical relevance to the solution of many of the problems associated with the transport industry and with its activities. (White and Senior 1983: 207)

Here and elsewhere, transport geographers explain both transport systems (roads, rails, terminals, etc.) and the actual act of moving, with reference to the measurable difference between locations (expressed as "place utility") and to more general spatial patterns. Movement, then, is simply the outcome of rational choices involving the comparison of one location with another. Only the occasional motorist, cruise ship passenger, or trainspotter is allowed to conduct mobility for its own sake. These, needless to say, are beyond the reach of the scientific approach.

Transport geography is but one example of the wider interest in movement and process within spatial science. Indeed, all kinds of movements have been described in much the same way as transport is discussed above. One leading textbook from 1972 accounts for movements in the following way:

> Point-to-point moves are basic because all moves can be reduced ultimately to this scale. Consider point-to-point moves in the light of the modified principle of *least net effort* and returning maximum net benefits (that is benefits minus costs are at a maximum). In an ideal, undistracted, unencumbered movement, a person wastes no time going directly from A to B. (Abler *et al.* 1971: 240)

The idea of "least net effort" is a key notion in the spatial scientific approach to moving things and people. Moving is "effort" and effort is something which we constantly seek to reduce by producing better spatial arrangements of things – arrangements that require less movement (effort). Note how the "undistracted, unencumbered movement" from A to B is described as ideal. Any deviation from this is labeled as dysfunctional – as evidence for either irrationality or a system of movement that does not work efficiently.

As with the earlier exceptions made for trainspotters and the like, Abler *et al.* note the always awkward exceptions to general rules of movement:

> We stress *net* effort to emphasize that the very movement process itself may carry benefits at the same time that costs are incurred. A commuter sitting in a traffic jam inhaling gasoline and carbon monoxide fumes pays a high cost for his trip, but he also has relative peace and quiet twice a day, a radio to listen to, and the feeling that for a while at least he is the boss. If his job were to move next door to his house he would probably move. (Abler *et al.* 1971: 253)

This is but one example of how movement is often a secondary geographical fact made necessary by the arrangement of primary considerations of space and location. So despite Crowe's call for geographers to focus on "men and things moving" rather than the nodes or networks that provide the material infrastructure for such movement, geographers who worked obsessively on movement continued to relegate movement to something logically secondary to the arrangements of space and place. This is why the apparently throw-away references to forms of mobility that do not fit the various models put forward by spatial scientists catch the attention. The "motorists who simply drive into the country, passengers on cruise liners and 'railfans'" (White and Senior 1983) and the commuter enjoying "relative peace and quiet twice a day, a radio to listen to, and the feeling that for a while at least he is the boss" (Abler *et al.* 1971) are, in fact, fascinating. Turning these experiences into a footnote is a result of thinking of movement as a cost. The search for generalization is made clear in the following way:

> 'Why did Jones go downtown to work?' Rephrased and generalized, this question becomes: 'Why do people make a journey to work?' By answering the second question satisfactorily, we discover a principle applying not only to Jones, but also to anyone else who travels daily to a job. Particularistic knowledge of a unique event is practically useless by itself as a contribution to wider understanding. (Abler *et al.* 1971: 196)

A similar point is made in another spatial science account of the geography of movement:

> ... we have tried to provide an exposition of the underlying regularities and processes that may lead to an understanding of the nature of human movement. For this reason, we have excluded modes and regions. Movement in both Asia and North America can be adequately treated in the same framework. Our examples are taken mostly from the American context, but alternatives can be rather easily substituted. In any event, this point may be irrelevant since examples are meant to support generalizations; hence whether a trip for medical attention, for example, is to a witch doctor or to a medical complex is totally immaterial. (Lowe and Moryadas 1975: npn)

And it is not only human movement that is generalizable. Movement in general is often portrayed as something which unites social and physical science, human and physical geography. In *Spatial Organization: The Geographer's View of the World*, Abler *et al.* describe a number of ideal movements. These ideal movements include movements from an area to a line or an area to another area, for instance. They are descriptions of patterns. Movement from an area to a line describes water flowing off of a roof and into a gutter and commuters from a suburb entering a highway. Not only are differences between humans erased in this, but differences between human movement and movement in the natural world:

> When water runs off a roof and into a gutter during a rainstorm the dimensions of the moves are the same as those when animals come out of a forest to drink at the river; or when commuters leave their garages for the street; or when soil is wasted through sheet erosion into ditches, gullies, and stream beds. In all cases something moves from an area to a line with least net effort expended. (Abler *et al.* 1971: 251)

Movement was at the center of the quantitative revolution in geography. Even location models (such as central place theory, industrial location models, or the isolated state) had the notion of equal access to mobility, and least net effort, at the center of them. But the models and laws developed to deal with movement, **spatial interaction theory** and **gravity models**, for instance, never really met the criteria of strict science. For all their apparent complexity (and some volumes of spatial science are full of equations well beyond the mathematical capabilities of this author), the conclusions of such theorization and modeling were startlingly simple. The main conclusion was that volume of movement over distance declines as some function of distance. Things further away are less likely to interact than things close together. Big places have a bigger influence over a wider range than small places. All neatly summed up by Zipf's inverse-distance law – migration between two centers is proportional to the populations of the centers and inversely proportional to the distance between them (Zipf 1949). A related concept is Waldo Tobler's first law of geography (presumably somewhat tongue in cheek) that "Everything is related to everything else, but near things are more related than distant things" (Tobler 1970). But despite the apparent simplicity of these ideas, such "laws" prove to be less than helpful. As Harvey has indicated, they do not allow us to predict anything very successfully and, on the ground, are specific and regionally variable: "Although the formulation provides a good fit in a large number of circumstances, the parameters fluctuate wildly. Thus the rate of change in interaction over distance varies over place, according to time, and according to social characteristics in the population" (Harvey 1969: 110–111). In the end, geographers need to do empirical research on the ground, in a particular place, in order to find out how this rule operates.

Quantification and Physical Geography

The call for quantification transcended subdisciplinary boundaries. Both Schaefer and Bunge made it clear that the best kinds of geographical law were the ones that encapsulated the whole field of geography:

What is new is the sudden great acceleration of spatial cross-fertilization at the theoretical level with opportunities appearing on every hand. For instance, are not animals located according to central place principles with the animal food chain providing the hierarchy? Why cannot portions of Ullman's concepts dealing with exotic and dioric streams be applied to highways? Examples can be found in the most common everyday experience. Is not the problem of raking the lawn similar to the problem of urban arrangement? Leaf piles can be identified with minor cities, compost piles with major cities, areas of leaves with market areas, etc. (Bunge 1962: 27)

Of course, regional geography had always maintained connections between human and physical geography. A total account of a region would typically start with descriptions of soil types, topography, and climate. But Bunge and others had something else in mind. They were seeking to replace the focus on "things" such as regions with a focus on spatial relations. Schaefer insisted on the importance of relations between "things" as the focus of a new geography. This "spatial logic," Bunge insisted, would allow statements to be made which could include such things as the arrangement of cities and the raking of lawns:

Of course, caution must be used so that not too literal a comparison is made between different branches of geography, but by skillful selection of the underlying spatial logic of different branches, the power of geographic concepts begins to unfold. Geography lends great efficiency to spatial studies precisely because of the repetitive spatial situations found in the observable world. (Bunge 1962: 32–33)

It had not just been human geographers who were trying to make the discipline more scientific and more quantitative. Physical geographers had too. One of the earliest formulations came from the geomorphologist Arthur Strahler, in a call for a more scientific accounting of landforms. "For more than half a century," he wrote, "the study of landforms in North America was dominated by an explanatory-descriptive method of study used by W. M. Davis and his students" (Strahler 1952: 924). This needed to be challenged, he went on, by making geomorphology properly scientific and at the same time more concerned with *process* (a call that reflects Crowe's concern both with quantification and with process and movement in human geography). As with human geographers, Strahler looked to other disciplines for examples and, as with human geographers, it was mathematics that seemed particularly appealing:

If geomorphology is to achieve full stature as a branch of geology operating upon the frontier of research into fundamental principles and laws of earth science, it must turn to the physical and engineering sciences and mathematics for vitality which it now lacks. (Strahler 1952: 924)

Mathematics, he argued, would allow geomorphologists to construct general models based on simple principles. This would be coupled with the traditional physical geography pursuit of fieldwork in order to gather data that might make the mathematical model more accurate:

… the geomorphologist may formulate, through a type of invention or intuition based upon the sum total of his experience, a relatively simple mathematical model which is a quantitative statement of some point of important general theory otherwise definable only in words, qualitatively. The establishment of such mathematical models may be regarded as the highest form of scientific achievement because the models are precise statements of fundamental truths. The two methods – empirical and rational – would tend to converge as time goes on and the fund of information grows. The statistical analyst cannot hope to derive quantitative relationships of general application from small samples because of their inherent variability, but, as his sample data increase and the influences of variables are isolated, his empirical equations tend to approach the status of general laws. New knowledge of the observed influences of variables in turn results in keener deduction on the part of the analyst who is formulating his general mathematical laws by intuitive, deductive mental processes. (Strahler 1952: 936)

Strahler sums up his plea for a more scientific geomorphology in the following way:

> In summary, the proposed program for future development of geomorphology on a dynamic-quantitative basis requires the following steps: (1) study of geomorphic processes and landforms as various kinds of responses to gravitational and molecular shear stresses acting upon materials behaving characteristically as elastic or plastic solids, or viscous fluids; (2) quantitative determinations of landform characteristics and causative factors; (3) formulation of empirical equations by methods of mathematical statistics; (4) building of the concept of open dynamic systems and steady states for all phases of geomorphic processes; and finally (5) the deduction of general mathematical models to serve as quantitative natural laws. The program is vast and qualified investigators few, but we are already a half-century behind if development is to be measured against chemistry, physics, and the biological sciences. The need for rapid dynamic-quantitative advances is, therefore, all the more pressing. (Strahler 1952: 937)

This is as clear a statement as any of the perceived need for a more scientific and more quantitative geography in the 1950s. The formulation Strahler suggests includes the importance of numbers – of quantification (rather than the description or elaborate hand-drawn diagrams favored by Davis), the use of statistical equations (Strahler suggests simple forms of regression analysis), concept building, and finally the development of "mathematical models to serve as quantitative natural laws." To finish things off, Strahler draws the reader's attention to the hard natural sciences and their considerable achievements in this kind of research and analysis. There is little difference (except in the details) between this argument and those being made by human geographers such as Bunge and Harvey.

On the whole, and with some notable exceptions, physical geographers have been reluctant to embrace the philosophical aspects of geography. One often-repeated quotation is worth repeating again. "Whenever anyone mentions theory to a geomorphologist he instinctively reaches for his soil auger" (Chorley 1978: 1). There are, of course, a multitude of theories in physical geography. Theories about climate systems or slope formation, for instance. But this is not a place for the discussion of such specific theories. Here we are more interested in general theoretical perspectives and it is these that are unlikely to find a warm embrace in the arms of physical geographers. Nevertheless, as with all intellectual inquiry, physical geography is based on implicit theoretical standpoints. The broad principles of positivism fit well with physical geography. Almost all physical geographers would subscribe to the philosophical view that there is an external reality that is observable. Most would be comfortable with the process of generalizing from multiple observations and moving toward the formulation of law-like statements. Many would also be happy to describe their methods as quantitative. Few, however, would subscribe to the strictest definitions of positivist philosophy.

The broad philosophical approach that finds most favor in the work of those geographers who have written about philosophy and physical geography is **critical rationalism** (Haines-Young and Petch 1986; Gregory 2000; Inkpen 2005). This is an approach associated with the critic of logical positivism, Karl Popper. Popper was happy with most of the ontological and epistemological standpoints of logical positivism but believed that a theory could never be verified. Instead, he advocated a process of "falsification" whereby theories would be assumed to be correct until they were proved wrong. Even if a theory appears to be "true" for years on end, we can never be sure it is correct. For many centuries scientists believed that the earth was at the center of the universe and observable experience appeared to confirm this. Eventually, of course, we learned that it is not, and a better theory took its place. While positivists believe in a perfect fit between theory and the world, critical rationalists can only believe in a provisional fit between theory and the world that might one day have to be thrown out. Popper denies that it is ever possible to show that a theory reflects reality accurately.

Critical rationalists also have little faith in the **inductive** process that is central to positivist philosophies of science. An inductive process is one that starts with simple observations and accumulates them in order to develop theories and, eventually, laws. Many physical geographers recognize that there is no such thing as a pure observation that is not in some way theoretically informed

(Rhoads and Thorne 1994; Gregory 2000). The principle of induction, they recognize, is logically flawed as no accumulation of observations that confirm a theory will make it true. The theory-laden nature of observations and the impossibility of a truly inductive theory construction enabled the general acceptance of *deductive* theory. The deductive process starts with a logically consistent theory and then looks to empirical evidence to test its adequacy. Ideally there might be a set of theories competing to explain a particular theory and these will gradually be eliminated until the best theory is left. Inkpen summarizes Popper's critical rationalist view in the following way:

> The truth-value of a theory increased the more testing it survived, but it never became true. This rather unsatisfactory end point for a theory implied that key scientific ideas such as the law of gravity were not necessarily true, just unfalsified. Their stability as cornerstones of scientific thought came from the confidence derived from their continued non-falsification, and the lapse in interest in trying to falsify them resulted from this confidence. (Inkpen 2005: 30)

Spatial Science in the Twenty-First Century

Spatial science, positivism, and quantification all came under significant theoretical attack (mostly from human geographers) from the late 1960s onward. There had always been some regional geographers who never accepted the claims of spatial science in the first place. But by the end of the 1960s, new theoretical directions, based in both a humanistic tradition and more radical (mostly Marxist) theory, began to reject spatial science for new reasons (see Chapters 6 and 7). By the early 1980s, it was quite hard to maintain a commitment to some pure form of spatial science, such was the weight of the onslaught against it. Many books on "theory" in geography are accounts of post-positivist theory as though spatial science has nothing to offer the student of geography in the twenty-first century. This is a mistake. There are many ways in which the traditions surrounding spatial science continue to be important. Many geographers still practice spatial science in one form or another. There are whole journals dedicated to its continued existence (see for instance, *Geographical Analysis: An International Journal of Theoretical Geography*). Some departments continue to be marked by their interest in quantitative methods.

It is also the case that the use of quantification in geography has moved away from the desire to produce law-like generalizations and towards more instrumental and often critical purposes. Many of the most vocal critics of both positivist philosophy and quantification have been critical geographers, including Marxists, Feminists, and Post-structuralists. They have objected to both the sense of godlike objectivity that quantification can be guilty of and the alleged political uselessness of what can seem to be overly elaborate forms of description that are numerical rather than textual. Some critical geographers, however, have continued to argue for the instrumental usefulness of quantitative approaches to important social problems. Tim Schwanen and Mei-Po Kwan, for instance argue that not all quantifiers are positivist and that the use of quantitative methods is not, therefore, subject to the flaws of positivism. They point to **feminist empiricism** where critical feminist scholars frequently used quantitative techniques to show how disregarding women in geographical analysis is conceptually and empirically flawed, as well as work in critical GIS where advanced computer-based geographical information systems are used to address issues of inequality and oppression (Schwanen and Kwan 2009). The use of quantitative methods, they argue, "makes their arguments more difficult to dismiss as irrelevant, lacking scientific rigor, or even unscientific by those served by neoliberalism, patriarchy, and other systemic inequalities" (Schwanen and Kwan 2009: 461–462).

Quantitative research in health geography is one area where quantitative methods are used to address significant social problems. Niamh Shortt and colleagues, for instance, explore the links between tobacco and alcohol used and social deprivation by accounting for the availability of tobacco and alcohol through sales outlets in Scotland (Shortt *et al.* 2015). They accessed the addresses of every tobacco and alcohol outlet in Scotland and used quantitative techniques and GIS to create measures of outlet density. This was statistically compared with the Scottish Government index of multiple deprivation in each official "data zone" – areas with more-or-less 817 people living in them. The researchers specifically focused on income deprivation and were able to show with confidence that the highest densities of outlets were in areas of highest income deprivation.

Another example of the use of quantitative methods to address important questions is provided by Elvin Wyly and C.S. Ponder in their analysis of the links between gender, age, and race in the allocation of risky and often high interest subprime mortgages in the years leading up to the economic crash of 2008 (Wyly and Ponder 2011). Wyly and Ponder reflect on the difference between anecdotes – stories told by individuals – and "data" – which is most often quantitative in nature. Anecdotes, they note, are generally ignored when they come from members of marginalized and oppressed groups in society. This was the case when stories emerged about predatory and discriminatory lending practices connected to subprime mortgages in the United States. Here is one example.

In 1997, a 71-year old African American woman named Veronica Harding needed money for repairs on her rowhouse in North Philadelphia, and wound up with a $35,000 loan from a firm called American Mortgage Reduction, Inc (AMR). The loan carried an interest rate of 11.4 percent, settlement fees of $4,400, a broker's fee of $3,500, a $2,815 insurance policy financed at an interest rate of 22.5 percent, and a balloon payment of $32,000 due in 27 years. The AMR loan was only the most recent mortgage for Harding – she had taken out 14 over the previous dozen years – and she explained that "brokers usually worked out a loan at her kitchen table. … 'They make it so easy. … They tell you they are going to pay off all of your bills. And then they give you a check. But a couple of months later you are in more debt than before.'" (Wyly and Ponder 2011: 530)

A story such as this does not count as "data" in quantitative analysis and is therefore, easily dismissed as not having the same weight as something that can be quantified and generalized. Just because this happened to Veronica Harding, it might be argued, it does not follow there was a problem with subprime lending. In their paper, Wyly and Ponder do not dismiss these "anecdotes" but, rather, explore them by applying quantitative techniques to address the hypothesis that "discriminatory impacts in the subprime boom were greater among African Americans compared to otherwise similar non-Hispanic Whites, more severe among African American women, and even more pronounced among older African American women" (Wyly and Ponder 2011: 537). In other words, Wyly and Ponder use quantitative methods to explore important questions of intersectional discrimination with significant social justice implications. Specifically, they use two existing data sets, the broad but shallow Home Mortgage Disclosure Act (HMDA) data set, and the much narrower but deeper National Mortgage Data Repository (NMDR) which only includes records of about 600 mortgages but does include data on age, gender, and race. By comparing these two data sets and using regression analysis they were able to provide what they refer to as "circumstantial links" between race, gender, and age in provision of subprime mortgages.

Single female subprime borrowers outnumber single males as well as couples among non-Hispanic African Americans, and not for any other racial/ethnic group. Single African American women are five times more likely than non-Hispanic White couples with the same incomes and loan-to-income ratios to wind up with subprime loans; disparities persist even when single Black women are compared with single White women and African American couples. (Wyly and Ponder 2011: 559)

Importantly, Wyly and Ponder return to the issue of anecdotes they started with to suggest that perhaps the anecdotes should be taken more seriously as there is "something representative and generalizable from stories" told by victims of predatory mortgage lending and that too often "methodological conservatism abets ideological conservatism" (Wyly and Ponder 2011: 559). Wyly and Ponder's conclusions are far from conservative, suggesting "structural legal changes" to address "systemic inequalities" and the need to re-evaluate "the rights of home as distinct from the rights of profit, speculation, and deception" (Wyly and Ponder 2011: 560).

The use of quantitative methods as part of a toolkit for critical geography is not the only way in which quantification in geography has changed. While the origins of spatial science saw quantitative methods go hand in hand with positivism and a desire to achieve law-like generalizations that could be applicable anywhere, more recent quantitative work has dropped the urge to generalize and become more place-specific in its applications. A. Stewart Fotheringham and Chris Brunsdon have discussed the emergence of local forms of spatial analysis that use quantitative techniques to explore the specificity of particular locations, rather than as part of a search for universal truths (Fotheringham and Brunsdon 1999). They helpfully outline one of the problems with global forms of spatial analysis that attempt to erase the locally or regionally specific.

> Models of spatial processes and methods of spatial analysis have usually been applied at a global level, meaning that one set of results is generated from the analysis and these results, representing one set of relationships, are assumed to apply equally across the study region. Essentially, what is being undertaken in a global analysis, but is rarely acknowledged, is the generation of an "average" set of results from the data. If the relationships being examined vary across the study region, the global results will have limited application to specific parts of that region and may not, in fact, represent the actual situation in any part of it. Calibrating a global model is therefore akin to being given the information that "the average precipitation in the United States last year was 32 inches." This is a "global" statistic in that it provides information about the study area in general but not necessarily about any specific part of it. Consequently, it is of little use if precipitation does vary locally. (Fotheringham and Brunsdon 1999: 341)

The different forms of "local analysis" that Fotheringham and Brunsdon introduce focus on variation in spatial relationships rather than the establishment of law-like certainties with global application. Quantification in geography, in other words, has moved a long way from its positivist ambitions to produce universal laws and returned, in some sense, to the theme of areal differentiation that was at the center for earlier forms of regional geography (see Chapter 4).

One form of "local analysis" in quantitative geography has been the exploration of the **neighborhood effect**. The neighborhood effect posits that the specific characteristics of a local area have effects on individual and group forms of behavior. The work by Shortt and her colleagues on tobacco and alcohol use in Scotland demonstrates a kind of neighborhood effect (Shortt *et al.* 2015). The neighborhood effect is not entirely a spatial science concept and has been explored through a number of theoretical lenses. But at its heart is the familiar hypothesis that things close by have a greater influence than things further away. People, it is supposed, are likely to make decisions based on the values and activities of their neighbors rather than, for instance, rational cost–benefit analyses. But even the seemingly place-sensitive notion of the neighborhood effect is based on a largely uncritical reading of ideas like place and neighborhood. Critical geographers have long theorized places and neighborhoods as far more than spatially bounded areas with largely homogenous characteristics. They are made as much from what moves between them as what occurs within them. Mei-Po Kwan has argued for a more mobile sense of environmental influences of behavior pointing out that people generally only spend part of their day where they live and are mobile on a daily basis. To quantitatively account for effects on behavior therefore, it would be necessary to follow time–space trajectories (journeys), perhaps using GPS data, rather than confining analysis to an always artificial "neighborhood."

> Because individual exposures to environmental influences are determined by the interactions between environmental influences and individual mobility, such exposures could face different types of

contextual uncertainties. For example, individual exposure to air pollution is determined by physical contact between pollutants and humans, and both vary or move over space and time. (Kwan 2018: 1483)

Contextual influences, Kwan tells us, are much more complicated than what can be accounted for in a contiguous area. Kwan's critique of the neighborhood effect points towards a wider critique of even "local" forms of spatial analysis as too easily based on naïve assumptions about the spatial division of people's lives. Returning to Shortt *et al.*'s examination of the links between areas of multiple deprivation and alcohol and tobacco outlets, for instance, Kwan might note that people who live in these areas do not necessarily work in them. People travel through them, into them, and out of them, and that too might have an influence of alcohol and tobacco use.

Conclusions

Despite the decades of critique of spatial science since the 1970s, the legacy of the quantitative revolution is implicit in pretty much all of human geography today. Before the 1950s, it would have been almost impossible to write a book on theory in geography. Geographers, generally, simply did not think of what they did as "theory." Bunge's *Theoretical Geography* and Harvey's *Explanation in Geography* were two books which explicitly argued for the importance of theory in geography. Their notion of what constituted theory was itself a product of positivist thinking and was therefore fairly narrow and unrecognizable to contemporary human geographers who write about theory. The second legacy of spatial science was the development of a focus on quantitative methods that has remained strangely hegemonic in a discipline where very few people actually practice them. When I was taking my PhD in Wisconsin in the late 1980s, we had to take a "methods requirement." The choice was statistics or a language. Physical geography colleagues made the case for scuba diving. Human geographers busy reading literary theory wanted to take semiotics but these were forbidden. Statistics were king of the methodological castle. In Britain, the training requirements of the Economic and Social Research Council (who fund most human geography PhDs) demand fairly unequivocally that students are trained in quantitative methods despite the fact that so few of them are ever likely to use them. Generations of human geography PhD students still have to learn statistics (again) even if they are never going to use them. The effects of positivism and quantitative social science still resonate in the institutional arrangements for doctoral training. The third legacy is an interest in space and the spatial. Harvey, Bunge, Haggett, and others involved in the promotion of spatial science were keen to move away from a focus on regions (or areal differentiation) and to refocus on spatial relations. "Space" became an important part of the geographical theoretical vocabulary. Although there have been many developments in the ways in which space has been theorized (and spatial science perspectives on space can seem quite naïve in hindsight), the discipline owes much to spatial science for the focus on space as a geographical phenomenon. The fourth legacy is a focus on movement and process. Crowe's early call for a modern progressive geography asked geographers to move away from their obsession with the topography of "things" and to look, instead, at the dynamic nature of human space relationships. Spatial scientists followed Crowe's lead and developed a considerable body of work on movement and process. It is only recently that this focus has been rediscovered.

Recent decades have also seen the rise of what Elvin Wyly calls "the new quantitative revolution" (Wyly 2014). This is less a revolution in geographic thought and more a revolution in the everyday world we inhabit. Wyly's paper was written in response to a paper by Ron Johnston and colleagues that was, in part, a critique of the first edition of this book – a critique that has led to many of the paragraphs you have just read (Johnston *et al.* 2014). Johnston *et al.*, among other things, insisted on the importance of education in quantitative methods within geography so that future geographers and future citizens would be equipped to deal with a world that is increasingly calculated and quantified. A world of "Big Data." Wyly cautions that the "new quantitative revolution" of Big Data is driven by neo-liberal political economy where the very idea of (always quantitative) data is used to support elite groups and their program of "evidence-based" policy – a program it is ever more difficult to be outside of. Wyly argues that "[t]he infusion of competitive market metrics, technocratic instrumental rationality, and neoliberal axioms of consumer choice are transforming education at an accelerating pace" (Wyly 2014: 28) which means that the logic of quantification has successfully invaded every nook and cranny of life in higher education where our "performance" is often reduced to metrics – often to simple numbers.

> Big data give us a quickly expanding, shallow view of the vast horizontal landscape of the desert of the present real, with each new technological advance accomplishing new kinds of devalorization of past generations of human knowledge. The speedy exponential cascade of today's digital vivisection of events large and small yields infinitely changing measures of attention and impact, as in the current enthusiasm for Klout scores and global speeds measured as tweets per minute; anything from the past that cannot be digitized, or that is disallowed from indexing on a server's robots.txt file, becomes another Anaximander fragment. (Wyly 2014: 28)

This quantification of almost everything in teaching and research is nested within a wider world where numerical data and hidden algorithms are ubiquitous in work, leisure and pretty much everything else. The question for geographers, as well as fellow academics in other disciplines, is how much our embrace of quantitative methods allows us to critically engage with this Big Data world, and how much if means we become part of it.

References

* Abler, R., Adams, J., and Gould, P. (1971) *Spatial Organization: The Geographer's View of the World*, Prentice Hall, Englewood Cliffs, NJ.

* Bunge, W. W. (1962) *Theoretical Geography*, Royal University, Lund.

Chorley, R. J. (1978) Bases for theory in geomorphology, in *Geomorphology: Present Problems and Future Prospects* (eds C. Embleton, D. Brunsden, and D. K. C. Jones), Oxford University Press, Oxford, pp. 1–13.

Chorley, R. J. and Haggett, P. (eds) (1967) *Models in Geography*, Methuen, London.

Christaller, W. (1941) *Die Zentralen Orte in den Ostgebieten und ihre Kultur- und Marktbereiche. Struktur und Gestaltung der Zentralen Orte des Deutschen Ostens, Teil 1*, K. F. Koehler Verlag, Leipzig.

Christaller, W. (1966 [1933]) *Central Places in Southern Germany*, Prentice Hall, London.

Crowe, P. R. (1938) On progress in geography. *Scottish Geographical Magazine*, 54, 1–18.

Fotheringham, A. S. and Brunsdon, C. (1999) Local forms of spatial analysis. *Geographical Analysis*, 31, 340–358.

Gregory, D. (1978) *Ideology, Science and Human Geography*, Hutchinson, London.

Gregory, K. J. (2000) *The Changing Nature of Physical Geography*, Arnold, London.

* Haggett, P. (1965) *Locational Analysis in Human Geography*, Edward Arnold, London.

Haines-Young, R. H. and Petch, J. R. (1986) *Physical Geography: Its Nature and Methods*, Harper & Row, London.

Hartshorne, R. (1939) *The Nature of Geography: A Critical Survey of Current Thought in the Light of the Past*, The Association of American Geographers, Lancaster, PA.

* Harvey, D. (1969) *Explanation in Geography*, Edward Arnold, London.

Hudson, R. (1990) Re-thinking regions: some preliminary considerations on regions and social change, in *Regional Geography: Current Developments and Future Prospects* (eds R. J. Johnston, J. Hauer, and G. A. Hoekveld), Routledge, London, pp. 67–84.

* Inkpen, R. (2005) *Science, Philosophy and Physical Geography*, Routledge, London.

Johnston, R. J. and Sidaway, J. D. (2004) *Geography and Geographers: Anglo-American Human Geography Since 1945*, Arnold, London.

Johnston, R., Harris, R., Jones, J, *et al.* (2014) Mutual mis-understanding and avoidance, mis-representations, and disciplinary politics: spatial science and quantitative analysis in (UK) geographical curricula. *Dialogues in Human Geography*, 4(1): 3–25.

Kwan, M. -P. (2018). The limits of the neighborhood effect: contextual uncertainties in geographic, environmental health, and social science research. *Annals of the Association of American Geographers*, 108(6), 1482–1490.

Lösch, A. (1954) *The Economics of Location*, Yale University Press, New Haven, CT.

Lowe, J. and Moryadas, S. (1975) *The Geography of Movement*, Houghton Mifflin, Boston, MA.

Mercer, D. (1984) Unmasking technocratic geography, in *Recollections of a Revolution* (eds M. Billinge, D. Gregory, and R. Martin), Macmillan, London, pp. 153–199.

Morrill, R. (1984) Recollections of the "quantitative revolution's" early years: the University of Washington 1955–65, in *Recollections of a Revolution* (eds M. Billinge, D. Gregory, and R. Martin), Macmillan, London, pp. 57–72.

Mountz, A. and Williams, K. (2023) Let geography die: the rise, fall, and "unfinished business" of geography at Harvard. *Annals of the American Association of Geographers*. https://www.tandfonline.com/doi/full/10.1080/24694452.2023.2208645?scroll=top&needAccess=true&role=tab.

Rhoads, B. L. and Thorne, C. E. (1994) Contemporary philosophical perspectives on physical geography with emphasis on geomorphology. *Geographical Review*, 84, 90–101.

Rossler, M. (1989) Applied geography and area research in Nazi society: central place theory and planning, 1933–1945. *Environment and Planning D: Society and Space*, 7, 419–431.

* Schaefer, F. K. (1953) Exceptionalism in geography: a methodological examination. *Annals of the Association of American Geographers*, 43, 226–249.

* Schwanen, T. and Kwan, M.-P. (2009) "Doing" critical geographies with numbers. *The Professional Geographer*, 61(4), 459–464.

Shortt, N. K., Tisch, C., Pearce, J., Mitchell, R., Richardson, E.A., Hill, S., and Collin, J. (2015) A cross-sectional analysis of the relationship between tobacco and alcohol outlet density and neighbourhood deprivation. *BMC Public Health*, 15, 1014. doi: 10.1186/s12889-015-2321-1.

Smith, N. (1987) Academic war over the field of geography: the elimination of geography at Harvard, 1947–1951. *Annals of the Association of American Geographers*, 77, 155–172.

* Strahler, A. N. (1952) Dynamic basis of geomorphology. *Bulletin of the Geological Society of America*, 63, 923–938.

Tobler, W. (1970) A computer movie simulating urban growth in the Detroit region. *Economic Geography*, 46, 234–240.

Ullman, E. (1956) The role of transportation and the bases for interaction, in *Man's Role in Changing the Face of the Earth* (ed. W. L. Thomas), University of Chicago Press, Chicago, IL, pp. 862–880.

Ullman, E. E. (1953) Human geography and area research. *Annals of the Association of American Geographers*, 43, 54–66.

Ward, S. V. (1992) *The Garden City: Past, Present and Future*, Spon, London.

White, H. P. and Senior, M. L. (1983) *Transport Geography*, Longman, New York.

Wilson, A. G. (1972) Theoretical geography: some speculations. *Transactions of the Institute of British Geographers*, 57, 31–44.

* Wyly, E. (2014) The new quantitative revolution. *Dialogues in Human Geography*, 4(1), 26–38.

Wyly, E. and Ponder, C. S. (2011) Gender, age, and race in subprime America. *Housing Policy Debate*, 21(4), 529–564.

Zipf, G. K. (1949) *Human Behavior and the Principle of Least Effort: An Introduction to Human Ecology*, Addison-Wesley Press, Cambridge, MA.

Chapter 6

Humanistic Geographies

Imagine you are hungry. You want to get some take-out food, fish and chips perhaps. Making sure you have your wallet, you leave your house, and make your way toward the fish and chip shop. But which one? Perhaps you think about the distance between where you are and the various fish and chip shops in the area that you are aware of. Perhaps you simply go to the nearest one. Fish and chips are pretty much the same wherever you go, so why waste energy. But perhaps not. Perhaps you know where the very best and crispiest batter is and the most succulent, not-too-greasy chips. Maybe it is a little further but surely worth it. And then again, what about the person that works behind the counter at one still further away. You like them and get to flirt for a few minutes. Perhaps you imagined it, but they appeared to flirt back last time you went. The fish and chips are dreadful, but the human interaction is first rate. Still further away is the new fish and chip shop. It is cool. It feels nice to be in. It has a nice vibe. Life is complicated. Humans are complicated. We are supposed to be marked out from other forms of life by our rationality but we also have hopes, dreams, and desires. We have imaginations.

What could a spatial scientist make of this choice? How could a spatial scientist deal with hopes, dreams, and desires, with the murky, messy world of imaginative humanity? The world of the spatial scientist is inhabited by a particular kind of imaginary person called a "rational being." In economics this figure is referred to as a "rational economic man." He always makes sensible decisions based on the careful weighing up of costs and benefits. He would definitely go to the closest fish and chip shop. But he is imaginary. Consider the following joke (always popular with my theory class): Two spatial scientists are stuck on a desert island. They have nothing but a year's supply of canned food that has been washed up by the waves. They have no can-opener. For several days they stare at the food. One says to the other, "what do you think we should do?" The other replies, "first, imagine we had a can-opener." This is (stereotypically) the world of the spatial scientist. A world in which we first imagine

Geographic Thought: A Critical Introduction, Second Edition. Tim Cresswell.
© 2024 John Wiley & Sons Ltd. Published 2024 by John Wiley & Sons Ltd.

something that is not true and then proceed from there. "All other things being equal" (they are not), "imagine the world is flat" (it is not). For a "real world" example of this kind of thinking, consider the following, taken from the introduction to an influential textbook on land use:

> Both the Puerto Rican and the Madison Avenue advertising man will be *reduced* to that uninteresting individual, economic man ... we shall assume that the city sits on a featureless plain ... what it does not have are such features as hills, low land, beautiful views, social cachet, or pleasant breezes. *These are undoubtedly important, but no way has been found to incorporate them into the type of theory that will be presented.* (William Alonso quoted in Ley 1980: 9)

Indeed, how can science be applied to creative, imaginative, thinking humans? Science likes numbers. Numbers have a beauty that is pleasing. They make the chaotic seem ordered. They give us answers. Allow me one more joke. Three geographers go out hunting. They spot a deer. One of the geographers, a humanist, asks to shoot first. He does and misses, 10 ft to the right. The Marxist geographer laughs and takes aim. He shoots, and misses, 10 ft to the left. Suddenly the third geographer, a spatial scientist, cheers loudly. "Why are you cheering?" her companions ask. "We got him," she explains.

During the 1970s, several sets of reactions to the excesses of spatial science emerged as human geographers began to question its deficiencies. One of these reactions became known as **humanistic geography**. Science, they argued, falls short when it is applied to creative, imaginative, thinking human beings. It may (and often does) work well when applied to the physical world, to rocks falling down slopes or sediments moving downstream, but it cannot account for the humanity of humans. Spatial science, it was suggested, suffered from the assumptions that came with a positivist world-view. The idea that there is a singular, measurable truth "out there" waiting to be discovered and explained that can be accessed by adopting a lofty, God-like, objective "view from nowhere" was dismissed. The notion that all-important things are quantifiable was questioned. Humanistic geographers were concerned with the lack of basic human meaning in the intricacies of spatial science. People, they argued, are not rocks or atoms. They present a different kind of problem. Human geography, they suggested, needed desperately to put humans back in.

In spatial science's search for general geographical "laws," the view from nowhere had obscured the view from somewhere. Or, more precisely, the singular view from nowhere, the God-like perspective of objectivity, had erased the multitude of somewheres that exist in actuality. The "general geography" of which Varenius wrote had overwhelmed his "special geography." Humanistic geographers were keen to rediscover and underline the importance of somewhere; of the specificities of particular places and of the nature of particularity in general. This set of arguments is, as we have seen, as old as geography. For every Ptolemy, searching for the measurable and universal laws of space, there was a Strabo or a Herodotus telling us about the specific qualities of place. And there is a paradox at the heart of this age-old discussion of the particular and the universal. There is, after all, nothing more universal than the particular. We all know the particular, the local, the place we are in. And because we all have this experience, we have a basis for relating to the experience of others, in different particular places. None of us, on the other hand, have ever experienced the universal. Somewheres are everywhere while everywhere is nowhere.

The emergence of humanistic geography as an identifiable approach with its own bundle of theories in the late 1970s can be explained as both a reaction to the kinds of research undertaken as spatial science (and **structuralism**) and as a positive affirmation of the importance of worlds of meaning and experience in the human relationship to the earth. Let us take each of these in turn.

Critiquing (In)Human Geography

A key reason for the emergence of humanistic geography was the belief that other approaches, including spatial science and various forms of structuralism (such as Marxism), had, sometimes willfully, erased the human from the human world. This was particularly forcefully argued by David Ley in his essay "Geography without man" (Ley 1980). He points to how both emerging spatial science and the structuralist anthropology of Lévi-Strauss actively sought to extinguish the human. The Chicago School sociologist Robert Park, in an attempt to develop a scientific "human ecology," had, for instance, written: "[R]educe all social relations to relations of space and it would be possible to apply to human relations the fundamental logic of the natural sciences" (Park 1936). Lévi-Strauss, in his search for universal structural codes that govern the logic of human culture, had declared his belief that "the ultimate goal of the human sciences is not to constitute man but to dissolve him" (Lévi-Strauss quoted in Ley 1980: 4). Both of these ambitions, Ley argues, remove real, fleshy, thinking people from the human sciences in an attempt to provide law-like codes for the beliefs and behavior of people. This, he suggests, represents the triumph of Durkheim and Comte's positivism against the interpretive enterprise of Vidal (see Chapter 4). Measurement, quantification, and observable "facts" had been victorious over the intangible and interpreted world of feelings and emotions. Geography had, thus, been dominated by the hexagons of Christaller and the study of spatial relations in which "the richness of human encounter with the environment was reduced ... to the single behavioural assumption of rational activity expressed in distance minimization in the journey to shop or trade" (Ley 1980: 6).

But it was not just spatial science that had erased the process of interpretation from geography and other social sciences. Similar processes could be seen in views of culture as a "superorganic" entity that molded and directed human action (Duncan 1980); in versions of structural Marxism that saw all cultural expression as a product (in the final instance) of an economic base (see Chapter 7); and in the processes of structuralism in anthropology and literature that saw all human expression and creativity as products of iron-clad structural laws. All of these ways of thinking, Ley argued, involved mysterious and largely abstract forces doing things to humans. Each of them depleted the capacity of people to act as agents.

A similar evocation of the simple postulates of spatial science and economic theory is made by Tuan in a rousing conclusion to his book *Space and Place: The Perspective of Experience* (Tuan 1977):

> The scientist postulates the simple human being for the limited purpose of analyzing a specific set of relationships, and this procedure is entirely valid. Danger occurs when the scientist then naively tries to impose his findings on the real world, for he may forget that the simplicity of human beings is an assumption, not a discovery or a necessary conclusion of research. The simple being, a convenient postulate of science and a deliberate paper figure of propaganda, is only too easy for the man in the street – that is, most of us – to accept. We are in the habit of denying or forgetting the real nature of our experiences in favor of the clichés of public speech. (Tuan 1977: 203)

The point of a humanistic geography, he goes on, is to "increase the burden of awareness" in the face of forms of thought that pay no attention to awareness whatsoever. *Space and Place*, one of the key texts in humanistic geography, asks us to consider the relationship between people and the world through "the perspective of experience." The "simple being" of science, Tuan suggests, has no experience as such and this is a considerable shortcoming. We need, he argues, to be in touch with the world through our bodies, through our senses and our emotions, rather than discarding these at the outset of our research as trivial or too subjective. It is no surprise, therefore, that another of the key figures in the emergence of humanistic geography, Anne Buttimer, should write that "from whatever ideological stance it has emerged, the case for humanism has usually been made with the conviction that there must be more to human geography than the *dance macabre* of materialistically motivated robots which, in the opinion of many, was staged by the post World War II 'scientific' reformation" (Buttimer 1993: 47).

This *dance macabre* of rational economic (and always geographical) people carried with it several dangers, according to humanists such as Ley, Tuan, and Buttimer. The assumption of the possibility of objectivity was erroneous and dangerous:

> The wrenching of knowledge from human interests has hidden the ideological content of scientific knowledge, that it is knowledge derived by a subject and for a subject. There is a clear need in the human sciences for the impenetrable appearance of objectivity to be demystified to illuminate the shifting anthropocentric presuppositions masked by the firm categories and rigorous procedures. (Ley 1980: 14)

Here was an early statement of the dangers of thinking that science carried with it the possibility of extracting *ourselves* from the possibility of knowledge production. The beauty of numbers on the page alongside the rigor of laboratories and instrumentation seemed to guarantee knowledge that was immune to both the subjective interests of the participants and the ideological interests of politics at work. "Categories" and "procedures" quite powerfully hid the fact that it was humans who were producing knowledge.

Simultaneously, the mask of "science" devalued the unquantifiable and mysterious world of consciousness and human action. By apparently removing subjectivity from the procedures of knowledge production, spatial science was able to declare neutrality and dis-interestedness. But it was more than this. It also explicitly and implicitly theorized human dreams, imaginations, beliefs, and consciousness as interference in the strict procedures of science. Finally, Ley argued, this could be equated with a "moral error" as: "Under the continuous extension of this formidable technical and managerial hegemony, even the expression of such basic human freedoms as free speech, free assembly and freedom to worship may be disqualified as incompatible with rational systems of control" (Ley 1980: 19). A similar point is made by John Pickles, quoting Gunnar Olsson, who suggested that humanism confronted the scientific procedure of predictive science with the observation that "this form of knowledge was morally questionable and that the applications it gave rise to, far from enriching the social world, resulted in further constraints on the freedom of action. 'What initially has presented itself under the disguise of humane methodology now appeared as crude, inhuman and power-ridden ideology. What our elders had told to be emancipatory we found out to be the opposite'" (Olsson quoted by Pickles 1987: 23). Others, such as Relph, made the argument that scientific methods were directly contributing to the production of real places in the planning process and that this had resulted in increasing "placelessness" in the landscapes of the developed world. Efficiency, rationality, and profit, Relph argued, combined to neglect truly human places in favor of mechanistic landscapes (think of Christaller and the planning of the Holocaust in Chapter 5) (Relph 1976).

Humanism, then, confronted science at epistemological, theoretical, and moral levels. This prefigures arguments made by theorists of **feminism**, **poststructuralism**, and **postmodernism** who have argued for the importance (theoretically, methodologically, and politically) of recognizing the **positionality** and situatedness of the production of knowledge.

What Is Humanistic Geography?

It would be wrong to think of humanistic geography as merely and only a response to, and critique of, spatial science. It certainly served that purpose, but its history is much deeper than that. It is, above all, a positive articulation of the relationship between people and the world built on its own set of philosophical inspirations. A foundational text on the history of western **humanism** is the *Oration on the Dignity of Man* (1486) by Pico della Mirandola in which it was claimed that "man" was responsible for himself and his own actions (Buttimer 1993). Man alone, he claimed, was able to descend to the animalistic and rise to the status of angels. His role in the universe was not fixed but largely up to him. Such a claim lies at the heart of humanistic geography. Humans are willful agents and not puppets of mysterious forces. They are imbued with intelligence, imagination, and consciousness and any truly *human* geography needs to foreground these rather than excluding them as worryingly unpredictable nuisances.

In fact, we have already encountered humanism in its broadest sense in the rediscovery of classical learning during the Renaissance. The geographies of Varenius or Kant, for instance, were in some sense humanist in that they recentered humanity in the search for knowledge. They displaced a God-centered view of the world. Indeed, the origins of properly modern philosophy, art, planning, architecture, economics, warfare, and a host of other fields of knowledge and practice grew out of a flowering of humanistic thought that placed humans at the center of learning in the fifteenth and sixteenth centuries. Even science, the seeming opposite of humanism in geography, grew out of this broad sense of humanism that emphasized the power of human reason rather than pre-ordained fate or destiny (Relph 1981; Cosgrove 1984). There is no little irony in a theoretical approach that opposed the overly scientific being labeled "humanism." This was recognized by Edward Relph:

> … humanistic geography opposes scientistic geography as humanism opposes the orthodoxies of scientism. The difficulty is that scientism is a direct extension of the humanist principles of free enquiry and rational thought, an extension in which these principles have become dogmatic and inflexible, and therefore deny themselves. In humanistic geography, humanism is challenging its own extension, not as an act of self-awareness but through the confusions that beset it and make it so difficult to know just what humanism stands for. (Relph 1981: 17)

While humanistic geography has its roots in the central claim that human geography should place humans at its center (this is what makes it "humanistic"), its direct lineage is more precise. There were several occasions over the course of the twentieth century where geographers argued for a realm of understanding that emphasizes the subjective and emotive side of geographical enquiry. Carl Sauer (1889–1975), for instance, a geographer who concerned himself mostly with the detailed analysis of the material landscape, found room to state that: "Having observed widely and charted diligently, there yet remains a quality of understanding at a higher plane which may not be reduced to formal processes" (Sauer 1996: 311). There is also a humanistic side to the evocations of region and "*genre de vie*" in the geographies of Vidal de la Blache and his followers in France that was later to inspire the humanistic geography of Anne Buttimer (Buttimer 1971).

Perhaps the most important precursor to the humanistic geography that became established in the 1970s was John Kirkland Wright. In an important presidential address to the Association of American Geographers in 1946, he proposed a geography of knowledge he called **geosophy**. In a world in which almost everything had been explored and mapped, he argued, we had yet to explore our internal geographies. He suggested that geographers need to know how people know their world. He argued for an exploration of their geographical imaginations. We should, he argued, heed the siren's voice of the unknown – the spell of the mysterious and the intangible. Subjectivity, he insisted, was an important domain of geographical enquiry (Wright 1947):

> Indeed, nearly every important activity in which man engages, from hoeing a field, or writing a book, or conducting a business, to spreading a gospel or waging a war, is to some extent affected by the geographical knowledge at his disposal. (Wright 1947: 14)

It does not appear that Wright had a particularly attentive audience for his presidential address. There was no great outpouring of geographical considerations of imagination or subjectivity in the following years. Indeed, it was not until 1961 that David Lowenthal took up Wright's challenge in a paper which developed the link between "personal geographies" and "world views" (Lowenthal 1961). We all, as Wright had previously suggested, have personal geographies, flights of fancy, and learned relations to the world around us. "The surface of the earth," Lowenthal wrote, "is shaped for each person by refraction through cultural and personal lenses of custom and fancy. We are all artists and landscape architects, creating order and organizing space, time and causality in accordance with our apperceptions and predilections" (Lowenthal 1961: 260). But, as this quote suggests, we also have shared or "cultural" worlds that we develop as we grow up and communicate with others. Not everyone, Lowenthal suggests, has this shared view. "The most fundamental attributes of our shared view of the world are confined, moreover, to sane, hale, sentient adults" (Lowenthal 1961: 244). Lowenthal goes on to list "idiots," "psychotics," "mystics," "claustrophobics," and children as among those unable to share in a geographical worldview. But some sense of a private world is important, and not beyond the scope of geographical enquiry:

> In each of our personal worlds, far more than in the shared consensus, characters of fable and fiction reside and move about, some in their own lands, others sharing familiar countries with real people and places. We are all Alices in our own Wonderlands, Gullivers in Lilliput and Brobdingnag. Ghosts, mermaids, men from Mars, and the smiles of Cheshire cats confront us at home and abroad. Utopians not only make mythic men, they rearrange the forces of nature: in some worlds waters flow uphill, seasons vanish, time reverses, or one- and two-dimensional creatures converse and move about. Invented worlds may even harbor logical absurdities: scientists swallowed up in a fourth dimension, conjurors imprisoned in Klein bottles, five countries each bordering on all the others. … If we could not imagine the impossible, both private and public worlds would be the poorer. (Lowenthal 1961: 249)

Lowenthal's paper considers the various ways in which the personal world, even the world of the fantastic, is connected to both worldviews and shared cultural norms. This focus on feeling, meaning, fantasy, and the subjective, while not wholly new to the discipline, certainly stood out at a time when human geography was still following a largely faithful regional model of describing in great detail the elements of the "real" world.

The voices of geographers like Wright and Lowenthal were largely ignored until the dominance of spatial science and the quantitative revolution pushed people into a more sustained critique in the mid-1970s. It was this period that saw the emergence of a distinct body of thought that acquired the label "humanistic geography." Key figures included Yi-Fu Tuan, David Ley, Edward Relph, Anne Buttimer, and David Seamon. As with any alleged "school of thought," there are many differences between the approaches and perspectives of these thinkers. They all share, however, a desire to put humans and human consciousness, feeling, thoughts, and emotions at the center of geographical thinking. They all insist that people are not puppets of mysterious forces but intentional beings with their own subjectivities. They all insist on human geography as the study of "people-in-the-world." This is an argument made, for instance, by Anne Buttimer:

> Neither humanism nor geography can be regarded as an autonomous field of inquiry; rather, each points toward perspectives on life and thought shared by people in diverse situations. The common concern is terrestrial dwelling; *humanus* literally means "earth dweller." (Buttimer 1993: 3)

A similar point is made by Donald Meinig when he suggests that humanistic geography is marked by a "self-conscious drive to connect with the special body of knowledge, reflection, and substance about human experience and human expression, about what it means to be a human being on this earth" (Meinig 1983: 315). The key here is "human being on this earth" – suggesting an analysis of the connection between humans and the world they inhabit.

The notion of dwelling and the associated term "home" is clearly at the center of the humanistic enterprise. Yi-Fu Tuan, in a short essay on the nature of geography, describes geography as the study of the earth as the home of people (Tuan 1991). Geographers obviously study the earth (most obviously in physical geography) and obviously study people (at least in human geography). But to Tuan the key term is "home." How do people make the earth into a home? Even physical geographers, he argues, stay close to the inhabited earth, the classical *ecumene*. Get too far away (too deep below the surface or too long ago) and they become earth-scientists and geologists. Geographers, he argues, have no interest in looking beyond the atmosphere of the earth. There is no geography of Mars or Jupiter. Geography is about the planet earth and what we have made of it. The notion of home is a deeply resonant one (Blunt and Dowling 2006). In the English-speaking world there is a clear difference between a house and a home. A house is a physical structure of some kind in which people live. To call a house a home denotes a sense of belonging and attachment. This is frequently used in advertising for stores such as IKEA (a home furnishings store) that declares that "home is the most important place in the world":

> Do you live in a house or a home?
> Are you in it for the money or the love?
> Is it perfect or is real … and still perfect?
> Do you look in estate agents' windows or do you look in your own window?
> And think "how lovely!"
> What do you put into it? Just money, or your life and soul?
> What will you get out? A quick profit, or everlasting memories?
> A house can always become a home. Love, not money, is what gives a home a soul.
> And a home's soul is not for sale.

Here, as in humanistic geography, the fully human is pitted against the rational economic actor. The house against the home. Making a house into a home means investing in it emotionally and aesthetically. It involves giving it meaning. It is this word "home" (not without its problems, as we shall see in Chapter 8) that most clearly conveys the geographers' task to Tuan. How do people make the world meaningful? This sums up the spirit of humanistic geography. One of the main ways people can be-in-the-world is through home making.

It was Tuan who first coined the term "humanistic geography" in a paper in 1974 (Tuan 1974). By 1978, an edited collection had been published with the title *Humanistic Geography: Prospects and Problems* (Ley and Samuels 1978). By the early 1980s, a series of key texts had been published in Anglo-American geography that defined the various approaches of humanists in geography (Relph 1976; Tuan 1977; Meinig 1979; Seamon 1979; Buttimer and Seamon 1980; Gold and Burgess 1982). Humanistic geography was and is both a critique of positivism and a positive set of theories based on the continental philosophies of meaning, **phenomenology**, and existentialism.

How is it possible to describe and explain people's **being-in-the-world**? One of the advantages of spatial science is that it has a clear and precise set of largely mathematical models through which the world can be "tested." Underlying this is a philosophy (positivism) that clearly states what counts as truth and facts and what does not. Despite its clarity, however, it has a hard time dealing with notions such as "being-in-the-world." If one person does not fear walking down an urban alley in the dark, this does not disprove the idea that dark alleyways are scary places. How we relate to the world and invest it with meaning is, by and large, not quantifiable or testable. Neither is understanding the meaningful world quite the same as describing places or regions in great detail, as an older tradition of regional geography might have done. "Being-in-the-world" is more than a series of facts. So what methodologies do humanists use?

The general answer is that humanists use qualitative methods. Among these are simple observation of the world around them and human intuition: looking at the world and thinking about it. Humanists tend to analyze their own experience of being-in the-world as a basis for interpreting the lifeworlds of others. Life is fieldwork for the humanist. But humanists also look for evidence of other people's being-in-the-world. One place this can be found is in artistic or cultural products such as paintings, novels, poems, and films. These are all examples of people attempting to convey facets of what it is like to be-in-the-world to others and thus represent evidence that has to be interpreted. Beyond this, some humanists are interested in undertaking qualitative fieldwork in which people are directly asked about their connections to, and experiences of, the world around them. Interviews, oral history, and participant observation are all methods that have been applied by humanistic geographers and that were rarely used before the mid-1970s by geographers (Ley 1974; Rowles 1980). It is a sign of the success of humanistic geography that qualitative methods are now commonplace and at the center of contemporary geographical scholarship (Cloke 2004; Gomez and Jones 2010).

Phenomenology and Existentialism

While geographers such as Buttimer and Cosgrove (Cosgrove 1984; Buttimer 1993) looked to Renaissance humanism for inspiration, the principal inspirations for humanistic geography came from the more recent philosophies of phenomenology and existentialism. These became very popular in the English-speaking world during the late 1960s and early 1970s (when many of the humanistic geographers were undertaking their postgraduate education). The novels and philosophical tracts of writers such as Albert Camus, Jean Paul Sartre, Martin Heidegger, Frederick Neitzche, Maurice Merleau-Ponty, and Franz Kafka were the treasured possessions of young scholars in the humanities and social sciences who passed them around coffee shop gatherings, accompanied by strong coffee and French cigarettes. These advocated a broad philosophy involving the central question of the meaning (and absurdity) of life. How does life become meaningful? they asked. The simple answer was that life was made meaningful through intentional and purposeful human activity. Existence precedes essence, as Sartre put it. Humans, not God or any other external forces, made their own meanings (Samuels 1978). In this sense, there is a direct link between these philosophies of meaning and the earlier proclamations of Pico della Mirandola concerning the ability of "man" (unlike any other being) to produce his own destiny.

While existentialism is a broad philosophy of human meaning that undoubtedly informed humanistic geographers, it is **phenomenology** that was most central to the work of humanistic geographers in the late 1970s and early 1980s (Tuan 1971; Relph 1976; Seamon 1979). Phenomenology is concerned with discovering what things really are – discovering their *essences*. In some ways it is more like a methodology than a philosophy. Philosophers such as Edmund Husserl, Martin Heidegger, and Maurice Merleau-Ponty wanted to develop ways of discovering the core essence of things through a process of "transcendental reduction." Transcendental reduction refers to the mental process of stripping away all that is unnecessary to make a phenomenon what it is and arriving at a kernel which is the essence of that thing (Pickles 1984). So if we ask, for instance, "what is a horse?" we then have to discount all the countless varieties of horse and come to some core sense of horsiness that all horses, however varied, share. Without this core essence the thing would no longer be a horse but would become something else – a donkey or a zebra perhaps. Now consider geographical "objects." A kind of question that had never been asked before in geography was "what is place?" Geographers had asked questions like "what is place X like?" or "how is place X different from place Y?" But the actual nature of place was assumed or taken for granted. Phenomenology asserts that it is important first to know the essence of what it is you are studying. Thus "what is place?" became a key question for a humanistic geographer. Phenomenological philosophers believed that all academic subjects (including the natural sciences) needed phenomenology as a starting point which tells us, in essence, what our subjects are about.

It is from phenomenology, and particularly from the work of Merleau-Ponty, that the notion of "being-in-the-world" arises. The hyphens in "being-in-the-world" symbolize another key term for phenomenologists, **intentionality**. The philosopher Edmund Husserl focused on the tricky issue of human consciousness and what it meant to be conscious. Consciousness, he argued, was always consciousness *of* something. There is no such thing as pure consciousness without something in the world to be conscious of. We are conscious of that large and scary dog over there. We are conscious of the gathering gray clouds or the way that sunlight plays on the surface of a river. We can even be conscious of immaterial things like love or talent. We can be conscious of consciousness. But there is always an "of" involved. This *ofness* of consciousness is described by the word "intentionality." This notion inspired some humanistic geographers to locate in this *ofness* a relation between thinking, feeling, imagining people and the world. Geography informed by phenomenology would not simply be about the world or about people but about people-in-the-world. And this being-in-the-world can be located in consciousness and in intentionality.

The uptake of phenomenology by human geographers is ongoing. Early uses of Husserl, Heidegger, and Merleau-Ponty were quite selective and tended to take small sections of the philosophy rather than an in-depth reading of its total implications (Pickles 1984). Recently, geographers have reengaged with the phenomenological tradition in the guise of, among other things, **nonrepresentational theory** (see Chapter 11) (Dewsbury 2000; McCormack 2003; Wylie 2005).

Space and Place

Humanistic geography grapples with many issues that have become central to human and, particularly, cultural geography. Issues around emotions and affect, the body and performance, for instance, can all find roots in the humanism of the 1970s. Similarly, contemporary methodological debates around qualitative and interpretive methods owe much to the pioneering work of humanistic geographers. But the most enduring legacy of humanistic geography is theoretical engagement with notions of space and place.

The notion of "place" is central to the theoretical development of humanistic geography. Spatial scientists might have used the word place in the sense of "central place theory." Here, place is **location**. It refers to an objective point on the earth's surface that can be simply described by using coordinates. It can also be described in relation to other locations by describing direction and distance. This location is 50 miles southwest of that location. "Place" adds layers to this notion of location. A place has a location but it is also a physical landscape with a particular shape and, perhaps most importantly for humanistic geographers, it has meaning. This attachment of meaning to place is often referred to as a **sense of place**.

Geographical definitions of place since the advent of humanistic geography in the 1970s tend to focus on the combination of location (an objective, definable point in space) and meaning (Tuan 1977; Agnew 1987; Cresswell 2004). Places are locations with meaning. This can be illustrated by the observation that latitude: 51° 30 18 N, longitude: 0° 1 9 W is a location, but London Docklands is a place. While they share the same objective position, London Docklands also includes Canary Wharf, a Docklands museum, office blocks, smart restaurants, and a hi-tech light rail line running through it. The Docklands also has a past. It was a place associated with the docks, with slavery, with a working-class population, and with centuries of immigration. Outside of the museum very little of this past is apparent. As well as being a location, then, place has a physical landscape (buildings, parks, infrastructures of transport and communication, signs, memorials, etc.) and, crucially a "*sense of place*." Sense of place refers to the meanings, both individual and shared, that are associated with a place. While this combination of location, landscape, and meaning is perhaps obvious in a settlement, it is less obvious in places at smaller scales. But even a favorite chair has a particular location (in front of the fireplace perhaps), a physical structure (worn armrests, wobbly legs), and meanings (it is where your father sat when reading stories to you as a child perhaps). Places are not necessarily fixed in place. A ship, for instance, may be shared for months on end by a crew of fishermen and become very much a home place while moving around. To say a place occupies a location is not the same as to say it is stationary. Wherever the ship is at any one moment it is still located somewhere on earth. Places, then, are particular constellations of material things that occupy a particular segment of space and have sets of meanings attached to them.

The term **place** has long been used by geographers but has a relatively recent history as a concept which has been explored for its own sake. Geographers have always been interested in places but not in "place." It is humanistic geographers who brought this to the fore in the 1970s when they insisted that geographers needed to pay attention to the subjective experience of people in a world of places (Tuan 1974; Relph 1976; Buttimer and Seamon 1980). To make human geography fully human, humanists argued, geographers needed to be more aware of the ways in which we bring a particularly human range of emotions and beliefs to our interactions with the physical world. Central to this awareness is the concept of place. As well as referring to things in the world (places), place describes a way of relating to the world. Key here is the idea of "experience." Experience, here, refers to the way we come to know the world through our senses – the way we sense the world around us. It refers to the process of knowing and being in the world rather than the alternative notion of experiences as particular segments of life – such as the experience of moving countries or buying a house. It is this notion of experience as bodily sensing (vision, sound, smell, touch, taste, etc.) that lies at the heart of the humanistic approach to place. Ideas such as "experience" were not in the vocabulary of human geographers in the early 1970s who had been constructing human geography as a "spatial science." Some broadly scientific geographers under the heading of **behavioral geography** had attempted to understand disaggregated individual acts through an understanding of psychological/cognitive processes (Gold 1983). But even under this heading, geographers had tended to think of people as objects and rational beings, or through the idea of relatively simplistic responses to local environments. Behavioral geography, despite its focus on individuals was still broadly positivist and determined to portray itself as scientific. Rational and behavioral beings were not "experiencing" the world and geographers studying them were, and are, certainly not interested in how they experience the world. To focus on sensual experience, therefore, was revolutionary. While the spatial scientists wanted to understand the world and the people in it objectively, in a way that equated people with rocks, or cars, or ice, humanistic geographers focused on the relationship *between* people and the world through the realm of *experience*. Leading humanistic geographer Yi-Fu Tuan writes: "[t]he given cannot be known in itself. What can be known is a reality that is a construct of experience, a creation of feeling and thought" (Tuan 1977: 9). Focusing on place, therefore, attends to how we, as humans, are-in-the-world – how we relate to our environment and make it into place. Place describes the central humanistic engagement with "being-in-the-world."

Relph, in his 1976 book *Place and Placelessness*, describes place, home, and roots as fundamental human needs: "to have roots in a place is to have a secure point from which to look out on the world, a firm grasp of one's own position in the order of things, and a significant spiritual and psychological attachment to somewhere in particular" (Relph 1976: 38). Here, place is tied up with the idea of rootedness and attachment. These are much more than functional goods but moral goods, infused with a positive glow. Relph, like other geographers of the time, was inspired by the writing of the German philosopher Martin Heidegger, and his notions of **dwelling** and *being there*.

Heidegger, throughout his career, had struggled with the nature of "being." To Heidegger, to be was to be *somewhere*. The word he used to describe this was *dasein* – or "being there." Note that this was not simply being in some abstract sense, as if in a vacuum, but being *there*. Human existence is existence *in the world*. This idea of being-in-the-world was developed in his notion of *dwelling*. A way of being-in-the-world was to build a world. Dwelling in this sense does not mean simply to dwell in (and build) a house, but to dwell in and build a whole world to which we are attached. Dwelling describes the way we exist in the world – the way we make the world meaningful, or place-like. Most famously, Heidegger used the image of a cabin in the Black Forest to describe both building and dwelling:

> Let us think for a while of a farmhouse in the Black Forest, which was built some two hundred years ago by the dwelling of peasants. Here the self-sufficiency of the power to let earth and heaven, divinities and mortals enter *in simple oneness* into things, ordered the house. It placed the farm on the wind-sheltered mountain slope looking south, among the meadows close to the spring. It gave it the wide overhanging shingle roof whose proper slope bears up under the burden of snow, and which, reaching deep down, shields the chambers against the storms of the long winter nights. It did not forget the altar corner behind the community table; it made room in its chamber for the hallowed places of childbed and the "tree of the dead" – for that is what they call a coffin there: the *Totenbaum* – and in this way it designed for the different generations under one roof the character of their journey through time. A craft which, itself sprung from dwelling, still uses its tools and frames as things, built the farmhouse. (Heidegger 1993: 300)

In this cabin everything seemed to have its place and the cabin sat almost organically in the natural world, linking the cosmological to the everyday. Here was the model kind of building and dwelling – a model kind of being-in-the-world. Obviously, it is not possible in a modern urban world for everyone to live like this, but the model that Heidegger describes becomes a way to thinking about the way we all dwell in the world. When we are confronted with a new environment, even the temporarily inhabited seat on a busy train, we do something to make it our own. We place a book on a table, or a bag next to us. Workers in open plan offices attempt to decorate the small piece of prefabricated "wall" that surrounds their computer with postcards. They might bring in their own mouse mat or even a pot plant. They try to make a little bit of the world more home-like. More like a cabin in the Black Forest.

Humanists are often compared to the older tradition of regional geography. Humanists, it is said, want to return to the particular and to retreat from the universal claims of spatial science. Humanistic work is **idiographic** while spatial science is **nomothetic**. The concern with place is given as evidence for this. Place as meaningful location is about the particular – about this particular place (not unlike the insistence on the ways of life in particular regions in the work of Vidal, for instance). This misses the point. In many ways humanistic geography is more general and more universal than spatial science. When Tuan or Relph write about place, they are not writing about Madison, Wisconsin, or Toronto. They are writing about place as an idea. They want to know what the essence of place, any place, is. What could be more universal than that? Or think about Anne Buttimer writing about the human spirit. Again, this is an ambitious stab at touching the universal and cosmological. Very few of the classic texts in humanistic geography are aimed at describing, interpreting, or explaining one particular place. In this sense, the work of humanistic geographers is far from idiographic and has been subsequently criticized by radical geographers for its universal and transhistorical ambitions.

Humanistic Geography's Afterlives

Literary geographies, geohumanities, geopoetics

Before the advent of humanistic geography, geographers generally had a limited set of ways of gathering information in order to conduct their research. Regional geographers could collect facts, observations, and measurements to construct their regional accounts. Quantitative geographers relied heavily on whatever was quantifiable and calculable. There were, however, a very small number of geographers who looked elsewhere for information and insight. This included geographers who looked to the arts, and particularly to literature, for access into geographical imaginations across time and space. One of the pioneers in this regard was J.K. Wright, who wrote early papers about the geographers of Homer and Dante (Wright 1924, 1926). Geographers had occasionally looked to the novel to provide accounts of regions in the past, or as a colorful decoration for more serious description. These included accounts of Thomas Hardy's Wessex (Darby 1948) or William Faulkner's American South (Aitken 1979). Under the banner of humanism, geographers revealed an increasing willingness to engage with creative writing – particularly the novel (Tuan 1978; Pocock 1981). This represents part of a wider willingness to engage with the creative arts in general, particularly painting (Cosgrove 1984; Daniels 1993). Humanistic geographers advocated a serious engagement with creative literature to uncover something of the connections between people and the world revealed in the creative process itself:

> Imaginative literature, a relatively small subset of the vast heterogeneous field of the printed word, has recently been espoused – "used" would be an inappropriate description – by a growing band of geographers seeking alternative perspectives and insights in the study of man–environment relationships. Disillusioned by an era of logical positivism, maybe shell-shocked by the quantitative revolution, perhaps rediscovering the literary heritage of geography – whatever the reason, the realm of literature has attracted increasing attention from our eclectic discipline. (Pocock 1981: 9)

Here, Douglas Pocock enlists literature as part of the struggle against "logical positivism" and the "quantitative revolution." Nothing could be further from the scatter plots and regression lines of spatial scientists, perhaps, than imaginative literature.

When a novelist (or poet) writes, they engage with the world. The novel they write is not a description of the world. Novels, on the whole, are not a good source for the mining of "facts." What they are, though, is evidence of the novelist's "being-in-the-world." Through the novelist's eyes we see the world anew by sharing, while we are reading, the novelist's position. While they may not be good sites for the discovery of facts, novels may be ideal places for the exploration of "truths." Or, as Pocock has put it: "the truth of fiction is a truth beyond mere facts" (Pocock 1981: 11). In the work of some humanistic geographers, there is an assumption that authors are special people. They are people with heightened powers of both observation and expression. Writers are those who are able to articulate powerful themes that most of us are incapable of expressing. There is certainly an inherent elitism in such claims and it contradicts the value placed on ethnographic methods by other humanistic geographers who pioneered talking to "everyday people" as a source of important insight.

To some (Meinig 1983), literature is a model for how geographers might consider writing. While there have been arguments aplenty about whether geography is a science, a social science, or part of the humanities, there have been few who have claimed it as an art. Yet this is one of the ways in which humanistic geographers (and some regional geographers before them) have used novels – as a model for geographical writing that is able to capture the feeling of landscape and place.

Since the heyday of humanistic geography in the 1970s and 1980s, the geographical engagement with creative literature has expanded. There is now a journal, *Literary Geographies*, that showcases both geographers engaging with literature and literary scholars thinking geographically. While this is clearly an example of work in the humanities, it has expanded well beyond a narrow definition of humanistic geography and includes many scholars informed by feminism, post-structuralism, and other theoretical traditions that will be explored in later chapters. Nevertheless, there are clear continuities between the pioneering work of humanistic geographers on "geography and literature" and current work on "literary geographies." A key figure in the field of literary geographies is Sheila Hones. Hones takes us beyond both the sense of literature as a record of the world and the use of literature as evidence for being-in-the-world. Instead, she asks us to think of the novel as an event, happening, or performance. Hones takes the relations between author, book, and reader seriously, relations that have their own literary geographies that allow literature to help constitute "real world" geographies that connect the author, text, and reader (Hones 2022). While the humanistic tradition of "geography and literature" was primarily concerned with what literature could provide geographers as evidence, "literary geographies" is equally interested in what geographical thinking might contribute to the study of literature and the way it works spatially. In this sense, Hones identifies literary geographies as an "interdiscipline" in which literary scholars have to not take geographical ideas such as "space" or "place" for granted and geographers have to stop taking literary ideas such as "author," "text," or "reader" for granted. Hones engages with a single novel, Culum McCann's *Let the Great World Spin* to "... explore a collaborative and interdisciplinary approach to the narrative spatiality of a work of contemporary fiction through a combination of theory and method in literary studies with theory and method in cultural geography" (Hones 2014: 3). Hones identifies three kinds of space for literary geography to focus in on. First there are the spaces of the text, perhaps the traditional domain of earlier geography and literature work. These include "its locations, distances, and networks" (Hones 2014: 8). The second "is the 'unending library' of intertextual literary space: in this case, the uncontained intertextual space that opens out from *The Great World* with every quotation McCann includes and every literary reverberation the reader senses" (Hones 2014: 8). Here, Hones is referring to the connections that the text of *The Great World* makes with other texts (and their locations, distances, and networks) that inform it and to which it alludes. The third is "the sociospatial dimension of the collaboration of author, editor, publisher, critic, and reader without which reading (and thus text) could not happen" (Hones 2014: 8). Here, Hones is referring to extra-textual spaces within which any text exists – the geography within which a text is written, produced, distributed, and read. Together, these take us a long way from the mining of novels for geographical facts or even the humanistic interest in creative writing as evidence of human relations to the environment.

The interplay of literary studies and geographical theory in literary geographies is indicative of a much wider interdisciplinary field of geohumanities (Cresswell and Dixon 2017). The field of geohumanities emerged in a quite intentional way as a project of the American Association of Geographers (AAG) to both boost the traditional role of geography as a humanities discipline and to increase interdisciplinary connections between geography and other humanities-based disciplines. A Symposium on Geography and the Humanities was held in Charlottesville, Virginia in 2007 that kickstarted a series of themed geography and humanities sessions at the annual conferences of the AAG. The Symposium and subsequent sessions led to two edited collections in 2011, *GeoHumanities: Art, History, Text at the Edge of Place* (Dear *et al.* 2011) and *Envisioning Landscapes, Making Worlds: Geography and the Humanities* (Daniels *et al.* 2011). This was followed in 2015 by the launch of a new AAG journal, *GeoHumanities: Space, Place, and the Humanities*. Both the books and the journal are notable for the range of disciplines represented in within them and on the editorial teams. Like literary geographies, geohumanities is an interdiscipline. Also, like literary geographies, geohumanities goes well beyond the remit of traditional humanistic geography at the same time as it clearly builds on the possibilities that were opened up in the 1970s. Some of the key players in the 2007 Symposium, such as Denis Cosgrove, Stephen Daniels, and J. Nicholas Entrikin were also scholars with at least one foot in the older traditions of humanistic geography. The edited books included essays some of the key figures in humanistic geography such as Yi-Fu Tuan and David Lowenthal as well as younger scholars clearly writing in a humanistic vein, such as Veronica della Dora. The geohumanities project continues the humanistic tradition of thinking of the earth as home, embraces newer forms of enquiry enabled by geolocatable technologies, pays attention to critical perspectives on space, place, landscape and other concepts that have emerged post humanistic geography and encourages a flowering of creativity in geography across artistic forms of representation. The "practices and curations" section of the *GeoHumanities* journal includes all kinds of creative geographical practice by geographers and others that fulfills Donald Meinig's 1983 hope that geography might one day properly be considered an art (Meinig 1983). There is a significant intersection between geohumanities and what are now called "creative geographies" – a field in which geographers are not simply interpreting the creative works of authors, artists, film-makers, and others, but are collaborating with them and undertaking self-consciously creative works themselves (Hawkins 2014).

One area of attention within geohumanities is geopoetics. The idea of geopoetics has its origins in 1979 in the work of Kenneth White and his International Institute of Geopoetics, which he founded 10 years later, in 1989. White described geopoetics as a reaction to the recognition that the earth is in danger and that poetics might have a role in confronting that danger. He saw geopoetics as a site where poetry could be inspired by science and vice versa. The geographer, Eric Magrane, developed this idea within geohumanities with an explicit "invitation for geopoetic texts and practices that draw on the work of poets as well as geographers, for an enchanted, earthy, and transaesthetic approach that moves to bring together contemporary poetics, particularly in the realm of ecopoetics, with critical human geography" (Magrane 2015: 87). Magrane draws on many influences including Alexander von Humboldt's interest in the aesthetics of landscape and Anne Buttimer's call for an approach that "involves poetics, aesthetics, emotion and reason in the quest for wiser ways of dwelling" (Buttimer 2010: 35). Magrane notes the increasing number of geographers working both creatively and critically with poetry, both as subject matter and as a mode of expression, as well as poets who work with a geographical/ecological focus. More than this, he outlines geopoetics as a kind of "geophilosophy." Geopoetics does not necessary involve poetry in the strictest sense, but more an approach that thinks of research in a way that is quite distinct from previous modes – and especially more scientific forms of geography.

> … a research agenda might not even begin with a question in the traditional social science sense, but with an immersion in a site that first pays close attention to the materialities and encounters of the site, and then intervenes in the site through a geopoetic form that is immanent to the site itself, one that is designed to enact, perform, comment on, critique, and, perhaps even recalibrate, the site itself. (Magrane 2015: 95–66)

Here, Magrane calls for a research process that starts from a sense of openness and attentiveness to a site or place – not a process driven by an instrumental question. The process then might lead to a different kind of result. Not necessarily a research paper but an intervention in the site, a performance, or, perhaps, a poem. Magrane's call resulted in an edited book featuring geographers, poets and others expanding on the idea of geopoetics and showing how, in practice, this new sensibility might actually work (Magrane *et al.* 2020). Both the book *Geopoetics in Practice* and the journal *GeoHumanities* are full of examples of geographers exploring the possibilities of geopoetic research processes and forms of expression (see Acker 2020; Jones 2022). In her contribution to *Geopoetics in Practice*, geographer and poet Maleea Acker juxtaposes (on left hand pages) poems and prose fragment vignettes of time she spent in Ajijic, Mexico with quotes from a range of geographers and others (on right hand pages). In this way a kind of dialogue is created between the authors that creates the possibility of sparks of insight on the part of the reader. In her words, "[t]he prose work sets the stage; the vignettes are my observations; the poems are the creative culmination of the study. I argue that, together, they offer a new way of doing geography within geopoetics" (Acker 2020: 132). In "05BH004 (1915–2019)" Samantha F. Jones, a physical geographer and also a poet, uses historical hydrometric data from a gauging station in Treaty 7 lands in what is now downtown Calgary, Canada, to form a set of constraints where each line of poem corresponds to a year of data between 1915 and 2019. The resulting poem is then written in a sinewy form that reflects the river. The poem, she writes, "is a case study in how data can be embedded directly into poetic form to perform process and create a poetic style that has potential to act as both a science communication tool and a method for data interrogation. This fusion of scientific data and the literary arts uses an interdisciplinary approach to bridge the gap between the sciences and the arts" (Jones 2022: npn).

Ethnographies of lifeworlds

If the arts were one realm where humanistic geographers looked for evidence of geographic imaginations, then another was in the lives of people themselves. While there were some exceptions, there are relatively few examples of geographers engaging with human subjects in the form of interview, participant observation or other ethnographic techniques before the 1970s. When J.K. Wright outlined the need for geosophy in 1947 – an exploration of the geographical imaginations of others – this opened up the possibility of both examining literature and the arts, as he had done 20 years earlier, and simply asking people about their geographies. Despite his provocation, the use of ethnographic techniques in geography was very rare before the 1970s. A properly human geography involved engaging with humans and their lifeworlds.

David Seamon directly applied his understanding of phenomenology to practical considerations of how people inhabit their worlds through bodily movement – which he defined as "*any spatial displacement of the body or bodily part initiated by the person himself*. Walking to the mailbox, driving home, going from house to garage, reaching for scissors in a drawer – all these behaviours are examples of movement" (Seamon 1980: 148). Seamon was keen to discover the essential experiential character of movement. Like spatial scientists before him he wanted to transcend specific examples and provide a general account of everyday movement. It was not so much what A or B did when they moved but how humans inhabit the world through movement. As a phenomenologist it was the "essence" he was after. He conducted a number of experiments with groups of people who were "interested enough in their own day-to-day contact with the geographical lifeworld to meet weekly for several months and share, in a group context, personal experiences relating to everyday movement and other related themes" (Seamon 1980: 151–152) and came to the conclusion that most everyday movement takes the form of *habit*. People will drive the same route to work and back every day without thinking. Occasionally people who have moved house find themselves going to their old house and only realize it when they arrive at the front door. People can habitually reach for scissors in the drawer while engaging in conversation. We can type without thinking. Such movements appear to be below the level of conscious scrutiny.

Seamon invokes the metaphor of dance to describe the sequence of preconscious actions used to complete a particular task such as washing the dishes. He describes this sequence of preconscious movements as a *body-ballet*. Throughout the day an individual will perform a series of these body-ballets to make up a *time–space routine*. In a particular place, such as a town center or outside the school gates when the school day is ending, people's time–space routines come together to form a *place-ballet* – a combination of different people's time–space routines which produces a sense of place. Seamon recounts some observations of the urban theorist Jane Jacobs to illustrate this notion of "place-ballet":

The stretch of Hudson Street where I live is each day the scene of an intricate sidewalk ballet. I make my own first entrance into it a little after eight when I put out the garbage can, surely a prosaic occupation, but I enjoy my part, my little clang as the droves of junior high school students walk by the center of the stage dropping candy wrappers … While I sweep up the wrappers I watch the other rituals of the morning: Mr. Halpart unhooking the laundry's handcart from its mooring to a cellar door, Joe Cornacchia's son-in-law stacking out the empty crates from the delicatessen, the barber bringing out his sidewalk folding chair, Mr. Goldstein arranging the coils of wire which proclaim the hardware store is open, the wife of the tenement's superintendent depositing her chunky three-year-old with a toy mandolin on the stoop, the vantage point from which he is learning the English his mother cannot speak. Now the primary children, heading for St. Luke's, dribble through to the west, and the children from P.S. 41 heading towards the east. (Jane Jacobs quoted in Seamon 1980: 160–161)

In this scene, as in many other scenes familiar to any one of us, the individual bodily routines coalesce into something larger than themselves in a fairly regular and predictable way. This produces a relatively enduring sense of place. This is not a sense of place produced through a world of cognition and meaning but through the repetition of activity – through practice.

Graham Rowles is a social geographer who uses a humanistic approach in his work. For many years he has been investigating the geographical experience of elderly people. An early example of this is his book *Prisoners in Space? Exploring the Geographical Experience of Older People* (1978). Rowles notes how much of humanistic geography appeared as other-worldly and esoteric – a world of armchair geographers who had little to say to the world of "clinical practice, and the formation of public policy" (Rowles 1980: 56), for instance. In contrast to this, Rowles was seeking to use the theoretical insights of humanism to directly inform the practice of how society related to the elderly. Rowles spent a great deal of time with five men and women ranging in age from 68 to 83. While he was with them (over a three-year period), he gathered "data" through "immersion within the participants' lifeworlds" pursuing "creative dialogue" and "mutual discovery." He was left with "reams of notes, photographs, and sketch maps; and many hours of taped conversations" (Rowles 1980: 57). These people inhabited a decaying area of a city, seemingly inhospitable to their aging bodies. Rowles expected that his research would focus on the impact of such an environment on their lives. He was surprised, therefore, when his participants spoke of other kinds of geographies instead:

> To my initial bewilderment, Marie seemed uninterested in these questions I had culled from extensive review and reflection on the literature of aging. She was reluctant to talk about physical restriction, reduced access to services, spending more time at home, problems of social abandonment, or fears of the future. Instead, as we sat in her parlor poring over treasured scrapbooks in which she kept a record of her life, she would animatedly describe trips she had taken to Florida many years previously. She would muse on the current activities of her granddaughter in Detroit, a thousand miles distant. She would describe incidents in the neighborhood during the early years of her residence. Blinded by preconceptions, I could not comprehend at first the richness of the taken-for-granted lifeworld she was unveiling. (Rowles 1980: 55)

Rowles suggests that his initial expectations were informed by a literature on aging that focused almost entirely on observable behavior (such as older people making shorter trips that were less frequent than younger people). Little research has focused on the "perceptual orientation" of older people, "their emotional and generally pre-reflective identification with places of their lives and the way in which life experience is incorporated within the constitution of a lifeworld" (Rowles 1980: 57). Through his three years of conversations, Rowles traces four aspects of what being-in-the-world is like for an older person. *Action*, he argues, becomes diminished as bodies become less capable within an immediate spatial context. Space is used more efficiently as people become less adept at using spaces they are not familiar with. Routine becomes more important. *Orientation* within the world becomes more focused on particular known spaces such as home. One of his participants, Stan, for instance, "knew the paths affording shade on a hot afternoon and safety on the icy days he dreaded. He was aware of the street crossings most hazardous during lunchtime traffic. He had internalized a series of detailed 'specific' linear schemata which facilitated his routinized movement under diverse environmental conditions" (Rowles 1980: 59). The lifeworlds of the elderly were also marked by a reservoir of *feeling* which attached themselves to particular locations. A place where one participant once danced became particularly important. Feeling was often more important than the physical condition of a place for the elderly. Finally, Rowles noted the role of *fantasy*. Elderly people, he suggests, inhabit places far away in time and space from the here and now, places associated with their past or with the worlds of their extended family. In this way immediate location is transcended.

These observations, Rowles argues, would not be possible without taking a humanistic perspective on the research. They could not be deduced from the immediately observable, measurable, and mappable. They only emerged after a process of immersion in the lifeworld of others that leads to what he calls "understanding." Understanding, he argues, is beyond that which is generalizable. It is a "deeper level of awareness which arises from drawing close enough to a person to become a sympathetic participant within her lifeworld and to have her integrally involved in one's own" (Rowles 1980: 68). To Rowles, understanding should be a "major goal in scholarly endeavor." It can be contrasted with *explanation*:

> By manipulating emergent themes in a search for more sophisticated *explanation*, there is the ever-present danger of falling into the reductionist trap of considering as legitimate only those aspects of experience which can be generalized. In so doing much of the richness of individual experience is cast aside. An understanding of person as subject – the author of a unique biography lived within a colorful lifeworld – is rejected. Ironically, as we progress toward more sophisticated explanation in developing a geography of growing old, our understanding may become progressively impoverished. Yet to the humanist geographer it is the understanding that is ultimately more important. (Rowles 1980: 69)

Geographers increasingly talked to people during the 1970s and 1980s. Most were informed by humanistic geography and the need to take the subjective lives of people seriously. David Ley explored the worlds of the "black inner city" in Philadelphia in a book length ethnography (Ley 1974). John Western used ethnography to chart the lived world of apartheid South Africa in *Outcast Cape Town* (Western 1981) and David Sibley conveyed the lifeworlds of British Gypsy/Travellers in *Outsiders in Urban Society* (Sibley 1981). Since 1980 it is no longer unusual for human geographers to engage with people and their geographical knowledge in research. Through the use of interviews, focus groups, participant observation, diaries, video logs, photography and a host of other increasingly creative methods, geographers have explored the lifeworlds of people other than themselves including people living "off the grid" in Canada (Vannini and Taggert 2015), urban explorers in London (Garrett 2013), and homeless people in Athens (Bourlessas 2018). While geographers using ethnographic methods have moved beyond humanism as it was in the 1970s, and often bring other critical theoretical lenses to bear in their work, there is no doubt that the long history of humanism in geography paved the way for all forms of qualitative methodology as practiced today.

Conclusion: Humanism Is Dead – Long Live Humanism

Humanistic geography was subject to critique almost as soon as it emerged. To positivist geographers (seeking explanation rather than understanding), humanism was merely subjective, untestable, a matter of opinion, as one piece of research could not be replicated or falsified. To others, its style was overly esoteric and abstract, producing what one critic called a "Mandarin dialect" (Billinge 1983). To more radical geographers, as we shall see, it did not pay enough attention to systematic and structural arrangements of power – of oppression, exploitation, and domination (let alone resistance to these). Humanists, they argued, inhabited a lofty plane of elitist assumptions drawn from high art, classic literature, and elite culture in general.

Despite these critiques, however, humanistic geography has had a profound influence on human geography in the past 30 years. In many ways its insights have become taken for granted, part of the very bedrock of the disciplinary common sense. Thematically, ideas surrounding space and place, the emotions and senses, the body, and the performance of everyday life continue to be important in human (particularly cultural) geography research (Cresswell 1996; Duncan 1996; Adams *et al.* 2001). Humanism (along with Marxism) was a key influence on the emergence of a reinvigorated cultural geography in the late 1980s (Cosgrove 1984; Jackson 1989; Daniels 1993; Duncan and Ley 1993). The critique of the notions of objectivity and the scientific method has continued in different guises. The humanistic insistence on the locatedness of knowledge in human consciousness has been central to this. This, in turn, has been important in the acceptance of a range of qualitative methods including participant observation, textual and visual analysis, and self-reflection which are now among the principal ways of conducting research in human geography.

Few geographers now refer to themselves as humanistic. This may be due to both the critiques of humanism from radical geographers more concerned with issues of power, domination, and resistance, and the fact that many of the lessons of humanism have become taken for granted. There are, however, some who continue to explicitly develop the notion of humanistic geography. Some of its original pioneers continued to write within a humanistic framework (Buttimer 1993; Tuan 1998; Cosgrove 2001). Others have acknowledged the continuing salience of a humanistic framework within their work (Entrikin 1991; Rodaway 1994; Pile 1996; Sack 1997; Adams *et al.* 2001). A wonderful example of work that straightforwardly continues the humanist tradition in geography is Veronica della Dora's *The Mantle of the Earth* (2021). In this book, she explores the metaphor of the mantle or veil (along with ideas of weaving and textiles) as they have traveled through history from Greek mythology, through medieval Christianity to the digital worlds of today. The earth's mantle is the thin surface crust – the very sphere that Tuan insisted was at the core of geography as the study of the earth as the home of people (Tuan 1991). At the center of her work, as with all humanistic geography, is the human imagination. The various mantle related metaphors that della Dora explores tells us something about how humans understood the worlds they lived in. They provide evidence for what it is, and what it has been, to be-in-the-world.

> Cloak, garment, vernicle. Stage, curtain, drape, veil. Tablecloth, carpet, surface. Web, tapestry, skin. Multiple incarnations of the earth's mantle metaphor speak of the unbounded imaginative power of the human mind and its continuous attempt to comprehend the planet we inhabit. Spanning three millennia of human history, all these metaphors share a basic matter of fact: our experience and knowledge of the planet are by necessity superficial. We dwell and move, struggle and thrive, live and die on the terrestrial crust or, perhaps more precisely, within the narrow bounds of the biosphere. Each variant of the mantle metaphor nonetheless reveals a different type of engagement with the terrestrial crust (or the biosphere): awe and respect in front of mystery; a desire for penetration and mastery; integration and interconnectedness. (della Dora 2021: 274)

Recently a "post-humanist" geography has arisen which, rather than being against humanism, seeks to take some of the insights of humanism into different orbits (McCormack 2003; Wylie 2005; Williams *et al.* 2019), ascribing forms of agency to animals and non-sentient objects (see Chapter 12). Post-humanists are troubled by the hubris of centering humans in the way proposed by Pico della Mirandola in the Renaissance. This centering of mankind is, afterall, what has led us to the geological era we now know as the Anthropocene – an era forever materially marked by the polluting excesses of humanity. Post-humanists also want to focus on the relationality of humans – the fact that humans are always parts of networks with living and non-living nonhumans that produce particular forms of agency that would be beyond the capacity of mere humans. Geographers and others have also begun to re-engage with phenomenology more directly in **postphenomenology** and **critical phenomenology** (Kinkaid 2021). Postphenomenology continues with the phenomenological interest in bodily experience but is critical of the way phenomenology in the mode of Merleau-Ponty or Heidegger overly values the singularity of human experience while relegating animals and objects to the status of mere objects. Post-phenomenologists thus have much in common with geographers in the "more-than-human" and relational modes explored in Chapters 11 and 12. They want to see intentionality and experience as the product of relations between humans and others rather than the capacity of singular knowing human subject. Eden Kinkaid has critiqued post-phenomenologists (like humanists in general before) for glossing over power differentials among and between humans. Critical phenomenology, Kinkaid argues, is actually a longstanding tradition that has:

> … described how worlds are experienced and embodied differently by subjects inhabiting forms of embodied difference, including gender, race, sexuality, and disability. By attending to how such subjects encounter worlds, these critical phenomenologies are not merely concerned with these subjects and their experience; rather, critical phenomenologies seek to illuminate how bodies, objects, spaces, and intersubjective worlds are (unevenly and differentially) composed. (Kinkaid 2021: 301)

Critical phenomenologists, Kinkaid tells us, explore the experience of being, for instance, Black, queer, or disabled. Kinkaid uses the insights of critical phenomenology to critique postphenomenology for its lack of attention to the differences between bodies that enter into relations with each other on a playing field that is far from even. In their keenness to get beyond the individual intentional human being and to embrace the importance of relationality between human and nonhuman bodies, post-phenomenologists have made the mistake of erasing important social differences that influence how bodies become part of networks of relations.

References

Acker, M. (2020) Lyric geography, in *Geopoetics in Practice* (eds E. Magrane, L. Russo, S. de Leeuw, and C. Santos Perez), Routledge, London, pp. 132–162.

* Adams, P. C., Hoelscher, S. D., and Till, K. E. (2001) *Textures of Place: Exploring Humanist Geographies*, University of Minnesota Press, Minneapolis.

Agnew, J. A. (1987) *Place and Politics: The Geographical Mediation of State and Society*, Allen & Unwin, Boston, MA.

Aitken, S. (1979) Faulkner's Yoknapatawpha County: a place in the American South. *Geographical Review* 69 (3), 331–348.

Billinge, M. (1983) The Mandarin dialect: an essay on style in contemporary geographical writing. *Transactions of the Institute of British Geographers*, 8, 400–420.

Blunt, A. and Dowling, R. M. (2006) *Home*, Routledge, London.

Bourlessas, P. (2018) 'These people should not rest': mobilities and frictions of the homeless geographies in Athens city. *Mobilities*, 13(5), 746–760.

Buttimer, A. (1971) *Society and Milieu in the French Geographic Tradition*, Published for the Association of American Geographers by Rand McNally, Chicago.

Buttimer, A. (1993) *Geography and the Human Spirit*, Johns Hopkins University Press, Baltimore.

Buttimer, A. (2010) Humboldt, Granö and geo-poetics of the Altai. *Fennia-International Journal of Geography*, 188(1), 11–36.

Buttimer, A. and Seamon, D. (1980) *The Human Experience of Space and Place*, St. Martin's Press, New York.

Cloke, P. J. (2004) *Practising Human Geography*, Sage, Thousand Oaks, CA.

Cosgrove, D. E. (1984) *Social Formation and Symbolic Landscape*, Croom Helm, London.

Cosgrove, D. E. (2001) *Apollo's Eye: A Cartographic Genealogy of the Earth in the Western Imagination*, Johns Hopkins University Press, Baltimore.

Cresswell, T. (1996) *In Place/Out of Place: Geography, Ideology and Transgression*, University of Minnesota Press, Minneapolis.

* Cresswell, T. (2004) *Place: A Short Introduction*, Blackwell, Oxford.

Cresswell, T. and Dixon, D. P. (2017) GeoHumanities, in *International Encyclopedia of Geography: People, the Earth, Environment, and Technology* (eds D. Richardson, N. Castree, M. F. Goodchild, A. Kobayashi, W. Liu, and R. A. Marston), Wiley Online Library. https://doi.org/10.1002/9781118786352.wbieg1169.

Daniels, S. (1993) *Fields of Vision: Landscape Imagery and National Identity in England and the United States*, Polity Press, Cambridge.

Daniels, S., DeLyser, D., Entrikin, J. N., and Richardson, D. (eds) (2011) *Envisioning Landscapes, Making Worlds: Geography and the Humanities*, Routledge, London.

Darby, H. C. (1948) The regional geography of Thomas Hardy's Wessex. *Geographical Review*, 38, 426–443.

Dear, M. J., Ketchum, J., Luria, S., and Richardson, D. (eds) (2011) *GeoHumanities: Art, History, Text at the Edge of Place*, Routledge, London.

Dewsbury, J. D. (2000) Performativity and the event: enacting a philosophy of difference. *Environment and Planning D: Society and Space*, 18, 473–496.

della Dora, V. (2021) *The Mantle of the Earth*, University of Chicago Press, Chicago, IL.

Duncan, J. S. (1980) The superorganic in American cultural geography. *Annals of the Association of American Geographers*, 70, 181–198.

Duncan, N. (ed.) (1996) *Bodyspace: Destabilizing Geographies of Gender and Sexuality*, Routledge, New York.

Duncan, J. S. and Ley, D. (eds) (1993) *Place/Culture/Representation*, Routledge, New York, p. 352.

Entrikin, J. N. (1991) *The Betweenness of Place: Towards a Geography of Modernity*, Johns Hopkins University Press, Baltimore.

Garrett, B. (2013) *Explore Everything: Place-Hacking the City*, Verso, London.

Gold, J. (1983) Behavioral and perceptual geography. *Progress in Human Geography*, 7(4), 578–586.

Gold, J. and Burgess, J. (eds) (1982) *Valued Environments*, Unwin Hyman, London, pp. 73–99.

Gomez, B. and Jones, J. P. (2010) *Research Methods in Geography: A Critical Introduction*, Wiley-Blackwell, Oxford.

Hawkins, H. (2014) *For Creative Geographies: Geography, Visual Arts and the Making of Worlds*, Routledge, London.

Heidegger, M. (1993) *Basic Writings: From Being and Time (1927) to The Task of Thinking (1964)*, Harper, San Francisco.

Hones, S. (2014) *Literary Geographies: Narrative Space in Let the Great World Spin*, Palgrave Macmillan, New York.

Hones, S. (2022) *Literary Geographies*, Routledge, London.

Jackson, P. (1989) *Maps of Meaning*, Unwin Hyman, London.

Jones, S. F. (2022) 05BH004 (1915–2019): generation of poetic constraints from river flow data. *GeoHumanities*, 1–5. https://doi.org/10.1080/2373566X.2021.1990783.

*Kinkaid, E. (2021) Is post-phenomenology a critical geography? Subjectivity and difference in post-phenomenological geographies. *Progress in Human Geography*, 45, 298–316.

Ley, D. (1974) *The Black Inner City as Frontier Outpost: Images and Behavior of a Philadelphia Neighborhood*, Association of American Geographers, Washington, DC.

Ley, D. (1980) Geography without man: a humanistic critique. University of Oxford, Research Paper 24.

*Ley, D. and Samuels, M. (eds) (1978) *Humanistic Geography: Prospects and Problems*, Croom Helm, London, p. 352.

*Lowenthal, D. (1961) Geography, experience, and imagination: towards a geographical epistemology. *Annals of the Association of American Geographers*, 51, 241–260.

Magrane, E. (2015) Situating geopoetics. *GeoHumanities* 1 (1), 86–102.

Magrane, E., Russo, L., de Leeuw, S., and Santos Perez, C. (eds) (2020) *Geopoetics in Practice*, Routledge, London, pp. 1–13.

McCormack, D. P. (2003) The event of geographical ethics in spaces of affect. *Transactions of the Institute of British Geographers*, 28, 488–507.

Meinig, D. (ed.) (1979) *The Interpretation of Ordinary Landscapes*, Oxford University Press, Oxford.

Meinig, D. (1983) Geography as an art. *Transactions of the Institute of British Geographers*, 8, 314–328.

Park, R. (1936) Human ecology. *American Journal of Sociology*, 42, 1–15.

Pickles, J. (1984) *Phenomenology, Science, and Geography: Spatiality and the Human Sciences*, Cambridge University Press, New York.

Pickles, J. (1987) *Geography and Humanism*, Geo Books, Norwich.

Pile, S. (1996) *The Body and the City: Psychoanalysis, Space and Subjectivity*, Routledge, London.

Pocock, D. C. D. (ed.) (1981) *Humanistic Geography and Literature: Essays on the Experience of Place*, Croom Helm, London, pp. 130–141.

* Relph, E. (1976) *Place and Placelessness*, Pion, London.

Relph, E. C. (1981) *Rational Landscapes and Humanistic Geography*, Croom Helm, London.

Rodaway, P. (1994) *Sensuous Geographies: Body, Sense, and Place*, Routledge, London.

Rowles, G. (1980) Towards a geography of growing old, in *The Human Experience of Space and Place* (eds A. Buttimer and D. Seamon), Croom Helm, London, pp. 55–72.

Sack, R. D. (1997) *Homo Geographicus*, Johns Hopkins University Press, Baltimore.

Samuels, M. (1978) Existentialism and human geography, in *Humanistic Geography: Prospects and Problems* (eds D. Ley and M. Samuels), Croom Helm, London, pp. 22–40.

Sauer, C. (1996) The morphology of landscape, in *Human Geography: An Essential Anthology* (eds J. A. Agnew, D. Livingstone, and A. Rogers), Blackwell, Oxford, pp. 296–315.

* Seamon, D. (1979) *A Geography of the Lifeworld: Movement, Rest, and Encounter*, St. Martin's Press, New York.

Seamon, D. (1980) Body-subject, time-space routines, and place-ballets, in *The Human Experience of Space and Place* (eds A. Buttimer and D. Seamon), Croom Helm, London, pp. 148–165.

Sibley, D. (1981) *Outsiders in Urban Societies*, St Martins, New York.

Tuan, Y.-F. (1971) Geography, phenomenology, and the study of human nature. *Canadian Geographer*, 15, 181–192.

Tuan, Y.-F. (1974) Space and place: humanistic perspective. *Progress in Human Geography*, 6, 211–252.

* Tuan, Y.-F. (1977) *Space and Place: The Perspective of Experience*, University of Minnesota Press, Minneapolis.

Tuan, Y.-F. (1978) Literature and geography: implications for geographical research, in *Humanistic Geography: Prospects and Problems* (eds D. Ley and M. Samuels), Croom Helm, London, pp. 194–206.

Tuan, Y.-F. (1991) A view of geography. *Geographical Review*, 81, 99–107.

Tuan, Y.-F. (1998) *Escapism*, John Hopkins University Press, Baltimore.

Vannini, P. and Taggert, J. (2015) *Off the Grid: Re-assembling Domestic Life*, Routledge, London.

Western, J. (1981) *Outcast Cape Town*, University of Minnesota Press, Minneapolis.

* Williams, N., Patchett, M., Lapworth, A., and Roberts, T. (2019) Practising post-humanism in geographical research. *Transactions of the Institute of British Geographers*, 44(4), 637–643.

Wright, J. K. (1924) The geography of Dante. *Geographical Review*, 14, 319–320.

Wright, J. K. (1926) Homeric geography. *Geographical Review*, 16, 669–671.

* Wright, J. K. (1947) Terrae incognitae: the place of the imagination in geography. *Annals of the Association of American Geographers*, 37, 1–15.

Wylie, J. (2005) A single day's walking: narrating self and landscape on the South-West Coast Path. *Transactions of the Institute of British Geographers*, 30, 234–247.

Chapter 7

Marxist Geographies

> There is an ecological problem, an urban problem, an international trade problem, and yet we seem incapable of saying anything of depth or profundity about any of them. When we do say something, it appears trite and rather ludicrous. In short, our paradigm is not coping well. It is ripe for overthrow.... It is the emerging objective social conditions and our patent inability to cope with them which essentially explains the necessity for a revolution in geographic thought. (Harvey 1973: 129)

Consider where you live. Or where your family lives. It is likely that somewhere nearby there is a place where wealthy people live. People more wealthy than you or your family. The houses will be bigger, the cars will be newer. People will dress differently, go to different places to eat. They may buy things in shops you do not even visit. They will take more regular holidays and stay in places beyond your budget. It is also likely that there is a place where people who are poorer than you live. There may be homeless people on the streets or in hostels. Houses may look run down. Street crime rates may be higher. Shops may be boarded up. It may be a place that lacks provision of services such as public transport. Somewhere near you there is likely to a place associated with environmental problems. Perhaps a company is releasing poisons into a river. Perhaps pollutants are leaching into the groundwater from illegal landfill. Perhaps the presence of a nearby busy road leads to higher rates of asthma. It is likely that these places are not surrounded by the homes of the wealthy.

Scroll through your newsfeed over your breakfast. As I am writing, the papers are full of stories of economic anxieties. Inflation is the highest it has been in the UK for over 20 years. Energy prices are at all-time highs. At the same time, wages are falling in real terms and a series of strikes are ongoing or planned. The very people who worked the hardest during the COVID lockdowns, the key workers, are the ones with the smallest pay rises. More and more people rely on soup kitchens for sustenance. Turning the page of my newspaper, I see images of wildfires across Europe as

---

temperatures reach record highs. In the UK, we are used to seeing these images in faraway places, not just to the east of London. There are continuing stories of desperate migrants seeking to enter the UK on small, unseaworthy boats crossing the English Channel. These are people desperate for a life which is better than the one they left. Meanwhile, there is news of the world's richest man, Elon Musk, buying Twitter, for 44 billion dollars. The likelihood is, if you are reading this book, that you are in a "developed" country – one with universities where they teach geographic thought and where you can buy or have access to textbooks. If so, in global terms, you are inhabiting the neighborhood of the wealthy. The global poor live elsewhere. They may be picking over your trash in enormous piles shipped out from our high-consumption economies. They may be taking apart electronic components from mobile phones that we replace every other year for a new model. They may be making your shoes or t-shirts for a fraction of even the minimum wage where you live. They might be dreaming of a life in your neighborhood.

There are important questions that arise from these observations. These are political and theoretical questions. Is this situation simply the way it is – a result of accident, luck, natural advantage, or disadvantage – something that is unavoidable even if a little disheartening? Or is it the necessary outcome of something systematic and unjust? Something called capitalism? Are the poor areas poor because the rich areas are rich? If the latter is true, is there something we can do to change it? Are environmental catastrophes just natural disasters or do they have something to do with the way we run our economies and societies? This leads to a second set of questions which go to the heart of geography. Should we, as academic geographers, be concerned with these issues? Should we be aiming to change things? And then, if your answer to these latter questions is yes, we have to ask ourselves why we have not done more. These are questions that were emerging in the late 1960s.

The late 1960s were a time of considerable turbulence in the world. The height of the Vietnam War was around the corner, the civil rights movement in the United States was hard to ignore, students rioted in Paris. It was a time when people were questioning much of what they were being told by those in positions of power and authority. Academics often found themselves at the center of arguments about what was wrong with the way the world was working and what a better world might look like. This was particularly true in France where celebrity academics such as Jean Paul Sartre found themselves at the center of hard-fought arguments. Geographers, for the longest time, seemed to contribute very little to questions of justice, equality, exploitation, and oppression. Spatial science, in particular, remained almost completely silent:

> . . . young geographers, propelled into a heightened state of social awareness by the events of the middle 1960s, noticed that the fine new methodology was being used only to analyze such socially peripheral matters as shopping behavior and the location of service centers. (Peet 1977: 10)

Humanistic geographers, while critical of spatial science's de-peopling of human geography, most often chose not to research and write about pressing social issues (but see Ley 1974; Western 1981). By the beginning of the 1970s, geographers were beginning to pay attention. Conferences, normally quite formal and disconnected affairs, were full of calls for increased "relevance." In 1968, *Antipode*, a journal for radical geographers of all persuasions, was formed at Clark University in Worcester, Massachusetts. It was in this context that Marxist geography emerged in the United States.

The Birth of Modern Marxist Geography

We have seen how geography and geographers have repeatedly been instrumental in the practice of governance, statecraft, and empire – from Strabo onward. The anarchist tradition of Kropotkin and Reclus had challenged this relationship but remained a marginalized tradition within the discipline.

It was in the late 1960s that some geographers, including some who had been key figures in the development of spatial science, began to draw on both the anarchist tradition and the writings of Karl Marx to produce a new radical geography that was an alternative to both spatial science and humanism. Geographers wanted to make a difference.

David Harvey is the central figure in the turn to Marxism. Harvey's book, *Explanation in Geography*, had been a key text for thinking theoretically about spatial science (Harvey 1969). By 1973, Harvey had become jaded about the point of research within the spatial science tradition. He looked around at the events happening in the world – "an ecological problem, an urban problem, an international trade problem" – and decided that the theoretical tools geographers were using were not up to the job. Geographers, it seemed, were irrelevant. "It is the emerging objective social conditions," he wrote, that led to "the necessity for a revolution in geographic thought" (Harvey 1973: 129). In 1973, he wrote this in the key text for early Marxist geography, *Social Justice and the City*, a remarkable turnaround from the contents of *Explanation in Geography* just four years earlier. The book is divided into two parts; the first – "Liberal Formulations" – cautiously considers issues of urban planning and income distribution in the United States. The second launches into a full-blown Marxist analysis. What links this remarkable book to the earlier one is an intense interest in the process of theory construction.

Harvey became a figurehead for a new Marxist, or Marx-inspired, geography. Other former quantifiers had been grappling with the need to apply geography to social and environmental problems. William Bunge, for instance, had also turned away from his calculations to lead "expeditions" in run-down areas of inner-city Detroit. He suggested that geographers needed to escape the confines of the academy and ask the inhabitants of the inner city how they could help. Opening up an office of geographers, he was soon helping local inhabitants with issues of rodent infestation and road safety. There was certainly radical intent in Bunge's work but it was not as part of any new systematic geographical theory – more a result of a feeling that geographers could not just stand around and watch (Bunge 1974).

The first few years of the 1970s has seen an unusual amount of soul-searching by geographers at the annual meetings of the Association of American Geographers – especially the meeting at Boston in 1971. The historical geographer Hugh Prince, not a noted radical, reported these new calls for "relevance" in the following way:

> Resolutions were passed inviting greater participation in the work of the Association by French and Spanish-speaking geographers, enlisting student representation on the Council, setting up an inquiry into the status of women in the profession and calling for an end to American military involvement in South-East Asia. Whatever may or may not be done to implement these directives, geographers have been reminded that, collectively and individually, they have responsibilities extending beyond their classrooms and libraries. (Prince 1971: 152)

There followed a heated debate about what it means to be relevant. Brian Berry, a spatial scientist, bemoaned the bleeding-heart liberals and hard-line Marxists (Berry 1972). David Smith, another commentator on the events of the 1971 AAG meeting, was busy constructing an alternative approach to relevant geography which might tackle social and environmental problems. This was known as a "welfare approach" (Smith 1971, 1973, 1977).

Marx-inspired radical theory is a very different thing than the kind of "theoretical geography" advocated by the spatial science versions of Bunge and Harvey. Perhaps the biggest difference is that Marxist theory makes no attempt to be politically neutral or objective. Harvey put it this way in his "Historical Materialist Manifesto": "Accept a dual methodological commitment to scientific integrity and non-neutrality" (Harvey 1996 [1984]: 105). This was put even more strongly by the Marxist geographer Richard Peet:

There is no such thing as objective, value free and politically neutral science, indeed all science, and especially social science, serves some political purpose; second, that it is the function of conventional, established science to serve the established, conventional social system and, in fact, to enable it to survive. (Peet 1977: 6)

Harvey put his analytical finger on the distinction between *fact* and *value* in orthodox science as a key to the problem. This distinction insists that facts are objectively removed from whatever values we may give to them. Separating facts from values is connected to the tendency to think of "objects as independent of subjects, 'things' as possessing an identity independent of human perception and action, and the 'private' process of discovery as separate from the 'public' process of communicating the results" (Harvey 1973: 11–12). Harvey identified this as a key problem with his earlier work in spatial science and, indeed, in the first half of *Social Justice and the City*.

The problem for Harvey, Peet, and others was that geographical theory had been almost entirely constructed in ways which served to reinforce the existing state of affairs rather than challenge it. The construction of geographical knowledge, they insisted, had to be understood as knowledge produced in relation to particular class interests and was therefore *ideological*:

The transformation from feudalism to capitalism in Western Europe entailed a revolution in the structures of geographical thought and practice. Geographical traditions inherited from the Greeks and Romans, or absorbed from China and above all Islam, were appropriated and transformed in the light of a distinctively Western Europe experience. Exchange of commodities, colonial conquest and settlement formed the initial basis, but as capitalism evolved, so the geographical movement of capital and labor power became the pivot upon which the construction of new geographical knowledge turned. (Harvey 1996 [1984]: 96–97)

Given that geographical knowledge had been produced under conditions of capitalist imperialism, they argued that a new way of thinking geographically was necessary – one which entailed a revolution in theory. The new Marxist geography would be a radical science that would expose the ways in which social problems, such as the formation of black "ghettos" in American cities, were related to deep social causes.

This revolution in thought obviously entailed a critique of existing arrangements of economy and power. Less obviously, it involved a critique of other attempts to ameliorate problems in society. Relevance was not enough. The liberal politics of "single issue" campaigns got in the way of more fundamental transformations in both theory and society:

During and after the [Vietnam] war, individuals and small groups of people broke off the issue-orientated liberal campaigns (antiwar, environment, appropriate technology, women's liberation, consumers etc.) and moved towards a deeper, more philosophical, radical politics. (Peet 1977: 8)

Harvey illustrates the need for a radical Marxist theory of space and the city by considering the shortcomings of the positivist use of von Thünen's theory of land use as applied to cities (see Chapter 5). The argument goes like this. A geographer wants to test the "truth" of von Thünen's theory and chooses to do so by testing it against the patterns of residential land use in Chicago. After much measuring and categorizing, it turns out that the theory works and is therefore correct. There are a few exceptions along the way that can be explained by Chicago's particular history and geography – one being racial discrimination in the letting and sale of property:

We may thus infer that the theory is a true theory. Thus truth, arrived at by classical positivist means, can be used to help us identify the problem. What . . . was a successful test of a social theory becomes for us an indicator of what the problem is. The theory predicts that poor groups must, of necessity, live where they can least afford to live. (Harvey 1973: 137)

Such an approach seems little more than a confirmation of the obvious. The theory is true but does that matter? The Marxist suggestion is that we need to rid ourselves of the conditions that make the theory true. "In other words, we wish the von Thünen theory of the urban land market to become *not* true" (Harvey 1973: 137). The Marxist approach is to dig deeper and develop a theory of the conditions that make it possible for such a theory to be true and then to challenge those conditions – in this case to replace a private competitive housing market with a socially controlled urban land market. Harvey is insistent that finding the solutions to "the ghetto" is not more studies confirming that they exist – we know they do. Neither is it a kind of ethnographic "emotional tourism" that humanists may have favored whereby we "live and work with the poor 'for a while' in the hope that we can really help them improve their lot" (Harvey 1973: 145). These approaches, Harvey tells us, are "counter-revolutionary." They militate against a real transformation in thinking that will produce a revolution in geographical thought. This "revolutionary" approach is located squarely in the realm of theory – "We are academics, after all, working with the tools of the academic trade. As such, our task is to mobilize our powers of thought to formulate concepts and categories, theories and arguments, which we can apply to the task of bringing about a humanizing social change" (Harvey 1973: 145). Harvey's ambition is considerable, for he aims to produce "such a superior system of thought when judged against the realities which require explanation that we succeed in making all opposition to that system of thought look ludicrous" (Harvey 1973: 146).

Historical Materialism: An Introduction

Before accounting for some of the ways in which geographers have developed Marxist geographical theory, it is necessary to cover some basic Marxist ideas. At the heart of the construction of Marxist theory is a belief that it is possible to get below a world of appearances to a deeper "reality." This reality is often referred to as the "last instance" that helps to explain everything else. Marxists consider Marxist theory to be "scientific" – indeed, more scientific than conventional science – in that the method of dialectical materialism allows practitioners to recognize what Marxists refer to as "objective conditions" operating below the level of mere ideology.

The bedrock explanatory foundations that many Marxists use to account for the world lie in the world of economic production. Perhaps underlying all other Marxist theories is his theory of history – referred to as **historical materialism**. In very simple terms this theory states that human history will inevitably progress through a series of stages from feudalism to capitalism and on to communism. The reason for this inevitable progression is that each stage in history is marked by a set of relations between people in order to produce things. Under feudalism this is the relationship between lords and the serfs who work their land. Under slavery it is the relations between master/ owners and slaves. Under capitalism it is the relation between capitalists and workers. These arrangements are called **relations of production**.

Another key part of the theory of historical materialism is the notion of an economic base and an ideological superstructure. This is called the **base–superstructure model**. It is drawn from a passage in Marx's *Preface to a Contribution to the Critique of Political Economy*, originally written in 1859:

> In the social production of their lives men enter into relations that are specific, necessary and independent of their will … The totality of these relations of production forms the economic structure of society, the real basis from which rises a legal and political superstructure…. The mode of production of material life conditions the social, political and intellectual life-process generally. (Marx 1996: 159–160)

A lot depends here on what Marx meant by the word "conditions," but the crudest form of the model argues that the economic base determines the superstructure. This means that beliefs and cultural forms of expression are always (or often) a reflection of the economic conditions that produce them and only change when the economic base changes. There have been many arguments about the passages where Marx defines this model. These arguments surround what the contents of the base and superstructure are and how one relates to the other. While the economy is often used as shorthand for the base, Marx wrote of a mixture of **productive forces** and relations of production. Productive forces include such things as machinery and know-how as well as natural resources. Marx's definition of the superstructure was quite limited but has been defined as including religion, belief, political propaganda, and more broadly as "culture." Thus culture is seen as a reflection of the economic system, something that both reflects and justifies it. In this view things as varied as organized religion, the *New York Times*, cricket, and *The X Factor* are reflections of a capitalist mode of production and serve to justify it and ensure its survival. A capitalist culture is a reflection and justification for a capitalist economic system. Only by changing the economic system would it be possible to change the culture. All meaningful resistance, therefore, needs to focus on the economic realm.

The final fragment in this brief summary of historical materialism is the mechanism that leads to historical change between different modes of production – a mechanism that leads to a transformation of the relations of production from, say, feudalism to capitalism. In Marxist theory this happens when the relations of production fetter the productive forces. History changes when the social arrangements that allow the economy to work are no longer the most efficient ones and can be seen to actively prevent improvements in production. Slavery, for instance, was no longer an efficient system for production in the United States when know-how and technology (productive forces) had developed to such a point that more could be produced with fewer laborers. Similarly, the feudal arrangements of attaching peasants to land held by lords became inefficient in early modern Europe, leading to capitalist relations between capitalists and workers. It was Marx's belief that, at some point, the relations between capitalists and workers would also fetter the productive forces under capitalism and a new arrangement would come into being.

The key for Marxist geographers was to identify the role that geography might play in this theory of history. This was formulated as a question concerning the role that space plays in capitalism.

The Production of Space and Uneven Development

It was Marxist geographers who first began to import the insights of social theory into the discipline. This was not theory as a spatial scientist such as Bunge might conceive of it. It was (and is) a theory that attempts to intervene in society in a critical way – not simply to understand the world, as Marx might say, but to assist in changing it. Marxist geographers sought to account for the ways in which space was produced. Space, they argued, was not simply the backdrop for things to happen, neither was it a universal category of existence. Rather space was produced through the actions of people. In capitalist societies this means that space is the result of a particular set of relations of production (between capitalists and workers) and the way in which these relations are used to transform nature in order to create profit.

For most of geography's history, space (and other geographical concepts such as place or region) had been thought of as a reflection of aspects of nature, culture, and society. In this sense, space was a product on a metaphorical conveyor belt and it was other things (society, politics, culture, etc.) that were doing all the work. Both regions and spatial patterns had been seen as *outcomes* of other processes. Critical geographers in the late 1970s and onward began to find this view of space too limited and inert. Some began to argue that space played a more active role in the production of society, effectively reversing the argument. Marxist geographers have made some of the most incisive theoretical contributions to this view of active space.

One way in which geographers have engaged with the writings of Marx (and Engels) is to think about what geographers might add to Marxist theory. Marx was not a geographer and did not spend a lot of time discussing things like space and place. He had other things on his mind. Geographers such as David Harvey, Neil Smith, and Milton Santos have developed very detailed theoretical accounts of the production of space under capitalism and "uneven development." In the mid-1970s, when Harvey was undertaking his conversion to Marxism, the Brazilian geographer, Milton Santos, was already engaging on a Marx-inspired rethinking of concepts such as space and territory from the perspective of a scholar located in what is conventionally known as the periphery – Brazil, but also in exile in Europe and Tanzania. Santos noted how Harvey and others were achieving success in applying Marx to the structures of the city, but this should be followed, he suggested "by similar studies of externalities, or the integral nature of space" (Santos 1974: 3). In expanding the geographical exploration of the nature of space, Santos continued, geographers could develop a Marxist understanding of geographies of development and underdevelopment on a global scale.

> If one regards space as a whole, then the artificial distinctions between "economic space" and "geographic space" could be abolished (. . .). One could be concerned with world space as a whole, and not just an aristocratic space, where the only flows studied are those of giant firms and persons of leisure. This would produce a true geography of poverty, i.e., a world geography where wealth and poverty are not treated as separate entities, but as complementary parts of a single reality. (Santos 1974: 3–4)

Just as the wealthy areas of cities depend on the areas of poverty, so "developed" countries in the world are logically and materially dependent on underdeveloped countries. There is, in other words, a world spatial system that includes within it "underdeveloped space." Santos highlights three points central to understanding the role of space in processes of development and underdevelopment. The first is that forces of modernization spread out unevenly from the centers of the system to the periphery producing variations within and between countries. The second is that forces at the center have an active role in deciding where and how peripheral spaces are reached and that this process is dictated by the need for maximum productivity for capitalism. The third is that "forces emitted from the centers ('poles') change their significance as they reach the periphery" (Santos 1974: 4). In other words, concepts change as they move. Something at the center is different when it arrives at the periphery. For this reason, Santos insists, in a way that prefigures decolonial thought (see Chapter 13), that:

> "Underdeveloped space" has a specific character; the priority of importance varies, even if the same forces are involved; because their combinations and results are different. This is something which Western geographers have had great difficulty in understanding. Why should we not then rally expertise from the underdeveloped countries themselves: to develop theories that would make sense to them both as geographers and as citizens? At the moment, "official" geography operates as though the West had a monopoly of ideas. (Santos 1974: 4).

Santos' exploration of the role of space in processes of development and underdevelopment prefigured and complemented work on space and its role in capitalist processes back in the metropolitan "poles" of the Global North – arguments that put geography at the center of explanations of how capitalism manages to creatively overcome an almost constant sense of crisis.

A key point in Marx's account of capitalism is that it is subject to periodic crises due to a number of contradictions which are inherent to its workings. In other words, crisis is integral to capitalism. Indeed, most generations experience a crisis of one kind or another when capitalism falters. When I wrote the first edition of this book, we were experiencing the collapse of banks and the beginning of what became known as "austerity" in the UK with massive cuts to public spending. This era of austerity lasted almost to the present day, interrupted only by massive public spending during the COVID pandemic. Today we have another round of crises including rapidly escalating global warming and an ongoing pandemic, both of which can be directly linked to the workings of a capitalist economy. The question is, how does capitalism prove to be so resilient if these crises keep happening? Neil Smith believes that the answer lies in geography:

> The point is that uneven development is the hallmark of capitalism. It is not just that capitalism fails to develop evenly, that due to accidental and random factors the geographical development of capitalism represents some stochastic deviation from a generally even process. The uneven development of capitalism is structural rather than statistical . . . uneven development is the systematic geographical expression of the contradictions inherent in the very constitution and structure of capital. (Smith 1991: xiii)

Think about the world for a minute and it becomes apparent that "development" varies across space at all scales. This was the lesson of Santos. Think about cities. Almost any city has well-maintained, smart areas where wealthy people live and run-down, poorly connected areas where poorer people live. There are areas of multiple deprivation and there are enclaves of extreme wealth. Then think about the national scale. In the United Kingdom, there has been a persistent gulf between an affluent and middle-class south and a poorer, working-class north – a distinction that has led to the language of "levelling up." Naturally this is not uniform (there are wealthy parts of the north) but it has generally been a supportable claim. In the United States, many of the states of the south have been marked by persistent poverty. Infant mortality rates in some parts of Mississippi match those in Bangladesh. Elsewhere people live in extravagant wealth and comfort. On a global scale we talk about the global north and south – developed nations and the developing world. We might also note how these change over time. They are not inevitable. Marxist geographers explain this through a process of **uneven development** which is inherent to the workings of capitalism.

For capitalism to work, it has to be possible to produce profit. Capital is invested for this reason. Some capital is invested in things that are relatively immobile – things like factories and machinery. These form part of the structure of real working places around the world. These fixed parts of the landscape are built to serve capital but, later, can become fetters on the further development of productive forces and profits start to fall:

> The produced geographical landscape constituted by fixed and immobile capital is both the crowning glory of past capitalist development and a prison that inhibits the further progress of accumulation precisely because it creates spatial barriers where there were none before. The very production of this landscape, so vital to accumulation, is in the end antithetical to the tearing down of spatial barriers and the annihilation of space by time. (Harvey 1996 [1975]: 610)

Perhaps the key idea here is the notion that space is part of the productive forces that form the bedrock of a mode of production such as capitalism. Just as the development of new technologies (such as the steam engine or the conveyor belt of mass production) allows increases in production and profit, so arrangements of space can facilitate an increase in profit. For this reason, space is a fundamental part of the equation of capitalism (or any other mode of production) rather than a mere afterthought or effect:

> Revolutions in the productive forces embedded in the land, in the capacity to overcome space and annihilate space with time, are not afterthoughts to be added on the final chapter of some analysis. They are fundamental because it is only through them that we can give flesh and meaning to those most pivotal of all Marxian categories, concrete and abstract labor. (Harvey 1996 [1975]: 611)

When a company invests in a place it may make considerable profits for a period of time before competitors move in and undermine that process. Eventually a company may relocate to a place with new opportunities or lower wages. The original place then enters a period of decline. Capitalist companies, the argument goes, are always looking for an ideal geography that allows for the production of maximum profit, and one way out of a crisis is to produce a new geography. This is called a **spatial fix**. This process means that the places that are doing well and the places doing less well are constantly shifting (Harvey 1982). It means that it is necessary for capitalism for there to be places that are less well developed than others. "Uneven development" is the general term for this process. Geography allows capitalism to overcome crisis. We are familiar with these stories. We know about companies that relocate to eastern Europe or Asia in order to exploit cheaper labor. We know that the towns of the British north were once the heart of the industrial revolution and producers of great wealth. The United States was once a marginal set of underdeveloped colonies for the British Empire.

Uneven development is one instance of a more general process we might call "the production of space." This phrase is most often associated with the French theorist, Henri Lefebvre. Lefebvre has had an enormous influence on radical geography since the late 1980s when his work appeared in English and was picked up by geographers such as Derek Gregory, Edward Soja, and David Harvey. Lefebvre's book *The Production of Space* was published in English in 1991. In it, Lefebvre takes on the notion that space is a geometric given – an abstract and absolute thing – and suggests instead that space is *produced* by capital and capitalism. Indeed, any mode of production, he argues, produces its own space:

> . . . every society – and hence every mode of production with its subvariants (i.e. all those societies which exemplify the general concept) – produces a space, its own space. The city of the ancient world cannot be understood as a collection of people and things in space; nor can it be visualized solely on the basis of a number of texts and treatises on the subject of space, even though some of these ... may be irreplaceable sources of knowledge. For the ancient city had its own spatial practice: it forged its own – *appropriated* – space. (Lefebvre 1991: 31)

In Lefebvre's work, space becomes central to an understanding of any mode of production or any society. Here he illustrates this point with ancient cities that did not simply exist in space but produced their own space through particular forms of spatial practice.

Lefebvre constructs his argument around a conceptual triad which describes and explains the different ways in which space is produced and reproduced. The key terms are *spatial practices*, *representations of space*, and *spaces of representation*:

1. *Spatial practice*, which embraces production and reproduction, and the particular locations and spatial sets characteristic of each social formation. Spatial practice ensures continuity and some degree of cohesion. In terms of social space, and of each member of a given society's relationship to that space, this cohesion implies a guaranteed level of *competence* and a specific level of *performance*.
2. *Representations of space*, which are tied to the relations of production and to the "order" which these relations impose, and hence to knowledge, to signs, to codes, and to "frontal" relations.
3. *Representational spaces*, embodying complex symbolisms, sometimes coded, sometimes not, linked to the clandestine or underground side of social life, as also to art (which may come eventually to be defined less as a code of space than as a code of representational spaces) (Lefebvre 1991: 33).

Lefebvre's work has been used by a diverse array of geographers to make a number of sometimes conflicting arguments (Harvey 1989; Shields 1999; Soja 1999; Merrifield 2006). One early interpreter was David Harvey. To Harvey, the world of spatial practices is one thoroughly infused with the requirements of capitalism. They are, he argues, "a permanent arena for social conflict and struggle. Those who have the power to command and produce space possess a vital instrumentality for the reproduction and enhancement of their own power. Any project to transform society must, therefore, grasp the complex nettle of the transformation of spatial practices" (Harvey 1989: 261). *Spatial practice*, to Harvey, is the world of space as *experienced* – it represents the actual material landscape we live in every day – the buildings and institutions as well as their arrangement in relation to one another and their connections through the flow of goods, money, and people. Lefebvre declares that spatial practice "embodies a close association, within perceived space, between daily reality (daily routine) and urban reality (the routes and networks which link up the places set aside for work, 'private' life and leisure)" (Lefebvre 1991: 38). In other words, there is a relationship between how we use space and the material reality of that space. We reaffirm spaces by using them as we are supposed to and so everyday life goes on. Think of it on a small scale. We enter a library, the library is quiet, we remain quiet in accordance with the expectations of that space and so the kind of space a library is remains a quiet space.

Lefebvre defines *representations of space* as "conceptualized space, the space of scientists, planners, urbanists, technocratic subdividers and social engineers, as of a certain kind of artist with a scientific bent – all of whom identify what is lived and what is perceived with what is conceived. (Arcane speculation about Numbers, with its talk of the golden number, moduli and 'canons', tends to perpetuate this view of matters)" (Lefebvre 1991: 38). This view of space is very close to the kind of space associated with spatial science, central place theory, gravity models, and the like. A calculated space. To Lefebvre, this space tends to be the dominant and dominating space in a society. Finally, *spaces of representation* refer to *lived* and *imagined* space. It is space as experienced from the point of view of inhabitants or users (rather than planners or spatial scientists). It is very close to the humanistic conception of "place." Here there is room for resistance and subversion – graffiti, carnival, festivals, utopian dreams. They are, Harvey says, "social inventions (codes, signs, and even material constructs such as symbolic spaces, particular built environments, paintings, museums and the like) that seek to generate new meanings of possibilities for spatial practices" (Harvey 1989: 261).

Lefebvre illustrates the utility and meaning of his conceptual triad with an illustration that focuses on the body:

> Considered overall, social practice presupposes the use of the body: the use of the hands, members and sensory organs, and the gestures of work as of activity unrelated to work. This is the realm of the *perceived* (the practical basis of the perception of the outside world, to put it in psychology's terms). As for *representations of the body*, they derive from accumulated scientific knowledge, disseminated with an admixture of ideology: from knowledge of anatomy, or physiology, of sickness and cure, and of the body's relations with nature and with its surroundings or "milieu." Bodily *lived* experience, for its part, may be both highly complex and quite peculiar, because "culture" intervenes here, with its illusory immediacy, via symbolisms and via the long Judeo-Christian tradition, certain aspects of which are uncovered by psychoanalysis. The "heart" as *lived* is strangely different from the heart as *thought* and *perceived*. (Lefebvre 1991: 40)

Lefebvre insisted that "social space is a social product" (Lefebvre 1991: 26). When he was writing about space, it was the not the space of the physicist or of geometry but space that has been produced within capitalism. As such it is capitalist space. The central point of Lefebvre, Harvey, and Smith is that space and our conceptions of space are products of capitalism and that we have to understand the active role of space under capitalism to be able to critique and transform it. You cannot bring about new kinds of society without producing new kinds of (postcapitalist) space. Marxist theorists of space reject the view that space is merely a passive backdrop or stage for social relations and insist that space (and knowledge of space) has an instrumental role in the mode of production. In fact, space appears at every stage in the production of social reality. It is the context for production (everything has to happen in space), it is a tool in production (we use space to produce particular forms of social relation), and it is a product (capitalism produces its own spaces through processes such as uneven development). Space is suddenly everywhere and appears to have considerably more theoretical power than it did in spatial science.

The Production of Nature

The idea of the production of nature seems to be even more paradoxical than the production of space. While nature, in the western tradition, is most often thought of as something inevitable and beyond the human, "production" suggests deliberate human action. It is a word we most often associate with industry and the manufacture of products. How can something that is beyond the human be produced in this sense? This is the puzzle that is central to Marxist thought about nature. One starting point for all Marxist theory is the recognition that people are confronted with something called "nature" and that they have to transform it (through "labor") in order to survive. All the various ways of organizing this labor (feudalism, slavery, capitalism, socialism, etc.) are a historical progression of forms of productive relations that transform nature in theoretically more optimal ways. At the heart of Marxist thought, therefore, is the issue of nature. Here, as Eric Swyngedouw has observed, nature and society are conceptualized as separate things, with nature being progressively more contaminated by society (Swyngedouw 1999). Marxist theorists in geography and beyond had remarkably little to say about nature for a long time. As with most other theoretical approaches, nature was assumed to be "just there" – an innocent backdrop to the machinations of human history (Fitzsimmons 1989). In part this is because Marx did not spend a lot of time theorizing nature. Scholars looking for a Marxist view of nature had to piece together small fragments of Marx's texts to reconstruct an argument.

Neil Smith, a student of David Harvey, wrote *Uneven Development* in 1984. In this book, he paid an unusual degree of attention to what he called "The Production of Nature." He made the argument that nature as we know it is socially produced and that nature has not been "natural" for a very long time. He divides nature into a series of types. *First nature*, he argues, is the original, primordial nature which is conventionally thought of as that which is beyond the influence of humans. This view of nature has historically been the dominant one in western thought, where there has been a post-Renaissance tendency to think through a binary division between the human and the natural. Nature, in this iteration, is, by definition, that which is not human or social or cultural. Smith, however, goes on to define "*second nature*" as socially produced nature – a human world appearing as if it were nature. First nature is transformed into second nature under capitalism through processes of labor and capital. Nature is transformed for firms to make profits:

> But it is not just this "second nature" that is increasingly produced as part of the capitalist mode of production. The "first nature" is also produced. Indeed the "second nature" is no longer produced *out of* the first nature, but rather the first is produced *by* and within the confines of the second. Whether we are talking about the laborious conversion of iron ore into steel and eventually into automobiles or the professional packaging of Yellowstone National Park, nature is produced. In a quite concrete sense, this process of production transcends the ideal distinction between a first and second nature. The form of all nature has been altered by human activity, and today this production is accomplished not for the fulfillment of needs in general but for the fulfillment of one particular "need": profit. (Smith and O'Keefe 1996 [1980]: 291)

In this sense, there is little difference between a rural landscape and an urban one. While a rural landscape may appear closer to some notion of the natural, it is every bit as produced as a housing estate in Birmingham (see Figure 7.1). Indeed, some geographers have deployed a reverse argument to show how the city is thoroughly infused with the stuff of nature (Cronon 1991; Gandy 2002). David Harvey has suggested that New York City is just as natural as anything else (Harvey 1996).

Figure 7.1 Yellowstone (Old Faithful geyser) and New York City (Times Square at Night) – which is more natural? Source: Photo of Yellowstone's Old Faithful geyser by Colin Faulkingham. http://commons.wikimedia.org/wiki/File:Old_Faithful_Geyser_Yellowstone_National_Park.jpg (accessed May 29, 2012). Photo of Times Square at Night by Matt H. Wade, CC-BY-SA-3.0. http://en.wikipedia.org/wiki/File:Times_Square_1.JPG (accessed May 29, 2012).

Most of what we recognize as "nature" in the landscape is in some sense humanly produced. Most obviously, agricultural land is the result of massive labor even if it has often been thought of as a space away from the city where we can get in touch with nature. There is a long tradition of removing signs of labor from such landscapes. Most notably perhaps, owners of English country estates would move whole villages (inhabited by the very people who worked the land) in order to construct views that appeared to be "natural" from their front windows (Williams 1973). Even though rural land-scapes are obviously the product of labor, they have been represented (in painting or poetry, for instance) as natural space. At the extreme end of nature is the concept of "wilderness" – a state which is defined by its distance from urban or human worlds. But this too can be thought of as second nature – as produced. Wilderness is often deliberately protected – cut off as part of a national park by human legislation and enforcement. The idea of wilderness is also a human construct, part of a long tradition of thinking of parts of the world as "innocent" or as products of God. Smith considers Yellowstone National Park – an area specifically defined as wilderness and protected for this reason:

> These are produced environments in every conceivable fashion. From the management of wildlife to the alteration of the landscape by human occupancy, the material environment bears the stamp of human labor; from the beauty salons to the restaurants, and from the camper parks to the Yogi Bear postcards, Yosemite and Yellowstone are neatly packaged cultural experiences of environments on which substantial profits are recorded every year. The point here is not nostalgia for a pre-produced nature, whatever that might look like, but rather to demonstrate the extent to which nature has in fact been altered by human agency. Where nature does survive pristine, miles below the surface of the earth or light years beyond it, it does so only because as yet it is inaccessible. (Smith 1991: 39)

And even the most remote and seemingly natural parts of the world are affected by humanly induced global transformations such as acid rain or climate change. In many instances, areas described as wilderness are directly managed in order to appear as close to our conception of wilderness as possible.

Marxism's encounter with nature touched on a variety of issues on the interface between human and physical geography. Consider soil erosion. This is clearly a process that occurs in nature. We learn the processes of erosion early on in our geographical education. The main culprits appear to be water, ice, and wind. We also know that humans directly erode the landscape as we make human landscapes. What is not so obvious is why soil erosion might be a process that interests Marxists. To answer this, we can look to the work of what has become known as "political ecology." One of the groundbreaking works in this field is the Marxist-inspired *Land Degradation and Society* by Piers Blaikie and Harold Brookfield (1987). In this book, the authors think about soil erosion not as (primarily) a product of natural causes or direct human action but as a result of the organization of society – particularly under capitalism. Their argument is that we can think of soil erosion as a product of various forms of marginality – land that is only marginally valuable, ecosystems that are marginal in their ability to sustain a large population, and areas of the world that are marginal in the world economy and produce raw materials but are not justly rewarded in terms of income. In the end, soil erosion is a product of the structure of the global economy. Poor people attempt to extract as much as they can from the land to compensate for the fact that profit is constantly being extracted from their labor.

This approach stood in distinction to theories that used a Malthusian logic that too many people put too much pressure on the land and erosion resulted. One simple claim that such an approach would make was that a given area of land has a "natural" carrying capacity that meant that when population reached a certain level the land could no longer sustain it. This is known as the "critical population density." Blaikie and Brookfield denied this, pointing out that the productivity of a piece of land varied year by year and month by month and that new technologies have often resulted in increased productivity (Blaikie and Brookfield 1987). Productivity, in other words, is as much an effect of the system of land use as it is of "nature." The traditional Malthusian approaches suggested that:

> [O]ne need not bother with the internal structure of human populations (including ethnicity, gender, class, power relations, etc.), with internal cultural differences in resource use and technology, or with the surrounding world system of interpopulational relations. In effect, the message is that anthropological concerns – not to mention those of other social sciences – can be left out of the analysis. (Painter and Durham 1994: 251)

Radical, Marxist-inspired, approaches to soil erosion and the wider issues of land degradation insist that we have to look to social causes for a proper understanding of apparently "natural" processes. Approaches such as this gave birth to a wider (not necessarily Marxist) field of **political ecology** that continues to insist on the tight connections between the natural world (ecology) and the realms of the social, economic, and political (Robbins 2004).

Likewise we can look at the fluvial landscape as thoroughly entwined with the workings of social power (Swyngedouw 1999; Loftus and Lumsden 2008). Swyngedouw, for instance, examines the hydraulic landscape of Spain and argues both that the modern social history of Spain and the hydrological process need to be understood as thoroughly entangled:

> [V]ery little, if anything, in today's Spanish social, economic, and ecological landscape can be understood without reference to the changing position of water in the unfolding Spanish society. The hybrid character of the water landscape, or "waterscape," comes to the fore in Spain in a clear and unambiguous manner. Hardly any river basin, hydrological cycle, or water flow has not been subjected to some form of human intervention or use; not a single form of social change can be understood without simultaneously addressing and understanding the transformations of and in the hydrological process. (Swyngedouw 1999: 444)

Modern Spain, Swyngedouw tells us, has nearly 900 dams, with every river basin being humanly altered and engineered.

Sometimes second nature is more than just a product of particular arrangements of power. Harvey looks to sites of environmental pollution which disproportionately affect poor people and people of color in North America:

> Property values are lower close to noxious facilities and that is where the poor and the disadvantaged are by and large forced by their impoverished circumstances to live. The insertion of a noxious facility causes less disturbance to property values in low income areas so that an "optimal" lowest cost location strategy for any noxious facility points to where the poor live. Furthermore, a small transfer payment to cover negative effects may be very significant to and therefore more easily accepted by the poor, but largely irrelevant to the rich, leading to . . . the "intriguing paradox" in which the rich are unlikely to give up an amenity "at any price" whereas the poor who are least able to sustain the loss are likely to sacrifice if for a trifling sum. (Harvey 1996: 368)

Harvey illustrates this point with reference to a number of sites of environmental degradation, including the burying of waste on Native American reservations and the siting of hazardous landfill among Black populations in Alabama. Sometimes this leads to organized and effective resistance – as in the case of Love Canal, in Buffalo, New York, where the households of the poor were filled with noxious gases from an infilled canal, leading to serious health effects in local children in the 1970s (see Figure 7.2).

Figure 7.2   Abandoned parking lot in Love Canal. Source: Photo by Bufferlutheran. http://commons. wikimedia.org/wiki/File:Abandoned_parking_lot_in_Love_Canal.jpg (accessed June 28, 2012).

Radical Cultural Geography

In the 1980s and 1990s, Marxist theorizations of nature combined with a renewed interest in the idea of **landscape**. Cultural geographers inspired by Marxist thought looked to uncover the cultural meanings of nature to reveal what Olwig called "nature's ideological landscape" (Olwig 1984). In this work, the so-called "natural landscape" is no longer seen as a raw medium for human transformation (as it was in the work of Carl Sauer, for instance), but as itself a cultural and social production.

This work formed part of a transformation in cultural geography where geographers informed by Marx (and others) sought to reinvigorate the world of cultural geography which had been dominated, with little change in emphasis over 60 years, by the North American tradition of Carl Sauer and the Berkeley School. The people who would come to be known as "new cultural geographers" took elements of Marxist theory and mixed it with elements of the humanistic tradition to think about the ways in which the cultural landscape is produced and then contested. The key theorists these writers looked to were interpreters of Marx such as the British literary scholar Raymond Williams and the Italian philosopher Antonio Gramsci as well as the art theorist John Berger (Gramsci 1971; Berger 1972; Williams 1973, 1977). They were also influenced by the new and dynamic discipline of cultural studies that emerged in the Centre for Contemporary Cultural Studies in Birmingham (Hall and Jefferson 1976; Hall 1980). Between them, these theorists had taken elements of Marxism and created an active role for culture in the radical critique of society. This engagement included a thoroughgoing revision of the base–superstructure model at the heart of historical materialism.

Scholars such as Williams and Gramsci insisted on the link between the economic and the cultural but were not so sure about the notion of one being "determined" by the other. These writers saw the economic and the cultural as dependent on one another and argued that forms of struggle and contestation in the realm of culture could be just as important as struggle over the forces and relations of production. A key notion here was Gramsci's theory of **hegemony**. Central to Marx's account of the world is the domination (oppression and exploitation) of labor by capital. Hegemony is an account of the form that such domination takes. Rather than crudely dominating through the use of force, Gramsci argued, the ruling class can dominate through the construction of common sense. The ruling class produces ideas that come to be accepted by the dominated class, despite that fact that those ideas do not serve their best interests from an "objective" point of view. In short, the dominated class (the workers) comes to believe that the ideas of the ruling class will benefit them. We see manifestations of this in recent years when political pundits have asked why, for instance, poor (White) rural workers in the United States vote Republican when they would (objectively) benefit more from a Democrat program. Similar questions were asked about British working-class people voting for Margaret Thatcher's government in the 1980s. The Gramscian answer is that these people were convinced that the ideas of the right were common sense and might, in some way, benefit them. If this is the case, then it is clear that the realm of social practice that orthodox Marxism refers to as the "superstructure" is actually an important arena of social conflict and that the economic base was not necessarily where all the action was. This raised the possibility of a radical cultural geography (Cosgrove 1983, 1987) – a geography that embraced culture as a site of struggle.

The new cultural geography looked predominantly to the humanist and largely British wing of Marxist thinkers such as Raymond Williams and E. P. Thompson, both of whom had challenged the orthodoxies of hard-line historical materialism and the structuralism of much of European Marxism (Thompson 1963; Williams 1977). To Thompson, for instance, class was not so much a product of a particular objective position in the relations of production but a form of consciousness developed in the cultural sphere. To Williams, culture was the most important sphere of contestation between dominant and emergent cultural formations (Williams 1980). He saw the realm of culture as simultaneously a realm of production and labeled his ideas "cultural materialism." Here he is referring to Marx's dictum that "it is not the consciousness of man that determines their being, but their social existence that determines their consciousness" (Marx 1970: 21). This is a pithy assertion of the base–superstructure model which relegates culture to a secondary role in understanding society. To Williams, the problem with this formulation was not that it was too materialist, but that it was not materialist enough. It was wrongheaded, he argued, to think of one arena of social life – the economy – as "material" and another – culture – as immaterial:

> … it is wholly beside the point to isolate 'production' and 'industry' from the comparably material production of 'defence', 'law and order', 'welfare', 'entertainment', and 'public opinion' … the concept of the 'superstructure' was then not a reduction but an evasion. (Williams 1980: 93)

To Williams, a football game or an opera was every bit as material, every bit as productive, as a car factory or a bank.

These ideas were taken up by proponents of both cultural studies and the new cultural geography in the 1980s (Hall 1980; Jackson 1989; Kobayashi and Mackenzie 1989). Culture suddenly became an arena of the political where various forms of domination and resistance were enacted as the different "maps of meaning" of different social groups came into conflict (Jackson 1989). Its proponents were willfully eclectic in their promotion of aspects of the world that were ripe for enquiry. Cultural geography, they argued, should be:

> … contemporary as well as historical (but always contextual and theoretically informed); social as well as spatial (but not confined exclusively to narrowly-defined landscape issues); urban as well as rural; and interested in the contingent nature of culture, in dominant ideologies and in forms of resistance to them. It would, moreover, assert the centrality of culture in human affairs. Culture is not a residual category, the surface variation left unaccounted for by more powerful economic analyses; it is the very medium through which social change is experienced, contested and constituted. (Cosgrove and Jackson 1987: 95)

The influence of Williams is clear here – culture should not be reduced to the economy as a residual category but explored as an arena of both change and conflict. This led to a focus on forms of domination and resistance at the level of meaning and practice in a variety of cultural domains. It proved to be the main inspiration for my own doctoral work and first book (Cresswell 1996).

These clashes were not limited to class but rapidly included gender, sexuality, bodily ability, age, and a host of other aspects of identity. While Marxism formed part of the critical impetus for new cultural geography, it rapidly became just one of a number of broadly critical theoretical impulses taken up by cultural geographers, including feminism, poststructuralism, and postcolonialism as well as radical queer theory. To some, this was a process of liberation from Marxist orthodoxy, while others saw it as a dilution of cultural geography's critical potential (Mitchell 2000).

New cultural geography quickly delved into many aspects of geographical existence. Perhaps the most important has been the idea of "landscape." *Landscape* has been the term most often associated with the work of Carl Sauer and his colleagues and students at Berkeley. It had remained mostly unscrutinized by sustained theoretical inquiry. Marxist-inspired geographers Denis Cosgrove and Stephen Daniels used ideas from Marx as well as Raymond Williams, John Berger, and others to produce a more theoretical account of landscape as the product of distinct social formations and as a carrier of ideology. Under the influence of Sauer, landscape was seen as either "natural" or "cultural," with the cultural landscape, usually rural, being the result of the work of culture on the natural landscape. Landscape was predominantly seen as a material topography that could be mapped and accounted for (Sauer 1965). This approach had been brought into question by the Canadian geographer James Duncan who described this approach to culture as "superorganic" (Duncan 1980). By this he meant that culture, in traditional cultural geography, had appeared to operate at a level above and beyond human life. Interestingly, Duncan was equally critical of Marxism which he saw as another version of the superorganic which posited political–economic structures as motivations for human action.

In Britain, Stephen Daniels and Denis Cosgrove developed a new cultural geography that refocused attention on what Daniels called the "duplicity" of landscape:

> … landscape may be seen as a 'dialectical image', an ambiguous synthesis whose redemptive and manipulative aspects cannot finally be disentangled, which can neither be completely reified as an authentic object in the world nor thoroughly dissolved as an ideological mirage. (Daniels 1990: 206)

To Daniels, landscape was neither a pure material thing in the world (as in Sauer's geography) nor an ideological illusion. It hovered between these two realms and its very materiality made it simultaneously a powerful ideological construct. Just as a book or a TV show may be ideological (which is to say it may convey meanings which serve the interests of particular powerful groups in society), so a landscape conveys meaning in its very materiality – and these meanings convey the interests of the powerful. Think again of the English country house with all evidence of labor removed in order that the landlord may enjoy the view. It may appear as beauty and nature in a Romantic tradition, but that appearance hides the social processes that went into its production. Such ideas of beauty serve the powerful.

But it was not just particular landscapes that served the powerful. The very idea of landscape was at the heart of the emergence of merchant capitalists in places such as Venice and Flanders from the fourteenth century onward, where art was married to architecture, agriculture, cartography, navigation, and a host of other flourishing forms of knowledge that grew up around the rise of capitalism itself. This point was made forcefully by Cosgrove:

> Landscape, I shall argue, is an ideological concept. It represents a way in which certain classes of people have signified themselves and their world through their imagined relationship with nature and through which they have underlined and communicated their own social role and that of others with respect to external nature. (Cosgrove 1984: 15)

Landscape, both in the sense of a material topography and in the sense of a form of representation (painting), was born alongside the rise of property and ownership for a new class of capitalists whose power lay outside of the church and the aristocracy.

Key to the development of landscape as an idea and art form was the rediscovery of linear perspective. Paintings before the Renaissance were far from realistic – often depicting religious scenes with important figures completely overwhelming any sense of perspective. Landscape painting meant representing the land with some fidelity to its appearance – it is represented as it is seen from a particular point – the point of view of the observer. Cosgrove likens this way of seeing to ownership. By looking at a landscape, we become the most important subject – the owner of the view. Not surprisingly, therefore, landscape paintings frequently appeared in the property of the wealthy, confirming their double ownership of the material landscape and the view. Landscape represented the control over space of the new bourgeoisie (Cosgrove 1984, 1985). Landscape, and the techniques of perspective that developed during the Renaissance, was just as much part of emergent capitalism as profits and labor exploitation.

An important part of this process of ownership was the removal of all evidence of the production of landscape – the labor that went into making it (Williams 1973). This is a point that has been picked up by the Marxist cultural geographer Don Mitchell (Mitchell 1996, 2000). Mitchell's work has consistently carried forward the initial radical impulse of new cultural geography and particularly the critical work on landscape. As the new cultural geography evolved, Marxism became less and less important to its practitioners. It is certainly hard to find in the later work of Cosgrove and Daniels, for instance. Mitchell has provided a useful corrective to this by focusing on the landscapes of labor. Landscape, to Mitchell, is a product of human labor and is therefore a form of what Marx and Harvey have called "*dead labor*" – labor made concrete in a fixed form. This is as true in the agricultural landscapes of California produced by the backbreaking work of Mexican migrants as it is in the heritage landscape of Johnstown, Pennsylvania. In the former case, Mitchell shows repeatedly how the beauty of the Californian landscape is linked to the repeated act of migrant bodies bending over in fields and to the plates of strawberries that appear on tables across the United States (Mitchell 2001, 2003). In the latter case, he describes the landscape of steel-manufacturing plants that once employed tens of thousands of workers and are now closed down:

> In it can be read the history of the industrial development of the United States, and the wrenching decline that followed. But to see landscape in such terms is almost to see it as static: the landscape passively "represents" some history or another. In reality, the landscape itself is an active agent in constituting that history, serving both as a symbol for the needs and desires of the people who live in it (or who otherwise have a stake in producing and maintaining it) and as a solid, dead weight channeling change in this way and not that (there are, after all, only a few uses to which a defunct steel mill can be put). "Landscape" is best seen as both *a work* (it is the product of human labor and this encapsulates the dreams, desires, and all the injustices of the people and social systems that make it) and as something that *does work* (it acts as a social agent in the further development of a place). In that regard too, Johnstown's landscape is just like any other. (Mitchell 2000: 94)

Johnstown was also the scene of a flood caused by the catastrophic failure of a large dam. The flood killed over 2000 people. The dam had not been properly maintained and had been "improved" to provide a pleasure lake for the wealthy of the area. Visitors to the area now can experience the flood as heritage in the Flood Museum which emphasizes the rebuilding of the town and the continued success (for a while) of the steel industry.

In this new landscape much is missing. Mitchell urges us to think of the contestations of culture as "culture wars":

> Culture wars are often about how meaning is made manifest in the very stones, bricks, wood, and asphalt of the places in which we live. There were thus no plans to represent the history of strikes, the geography of violence, or the *politics* of deindustrialization in the Third Century Plan for the makeover of the land-scape into a National Cultural Park. By stressing *industrial* history – the history of development, innova-tion, and the mechanics of making steel – the Johnstown landscape would minimize the contentious past within which such developments and innovations take place. The *work* of the landscape – the role it was assigned by planners – was to represent a heroic history, not a history of conflict. (Mitchell 2000: 98)

While the flood erased broad swathes of the workers' landscape in Johnstown, the heritage land-scape has been equally savage. The reality and struggle of work are made to disappear, while the glory and success of the steel industry are underlined. In this way, landscape does the work of capital.

Black Marxism, Racial Capitalism, and Abolition Geography

Marxist geography takes many forms. It is often part of a hybrid where the Marxist elements are mixed with elements from feminism, poststructuralism, or any number of other theoretical and philosophical approaches. Marxism has also been a key starting point for elements of Black geogra-phies (see Chapter 15). In the political scientist Cedric J Robinson's classic text, *Black Marxism*, he develops the idea of **racial capitalism** based on earlier work of radical scholars in apartheid era South Africa as well as the writings of Karl Marx (Robinson 1983). In Marx, slavery appears as a feature of "primitive accumulation" and, as such, is peripheral to the development of capitalism. Slavery is figured here as both pre-capitalist and non-capitalist. Slavery appears as a stage on a his-torical trajectory towards capitalism. Robinson insisted that this was not the case. He argued that the form slavery took in the Americas – chattel slavery – was integral to capitalism itself. While Marx theorized feudalism as something which fettered the productive forces and thus had to end when capitalism was born, Robinson sees certain kinds of continuity of feudal relations (Lord-serf) into the plantation/slavery complex at the heart of capitalism.

> The social, cultural, political, and ideological complexes of European feudalisms contributed more to capitalism than the social "fetters" that precipitated the bourgeoisie into social and political revolutions. No class was its own creation. Indeed, capitalism was less a catastrophic revolution (negation) of feudalist social orders than the extension of these social relations into the larger tapestry of the modern world's political and economic relations. (Robinson 1983: 10)

The problem for Marxist political economy, Robinson argued, was that it was built on a particular space – the early industrial mills of England – spaces where Marx and Engels observed the emer-gence of a new working class – the proletariat. Marx's theories, like most theories, did not come from nowhere – they were formed in a particular historical and geographical context. If you were to theo-rize from a different place, then the theory itself would be different. What, for instance, would it be like to start from the plantation in the Americas rather than the mills of England? It is not as though the two sites were disconnected. Afterall, the cotton being used in the mills was frequently produced under conditions of slavery in the plantations.

Robinson, unlike Marx, insisted on the centrality of what he called "racialism" to capitalism. "Racialism" referred to the process of seeing social organization and hierarchies as natural through the logic of race which was itself believed (falsely) to originate in "nature." It was this set of beliefs that allowed processes of exclusion, enslavement, and exploitation to develop as distinctly European social products that included both internal oppressions within Europe (such as directed against the Irish) and externally in colonies outside of Europe. Workers are not, in Robinson's view, a homogenous body born out of the mills but, rather, a differentiated body where forms of exploitation are based on that differentiation. Race is a prime way in which this differentiation happens and its logic extends from slavery to uneven development between the Global North (figured as White) and the Global South (figured as Black and Brown). The racial nature of global uneven development, itself built on colonialism, drives contemporary labor migration and the tensions between the "White" working class and racialized migrant labor. The logic of the mill and the logic of plantation are combined in ways which demand a nuanced theoretical approach that includes race at its center. Enslaved Black people, Robinson insists, are every bit as much revolutionary subjects as Marx's proletariat.

In sum then, Robinson contends that we need to look to different spaces to gain an understanding of the central role of race in capitalism. We need to look both beyond the space of Europe and beyond the specific site of the English mill.

> However, it is still fair to say that at base, that is at its epistemological substratum, Marxism is a Western construction – a conceptualization of human affairs and historical development that is emergent from the historical experiences of European peoples mediated, in turn, through their civilization, their social orders, and their cultures. Certainly its philosophical origins are indisputably Western. But the same must be said of its analytical presumptions, its historical perspectives, its points of view. (Robinson 1983: 1).

While Robinson was not a geographer, his ideas have been taken up by geographers, including Ruth Wilson Gilmore. Gilmore, following Robinson, insists that racial capitalism is all of capitalism – that capitalism cannot be understood without noting the centrality of the logic of race. It is not just a case of adding race to already existing Marxist formulations but, instead, insisting that the invention of race is foundational to the origins, and ongoing project, of capitalism. Gilmore defined racism as the "the state-sanctioned or extralegal production and exploitation of group-differentiated vulnerability to premature death" (Gilmore 2007: 28). Further, racism works through a process of partition where some member of the human race are divided from other members of the human race, most often through reference to skin color. Gilmore argues that the invention of race and the process of partition under capitalism allowed some people (and some parts of the world) to exploit other people (and other parts of the world). Capitalism creates groups through a racialization process so that it can then connect them in ways that are profitable. Capitalism, she insists, controls who connects to who and under what conditions – it creates specific exploitative forms of relations with race at the center. This process is always geographical.

Ruth Wilson Gilmore puts her theoretical tools to work in her account of the Californian prison system in her book, *The Golden Gulag* (Gilmore 2007). She asks how it is that the United States has come to build the biggest prison system in the world, in the heart of the world's wealthiest nation. Further, she ask why prisons were being built at astonishing rates while crime rates were decreasing, and why Black and Latinx people are disproportionately imprisoned. Incarceration, at its most basic, is a geographical technique.

> Incapacitation doesn't pretend to change anything about people except where they are. It is in a simple-minded way, then, a geographical solution that purports to solve social problems by extensively

and repeatedly removing people from disordered, deindustrialized milieus and depositing them some-where else. (Gilmore 2007: 14)

Gilmore traces the massive expansion of the prison system in California and how the expansion depended on the creation of new crimes that specifically criminalized elements of Black life in the United States. Prisons, she argues, were thus not a simple response to an expansion of crime but, instead, a response to various kinds of crisis in capitalism that accompanied the contraction of the welfare state from the 1980s onwards. She refers to a fourfold surplus: a surplus of capital that used to be used for other parts of state infrastructure such as schools and hospitals, a surplus of idle crop-land caused by drought available at low prices, a surplus of labor from people who were excluded from the formal economy, and a surplus in "state capacity" as welfare was cut back leaving a work-force with specific expertise available for the prison industry. Through this argument, Gilmore reveals how prisons are relational entities, linked to other places where everyday life was being eroded by the withdrawal of state investment in things such as welfare support, education, health, and public transit. This erosion of everyday life in places associated with poverty and mostly Black and Brown people led, inevitably, to more crime, feeding back into the expansion of the prison land-scape. The construction of the carceral landscape can thus be seen as specific version of what Harvey called the "spatial fix." As Gilmore puts it "[p]risons are geographical solutions to social and eco-nomic crises, politically organized by a racial state that is itself in crisis" (Gilmore 2022: 137). Capital needed new a place to go, and prisons fulfilled that function.

Gilmore's work is intended to provide a critical account of the prison landscape informed by Black Marxism. But she does more than that. Like Marx, Gilmore's point is to change things. With this in mind she proposes and accounts for an **abolition geography** (Gilmore 2022). While this starts with the idea of abolishing prisons, it is far more than that – it is the abolition of the conditions that pro-duce prisons in the first place. The abolition movement is anti-racist, anti-capitalist, feminist, and environmentalist – it is a coalition of activists informed by a critical diagnosis of society. Abolition means erasing the conditions that make people believe that prisons are necessary.

The End of Capitalism (as We Knew It)?

It seems easier for us today to imagine the thoroughgoing deterioration of the earth and of nature than the breakdown of late capitalism; perhaps that is due to some weakness in our imaginations. (Fredric Jameson [1994] in Gibson-Graham 2006a: ix)

One of the most remarkable theoretical approaches developed in radical geography has been devel-oped by Julie Graham and Katherine Gibson under the name J K Gibson-Graham. Their book *The End of Capitalism (as we knew it)* is rooted in a Marx-inspired critique of capitalism but also by developments in feminism, queer theory, and poststructuralism. It is based on a deceptively simple premise: that capitalism has an outside – that it is not always and everywhere. Their argument is that the *notion* of capitalism has become hegemonic. That is to say that scholars on the left, as much as the right, assume, as common sense, that capitalism is already achieved and everywhere. There has been a long tradition of left-wing intellectuals looking for places around the world that present an alternative to capitalism. For most of the twentieth century, this was the so-called communist econo-mies of the Soviet Union and eastern Europe. More recently, attention has focused on Cuba, Nicaragua, Bolivia, or Venezuela where reasonably large-scale state-led economies have been enacted. As each of these alternatives collapsed or proved to be problematic, we arrived at a position where capitalism appeared to have no actually existing alternative. Almost anything could be imag-ined, it seemed, apart from the collapse of capitalism.

J K Gibson-Graham's move was to recognize alternatives to capitalism in surprising places, close to home. They sought to expose the notion that capitalism was everywhere triumphant as a myth:

> Why might it seem problematic to say that the United States is a Christian nation, or a heterosexual one, despite the widespread belief that Christianity and heterosexuality are dominant or majority practices in their respective domains, while at the same time it seems legitimate and indeed "accurate" to say that the US is a capitalist country? What is it about the former expressions, and their critical history, that makes them visible as "regulatory fictions," ways of erasing or obscuring difference, while the latter is seen as accurate representation? (Gibson-Graham 2006a: 2)

In some senses they are arguing that the *idea* of capitalism has been even more successful than the mode of production that it describes. Radical geographers and others had therefore fallen into a trap of seeing only capitalism wherever they looked, ignoring the fact that the economy is just as diverse and fragmented as anything else. Even in the United States, the center of the capitalist world, there are farmers' markets, co-operatives, alternative currencies, barter arrangements, and a host of other alternative spaces of economic activity:

> But why can't the economy be fragmented too? If we theorized it as fragmented in the United States, we could begin to see a huge state sector (...), a very large sector of self-employed and family-based producers (mostly noncapitalist), a huge household sector (again, quite various in terms of forms of exploitation, with some households moving towards communal or collective appropriation and others operating in a traditional mode in which one adult appropriates surplus labor from another). None of these things is easy to see or to theorize as consequential in so-called capitalist social formations. (Gibson-Graham 2006a: 263)

The problem that this kind of thinking produces, according to J K Gibson-Graham, is that radical theorists can only respond to a universal and all-inclusive capitalism with something equally all-encompassing and universal – a global revolution. By breaking capitalism apart, revealing its fractures and incompleteness, they reveal the apparent unity of capitalism as a delusional fantasy and all kinds of other possibilities emerge.

Once again, then, geographers are presented with a theoretical face-off between the general and the particular. Mainstream Marxist and neo-Marxist theory seek to speak in universals (in some ways they share this with spatial scientists), while Gibson-Graham's account of the end of capitalism approaches change as it occurs in particular places in the multitude of ways that are possible. Their economy is not a unified thing but a multiplicity of a "diverse economy" (Gibson-Graham 2006b).

Recall David Harvey's account of the workings of the global economy and the way he defined the responsibility of the radical intellectual in the early 1970s. He declared that concerned intellectuals need to produce "such a superior system of thought when judged against the realities which require explanation that we succeed in making all opposition to that system of thought look ludicrous" (Harvey 1973: 146). Harvey argued that we are academics and that our best work is theoretical work. And this theoretical work must be so strong and powerful that it makes alternative visions look "ludicrous." His answer to an all-encompassing capitalism is an all-encompassing theory. While it seems likely that Harvey and Gibson-Graham agree more than they disagree, Gibson-Graham's proposition is much less all-encompassing and based on work in place-based alternatives that actually exist rather than the armchair of the theorist:

> The objective is not to produce a finished and coherent template that maps the economy "as it really is" and presents (to the converted or suggestible) a ready-made "alternative economy." Rather, our hope is to disarm and dislocate the naturalized dominance of the capitalist economy and make a space for new economic becomings – ones that we will need to work to produce. If we can recognize a diverse economy,

we can begin to imagine and create diverse organizations and practices as powerful constituents of an enlivened noncapitalist politics of place. (Gibson-Graham 2006a: xii)

The difference between Harvey's radical geography of the early 1970s and Gibson-Graham's radical geography of the turn of the century is the quarter century of theorizing that happened between them and the insights of feminism (Chapter 8) and poststructuralism (Chapter 10) in particular. As we will see, both feminism and poststructuralism exhibited a distaste for what they referred to as "*metanarratives*." Difference and the particular once again came to the fore.

Conclusions

It has been impossible to do justice to the broad impact of Marxism on geography. This has been the merest outline of what interpretations of Marx, both direct and indirect, have offered geography in the past four decades. They have made consistent pleas for geography to help understand and transform a capitalist world. Geography in the late 1960s was an odd subject with little to say about very important issues facing the world. Marxists help to rectify that to some degree. At the theoretical level, Marxists were and are at the forefront of prompting us to take the central geographical themes of "space" and "nature" seriously – a process that has gone on in other theoretical traditions since then. In the realm of economic geography, there have been important insights and interventions into the processes of global development showing how developing countries have been tied to the fortunes of exploitative developed nations (Santos 1974; Slater 1977; Peet 1991). Similarly, there have been innovations in our understanding of the way that labor and industry work spatially at urban and regional scales (Massey 1974, 1984; Storper and Walker 1989).

Marxism continues to be an important influence in geography, not least in the flourishing of Black Marxism and abolition geographies. The years since 1968 have seen it come under attack, first by the spatial scientists whom Harvey and others so thoroughly critiqued and then by humanists who saw it as a denial of human free will. More recently, it has been attacked by feminists (because of its blindness to gender) and poststructuralists (for its faith in metanarratives). Despite all this, however, Marxism remains one of the most powerful sets of theoretical approaches to the world we live in. There are still ghettos in cities. Countries in the global south are still dependent on the global north. We are still subject to waves of crises. The global economic chaos of 2008 saw renewed interest in Marxism (and anarchism) as a diagnostic of the workings of capitalism. David Harvey and Ruth Wilson Gilmore's most recent books are subjects of debate well beyond geography as their different but connected analyses of the limits of capital or racial capitalism are seen to provide important truths about the times we are living in. As Harvey put it in 1973, there is an ecological problem, an urban problem, an international trade problem….

References

Berger, J. (1972) *Ways of Seeing*, Penguin, Harmondsworth.

Berry, B. (1972) More on relevance and policy analysis. *Area*, 4, 77–80.

Blaikie, P. M. and Brookfield, H. (1987) *Land Degradation and Society*, Methuen, London.

Bunge, W. (1974) Fitzgerald from a distance. *Annals of the Association of American Geographers*, 63, 485–488.

*Cosgrove, D. E. (1983) Towards a radical cultural geography: problems of theory. *Antipode*, 15, 1–11.

Cosgrove, D. E. (1984) *Social Formation and Symbolic Landscape*, Croom Helm, London.

Cosgrove, D. E. (1985) Prospect, perspective and the evolution of the landscape idea. *Transactions of the Institute of British Geographers*, 10, 45–62.

Cosgrove, D. E. (1987) Place, landscape and the dialectics of cultural geography. *Canadian Geographer*, 22, 66–72.

Cosgrove, D. E. and Jackson, P. (1987) New directions in cultural geography. *Area*, 19, 95–101.

Cresswell, T. (1996) *In Place/Out of Place: Geography, Ideology and Transgression*, University of Minnesota Press, Minneapolis, MN.

Cronon, W. (1991) *Nature's Metropolis: Chicago and the Great West*, Norton, New York.

Daniels, S. (1990) Marxism, culture and the duplicity of landscape, in *New Models in Geography* (eds N. Thrift and R. Peet), Allen & Unwin, Boston, MA, pp. 177–220.

Duncan, J. S. (1980). The superorganic in American cultural geography. *Annals of the Association of American Geographers*, 70, 181–198.

Fitzsimmons, M. (1989) The matter of nature. *Antipode*, 21, 106–120.

Gandy, M. (2002) *Concrete and Clay: Reworking Nature in New York City*, MIT Press, Cambridge, MA.

* Gibson-Graham, J. K. (2006a) *The End of Capitalism (As We Knew It): A Feminist Critique of Political Economy*, University of Minnesota Press, Minneapolis, MN.

Gibson-Graham, J. K. (2006b) *A Postcapitalist Politics*, University of Minnesota Press, Minneapolis, MN.

Gilmore, R. W. (2007) *Golden Gulag: Prisons, Surplus, Crisis, and Opposition in Globalizing California*, University of California Press, Berkeley, CA.

* Gilmore, R. W. (2022) *Abolition Geography: Essays Towards Liberation*, Verso, London.

Gramsci, A. (1971) *Selections from the Prison Notebooks* (eds and trans. Q. Hoare and G. Nowell Smith), Lawrence & Wishart, London.

Hall, S. (1980) *Culture, Media, Language*, Hutchinson, London, in association with the Centre for Contemporary Cultural Studies, University of Birmingham.

Hall, S. and Jefferson, T. (1976) *Resistance Through Rituals: Youth Subcultures in Post-War Britain*, Hutchinson, London [for] the Centre for Contemporary Cultural Studies, University of Birmingham.

Harvey, D. (1969) *Explanation in Geography*, Edward Arnold, London.

* Harvey, D. (1973) *Social Justice and the City*, Blackwell, Oxford.

Harvey, D. (1982) *The Limits to Capital*, Blackwell, Oxford.

Harvey, D. (1989) *The Urban Experience*, Johns Hopkins University Press, Baltimore, MD.

* Harvey, D. (1996) *Justice, Nature and the Geography of Difference*, Blackwell, Oxford.

Harvey, D. (1996 [1975]) The geography of capitalist accumulation, in *Human Geography: An Essential Anthology* (eds J. A. Agnew, D. Livingstone, and A. Rogers), Blackwell, Oxford, pp. 600–622.

* Harvey, D. (1996 [1984]) On the history and present condition of geography: an historical materialist manifesto, in *Human Geography: An Essential Anthology* (eds J. A. Agnew, D. Livingstone, and A. Rogers), Blackwell, Oxford, pp. 95–107.

Jackson, P. (1989) *Maps of Meaning*, Unwin Hyman, London.

Kobayashi, A. and Mackenzie, S. (eds) (1989) *Remaking Human Geography*, Unwin Hyman, Boston, MA.

* Lefebvre, H. (1991) *The Production of Space*, Blackwell, Oxford.

Ley, D. (1974) *The Black Inner City as Frontier Outpost: Images and Behavior of a Philadelphia Neighborhood*, Association of American Geographers, Washington, DC.

Loftus, A. and Lumsden, F. (2008) Reworking hegemony in the urban waterscape. *Transactions of the Institute of British Geographers*, 33, 109–126.

Marx, K. (1970) *A Contribution to the Critique of Political Economy*, Progress Publishers, Moscow.

Marx, K. (1996) Preface to a contribution to the critique of political economy, in *Marx: Later Political Writings* (ed. T. Carver), Cambridge University Press, Cambridge, pp. 158–162.

Massey, D. (1974) *Towards a Critique of Industrial Location Theory*, Centre for Environmental Studies, London.

Massey, D. (1984) *Spatial Divisions of Labour: Social Structures and the Geography of Production*, Macmillan, London.

Merrifield, A. (2006) *Henri Lefebvre: A Critical Introduction*, Routledge, New York.

Mitchell, D. (1996) *The Lie of the Land: Migrant Workers and the California Landscape*, University of Minnesota Press, Minneapolis, MN.

Mitchell, D. (2000) *Cultural Geography: A Critical Introduction*, Blackwell, Oxford.

Mitchell, D. (2001) The devil's arm: points of passage, networks of violence and the political economy of landscape. *New Formations*, 43, 44–60.

* Mitchell, D. (2003) California living, California dying: dead labor and the political economy of landscape, in *Handbook of Cultural Geography* (eds K. Anderson, S. Pile, and N. Thrift), Sage, London, pp. 233–248.

Olwig, K. R. (1984) *Nature's Ideological Landscape: A Literary and Geographic Perspective on Its Development and Preservation on Denmark's Jutland Heath*, G. Allen & Unwin, Boston, MA.

Painter, M. and Durham, W. H. (1994) *The Social Causes of Environmental Destruction in Latin America*, University of Michigan Press, Ann Arbor, MI.

Peet, R. (1977) *Radical Geography: Alternative Viewpoints on Contemporary Social Issues*, Maaroufa Press, Chicago, IL.

Peet, R. (1991) *Global Capitalism: Theories of Societal Development*, Routledge, New York.

Prince, H. (1971) Questions of social relevance. *Area*, 3, 150–153.

Robbins, P. (2004) *Political Ecology: A Critical Introduction*, Blackwell, Oxford.

Robinson, C. (1983) *Black Marxism: The Making of the Black Radical Tradition*, Zed, London.

Santos, M. (1974) Geography, Marxism and underdevelopment. *Antipode*, 6(3), 1–9.

Sauer, C. (1965) The morphology of landscape, in *Land and Life* (ed. J. Leighly), University of California Press, Berkeley, CA, pp. 315–350.

Shields, R. (1999) *Lefebvre, Love, and Struggle: Spatial Dialectics*, Routledge, New York.

Slater, D. (1977) Geography and underdevelopment. *Antipode*, 9, 1–31.

Smith, D. M. (1971) Radical geography: the next revolution? *Area*, 3, 53–57.

Smith, D. M. (1973) Alternative "relevant" professional roles. *Area*, 5, 1–4.

Smith, D. M. (1977) *Human Geography: A Welfare Approach*, Edward Arnold, London.

* Smith, N. (1991) *Uneven Development: Nature, Capital, and the Production of Space*, Blackwell, Oxford.

Smith, N. and O'Keefe, P. (1996 [1980]) Geography, Marx and the concept of nature, in *Human Geography: An Essential Anthology* (eds J. A. Agnew, D. Livingstone, and A. Rogers), Blackwell, Oxford, pp. 282–295.

Soja, E. W. (1999) Thirdspace: expanding the scope of the geographical imagination, in *Human Geography Today* (eds D. Massey, J. Allen, and P. Sarre), Polity, Cambridge, pp. 260–278.

Storper, M. and Walker, R. (1989) *The Capitalist Imperative: Territory, Technology, and Industrial Growth*, Blackwell, Oxford.

Swyngedouw, E. (1999) Modernity and hybridity: nature, regeneracionismo, and the production of the Spanish waterscape, 1890–1930. *Annals for the Association of American Geographers*, 89, 443–465.

Thompson, E. P. (1963) *Making of the English Working Class*, V. Gollancz, London.

Western, J. (1981) *Outcast Cape Town*, University of Minnesota Press, Minneapolis, MN.

Williams, R. (1973) *The Country and the City*, Hogarth, London.

Williams, R. (1977) *Marxism and Literature*, Oxford University Press, Oxford.

Williams, R. (1980) *Problems in Materialism and Culture: Selected Essays*, New Left Books, London.

Chapter 8

Feminist Geographies

Before considering the sometimes-rarified world of feminist "theory," consider the following. Worldwide, at least one in three girls and women are sexually abused in their lifetime. Every year, four million girls and women are trafficked for activities related to the sex trade (http://www.feminist.com/antiviolence/facts.html). In 1970, in the United States, women earned approximately 60% of men's annual wages. In 2008, the figure was 77% (http://www.iwpr.org/pdf/C350.pdf). In 2003, UK women did, on average, 3 hours of housework a day compared to 1 hour and 40 minutes for men (http://www.statistics.gov.uk/CCI/nugget.asp?ID=288). Around 1% of land in the world is owned by women. Women do 2/3 of the world's work for 10% of the income. Out of 37 million people living below the poverty line in the United States, 21 million are female (http://www.internationalwomens-day.com/facts.asp). This is a list that goes on and on. Sometimes, some of us, especially in the developed West, think we live in a more or less equal world. The fact of the matter is that we do not. In the workplace, women are still discriminated against and underpaid. They are less likely to be senior managers. At home they tend to do most of the work and many suffer all kinds of physical and emotional abuse. In public, many women still live fearful lives. The media still (perhaps more than ever) projects women as objects for the enjoyment of men. Feminism and feminist theory are rooted in this state of existence. It is a grounded political and theoretical approach that starts from the observation that the world is systematically skewed against women.

A central point of this chapter is to convince you that feminism is a powerful collection of ideas and practices that can tell us a great deal about the world we live in. They are ideas that should be embraced. They are ideas that are connected to and rooted in the power relations between men, women, and others in the real world. Part of that real world is the discipline of geography and that is where we will start.

Geographic Thought: A Critical Introduction, Second Edition. Tim Cresswell.
© 2024 John Wiley & Sons Ltd. Published 2024 by John Wiley & Sons Ltd.

Women and Geography

For much of its history, geography has been an overwhelmingly male-dominated discipline. Not only have the vast majority of academic geographers been men until very recently, the discipline itself has been associated with various versions of particularly manly masculinity rooted in a history of exploration and conquest. We have previously been reminded that in David Livingstone's *The Geographical Tradition* (Livingstone 1993): "only two women receive mention in half a millennium of Western geographies!" (Royal Geographical Society (with The Institute of British Geographers). Women and Geography Study Group 1997: 17). In fact, women have been doing all kinds of work in and around the discipline of geography for a long time (Maddrell 2009).

A lot depends on what we mean by "geography" and "discipline," of course. What counts as geography has, itself, been largely defined by men. Institutions such as the Royal Geographical Society have put a lot of effort, particularly in the early years, into policing what counts as geographical knowledge. Publishers choose whose work to publish in books. Journal editors and referees determine what counts as geographical knowledge and what does not. Often, the people in positions of power have been men. The same applies to the tight connection between "geography" and the "discipline." If all the geography that counts as valued is within the discipline of formal university-taught geography then this too will have been defined by men. The observation that there have not been many women of note in the discipline of geography, then, is only to say that there have not been many women of note in a very narrowly defined notion of what counts as geography that has been defined, regulated, and disciplined (for the most part, and until recently) by men. No surprise there then!

So, whose knowledge is relevant in an account of geographical theory and its history? This is not a question that has worried too many of the theorists we have already encountered. Feminists made it important. A significant amount of energy has been expended by feminist geographers in reconsidering the discipline's history, in recovering voices and ideas that have been submerged. Gillian Rose, for instance, has argued that there has been a process that has systematically erased the presence of women from accounts of the history of geography. It is not that women were not there, it is simply that we have not looked hard enough (Rose 1995).

If, as Rose argues, there have been women involved in the constitution of geographical knowledge but their contributions have been erased, then who were they and what kinds of knowledge did they produce? Issues such as this came to the fore in a spirited exchange between Mona Domosh and David Stoddart concerning the telling of stories about the history of geography. Stoddart has written a highly regarded account of the history of geography, *On Geography and Its History* (Stoddart 1986). In 1991, Domosh published a paper arguing for a "feminist historiography of geography" in *Transactions of the Institute of British Geographers* in which she argued for the possibility that women's voices in the history of geography could be recovered. Examples of the kinds of women who contributed to geographical knowledge but were not credited or recorded as such were female travelers who took advantage of their class positions, their colonial privileges, and their whiteness to travel widely and record observations as they went. These included the British traveler Mary Kingsley and the American traveler Isabella Bird. In making her argument, she also took Stoddart's book to task for its erasure of women in the history of the discipline (Domosh 1991).

As an epigraph to her paper, Domosh quoted the following words of Isabella Bird's:

Such as it is, Estes Park is mine. It is unsurveyed, 'no man's land', and mine by right of love, appropriation and appreciation; by the seizure of its peerless sunrises and sunsets, its glorious afterglow, its blazing noons, its hurricanes sharp and furious, its wild auroras, its glories of mountain and forest, of canyon, lake, and river, and the stereotyping them all in my memory. (Isabella Bird [1879] quoted in Domosh 1991: 95)

At the heart of Domosh's paper is the question of why accounts such as this are excluded from accounts of the history of geography. Stoddart had made a case for including the tradition of geographic exploration and expeditions in the history of geographical knowledge: but only certain kinds of expeditions that were clearly scientific and sanctioned by disciplinary institutions. If he was willing to open up the definition of geographical knowledge in this way, Domosh argues, why not open it wider still to the unsanctioned, "unscientific" accounts of "lady travelers"?

Part of the issue is the definition of what counts as scientific knowledge. What counts as science and as geographical science had been defined by men. The kind of knowledge they approved of was "objective knowledge." Objective knowledge is knowledge that does not apparently depend on the position of the observer. It is the kind of knowledge that is supposed to be true everywhere and for all people. We have seen how humanistic geographers rejected this kind of knowledge from one angle, in their critique of spatial science. But even they were not keen to really reflect too deeply on their subject positions in the course of their research. The passage from Isabella Bird is very different from the lists of observations and measurements that fully institutionalized male geographers brought back from their expeditions. It reflects her own subjectivity and is written in an impassioned way. As Domosh put it:

> The stories of women travellers are incredibly diverse, yet they share some common threads, one of which is their quite explicit recognition of the personal goals of their travels. The so-called objective discoveries of new places were not separated from the discoveries of themselves. (Domosh 1991: 97)

There is no doubt that the travels of women in the nineteenth century and the early twentieth century produced knowledge; it is just that the knowledge did not fit into the set of knowledge that was labeled "geography" at the time:

> The subjectivity of fieldwork that women could claim as their special contribution to geographic knowledge was . . . systematically taken out of the realm of scientific geography. The suppression of the subjective and the denial of the ambiguity of observation was part of the legitimation of the academy and the professionalization of the social sciences that occurred in the first half of the twentieth century. (Domosh 1991: 99)

Women were excluded and erased from the history of geography at a number of levels and for a number of reasons. First, the knowledge they produced tended to be outside of the institutional frameworks that sanctioned knowledge. Women, for the most part, were not allowed to join these organizations such as the Royal Geographical Society, or were discouraged from doing so. Women were said to be unsuited to the rigors of proper science and certainly the rigors of expeditions. Any knowledge they produced was destined to be the knowledge of outsiders. Second, the kinds of accounts that were written by women were disqualified as "not scientific" and too personal and "subjective." The kind of personal geographies that were produced by "lady travelers" simply "brought them abruptly against the confines of the world circumscribed by a male-defined language" (Domosh 1991: 99).

Domosh's account is over 30 years old. Perhaps now, accounts of the history of the discipline and its key ideas have taken the contributions of women more seriously? In an event to announce and discuss the publication of the 13-volume *International Encyclopedia of Human Geography* (Kitchin 2008), a mammoth project which attempts to provide "An authoritative and comprehensive source of information on the discipline of human geography and its constituent, and related, subject areas" (http://www.elsevierdirect.com/brochures/hugy/), the feminist geographer Linda McDowell pointed out that of 60 key figures in geography chosen for the "people" section, only three of them were women: Jacqueline Beaujeu-Garnier (1917–1995), Doreen Massey, and herself. This a result of five years of thought and effort – three women from centuries of geographic thought. Other recent books are slightly better. In the book *Key Thinkers on Space and Place*, for instance, 8 out of 52 of these "key thinkers" are women (Hubbard *et al.* 2004). Take a look at the department you are taught in or work in. Look at the website or picture board that shows the faces of the faculty and staff and compare it to the picture boards of undergraduates and postgraduates. There are certainly exceptions but, for the most part, I would guess that over 70% of the faces of faculty members are male. I would also guess that over 70% of the administrative staff are female. The closer you get to the undergraduate level, the closer the number gets to 50% (in the "Western" world at least). During my PhD, I was based in a department where only one tenured faculty member was a woman. I once worked in a department where for a long time there had been only one woman out of nearly 30 faculty members (in 1999). Things have certainly improved since 1999, but there is still a long way to go.

What Is Feminist Geography?

Feminism is not a theory as such but there are feminist theories. Pick up almost any book on feminist geography and you are likely to encounter a willful celebration of the diversity of feminisms and feminist geographies. The book *Feminist Geographies: Explorations in Diversity and Difference*, for instance, is co-written by members of the Woman and Geography Study Group of the Institute of British Geographers. After 200 pages of spirited discussion of aspects of feminist geography the final line is "Ultimately, then, we cannot agree on how we interpret the current relationship between feminist geography and human geography" (Royal Geographical Society (with The Institute of British Geographers). Women and Geography Study Group 1997: 200). One anonymous reviewer for the proposal for the first edition of this book had little problem with much of what I proposed but was convinced that feminism did not merit its own chapter.

A landmark text in the development of feminist geography is *Geography and Gender: An Introduction to Feminist Geography*. It was written by a collective of the Women and Geography Study Group of the Institute of British Geographers and published in 1984. The Study Group had only been in existence for a few years. Its introduction asks whether the research and teaching of the Study Group should be "feminist geography" or a "geography of women." At the time it was still commonplace for the word "man" to be used to mean all of humanity. Courses had titles like "Man's role in changing the face of the earth" or "Man and nature." The critical interventions of Marxist and humanistic geographers had done little to change this, despite their often radical intent. "Rational economic man" had been at the center of some spatial science, Marxists had foregrounded class in a gender-blind way, and humanists posited a universal man at the center of a meaningful world. The very first paragraph of *Geography and Gender* takes the discipline to task for this:

> We are presented, for example, with *man* as the agent of change in agricultural landscapes, *men* digging for coal (or being made redundant by the closure of the coal-mines), and the results of surveys in which *men*, as heads of households, have been asked for their opinions on recreation resources, transport needs or housing. We might, in fact, be forgiven for thinking that *women* simply do not exist in the spatial world. (Women and Geography Study Group of the Institute of British Geographers 1984: 19)

Now, nearly 40 years later, most journals, not just those with radical intent, will not accept papers that use the word "man" in this way. The instructions for authors of *Transactions of the Institute of British Geographers*, for instance, include the statement that "papers should be written in non-sexist, non-racist language" (http://www.wiley.com/bw/submit.asp?ref=0020-2754&site=1). The use of sexist language is just as wrong as putting a full stop in the wrong place.

The authors of *Geography and Gender* were making a case for taking the geographies of women seriously: they were "encouraging geographers to consider women's daily lives and problems as legitimate, sensible and important areas for research and teaching" (Women and Geography Study Group of the Institute of British Geographers 1984: 20). But they did not want geographers to simply "add women and stir." Rather they argued for an approach that would help us to understand why it was that women had been marginalized – they were arguing for a feminist geography:

> What we argue for in this book is not, therefore, an increase in the number of studies of women *per se* in geography, but an entirely different approach to geography as a whole. Consequently we consider that the implications of **gender** in the study of geography are at least as important as the implications of any other social or economic factor which transforms society and space. (Women and Geography Study Group of the Institute of British Geographers 1984: 21)

The authors of *Geography and Gender* outline a number of key issues for feminist geography, all of which continue to this day. One issue is that of **essentialism**. Essentialism is a kind of thought that ascribes characteristics to something as innate and unchangeable – as the essence of what something is. This kind of thought is often applied to genders. Women are more caring, men are more aggressive, etc. Feminist geographers have consistently argued that, in most instances, such characteristics are socially produced and not natural. It is for this reason that some feminists distinguish between sex (as a biological category) and gender (as a social construct).

In 1984, the concept of **patriarchy** was central to a radical feminist agenda. If gender is socially constructed and not natural, then it is theoretically possible for any number of characteristics to be ascribed to the biological facts of sex. Why not have the reverse of gender stereotypes that we have today – women as aggressive and rational and men as caring and emotional? A theory of patriarchy attempts to explain why gender has been socially constructed in the way it has:

> Patriarchy can be defined as a set of social relations between men which, although hierarchical, establishes an interdependence and solidarity between them which allows them to dominate women. Thus, although men of different ages, races and classes occupy different places in the patriarchal pecking order, they are united because they share a relationship of dominance over their women. (Women and Geography Study Group of the Institute of British Geographers 1984: 26).

In the same way that Marxists seek to transform a system of social organization called capitalism, so (some) feminists seek to transform patriarchy – a system which is based on the ability of men to control and define women's roles in the economy as well as how they are represented and defined in broadly cultural realms.

Radical feminists see patriarchy as working on many levels. In the economy, women's jobs are generally lower paid and less prestigious. In the family, the role ascribed to women is undervalued and unrewarded. Ideologically, women are represented as objects for male desire and on a more immediate level women are subjected to a range of demeaning behavior, from sexist jokes in the workplace to sexual harassment and rape. All of these combine to produce, just as they express, a world which is systematically patriarchal.

The writings of the literary critic Kate Millett, particularly her book *Sexual Politics*, have been central to the development of radical feminism. In *Sexual Politics*, Millett revealed the operations of patriarchy in examples of "great literature" written by men, including novels by D. H. Lawrence and Norman Mailer. Millett argued that all known societies were effectively ruled by men and that this form of power was more important and more universal than power based on class or race (Millett 1971).

The concept of patriarchy was given sustained attention by Jo Foord and Nicky Gregson in an important article in *Antipode* in 1986. They argued that patriarchy was the "common thread" for the emerging feminist geography – that whatever differences women experienced around the world, differences that depended on context, they were united by an experience of patriarchy (Foord and Gregson 1986). The central point of their paper was to unite disparate theorizations of male domination with one coherent theory. This was an unusual enterprise in the annals of feminist geography which, for the most part, has been characterized by a willing embrace of theoretical diversity and suspicion of overarching theoretical orientations which are often described as "masculinist." Many feminist geographers responded to their paper making exactly this point (for an account of this debate, see Johnson *et al.* 2000: 84–87).

Perhaps the first task of feminist geography was not primarily theoretical but political. Women needed to be accounted for in geography. As we have seen, it was (and still is) the case that women were/are underrepresented in the discipline. It was (and still is) the case that the work that has been done by women (both as formal academic geographers and in the wider sense of producers of geographical knowledge) has been either intentionally or carelessly not given full credit. Early work in an emergent feminist geography sought to both highlight and then correct these problems of "women in geography" (Zelinsky 1973; Monk and Hanson 1982; Zelinsky *et al.* 1982). It was argued that women needed to be added both to the practice of researching geography and to the content of that research. Quite reasonably, it was suggested that if more women were included in the practice of geography, then geographers would be more likely to take account of women as subjects of analysis:

> One expression of feminism is the conduct of academic research that recognizes and explores the reasons for and implications of the fact that women's lives are qualitatively different from men's lives. Yet the degree to which geography remains untouched by feminism is remarkable, and the dearth of attention to women's issues, explicit or implicit, plagues all branches of human geography. (Monk and Hanson 1982: 11)

The suggestion here is that if more women were involved in conducting formal geographic research, then geographers would be more likely to research issues that concern or, at least, include women. Monk and Hanson identified "sexist biases" in geographical research produced, for the most part, unintentionally. In their early call for feminism in geography, they reject the need for a separate stream of geographical research labeled feminist and, instead, suggest that a feminist perspective should and could become part of all strands of the discipline. The reason they gave for geography remaining "untouched" by feminism was the fact that, at the time, less than 10% of researchers in the discipline were women and that as knowledge is a social creation it is likely to reflect the perspectives and life experiences of those who are involved in its production. None of the dominant philosophical/theoretical perspectives of the time (positivism, Marxism, humanism) seemed particularly adept at noticing or emphasizing the geographical worlds of more than half of the world's population:

> In sum, most academic geographers have been men, and they have structured research problems according to their values, their concerns, and their goals, all of which reflect their experience. Women have not been creatures of power or status, and the research interests of those in power have reflected this fact. (Monk and Hanson 1982: 12)

As with any theoretical approach, feminist geography continued to evolve. The most recent key collection of feminist geography, *Feminist Geography Unbound* (Gökarıksel *et al.* 2021) arose from a feminist geography conference held in 2017 at the University of North Carolina in Chapel Hill. The context of this meeting included the recent election of Donald Trump as President in 2016 and the outpouring of right-wing policy that followed, including a "Muslim ban" that banned people from predominantly Muslim countries from entering the United States. This included several scholars who were due to attend the conference. Others declined to attend the event due to anti-trans legislation in North Carolina. They were, in other words, difficult and tense times to be engaging in critical feminist scholarship. The contents of the collection are notable for the diversity of authors including a wide array of different intersectional identities and concerns. Racism, transphobia, and Islamophobia were very much part of the feminist agenda – as well as sexism and patriarchy. One of the targets of the collection is what the editors call "feel good" (largely White) feminisms as displayed through the wearing of pink pussy hats during massive women's marches.

> We find ourselves in a moment when feminism is sold to as a balm, with T-shirts and baby clothes that proclaim "the future is female." What might feminism look like if we refuse to aspire to the solace of "lean in" and "feel good" feminisms of pink pussy hats and US flag hijabs that deny the violence of nationalism, capitalism, and imperialism? (Gökarıksel *et al.* 2021: 1)

The editors insist on remaining attentive to potential feminist complicity with forms of racism and imperialism and to resist "glossy, marketized, and neoliberal solutions: the woman who has it all, the #girlboss" (Gökarıksel *et al.* 2021: 1). One of the main changes in feminist geography, and feminism more broadly, represented within the collection is the problematizing of the gender binary and the increasing awareness of trans politics.

Gender and Geography

Arguably the most important concept in the lexicon of feminist geography and feminism more broadly is **gender**. In most classic definitions arising from feminism, gender is differentiated from "sex" by stating that gender is what society/culture makes of sex. Sex, in this definition, remains a biological given. Gender, then, is a set of socially produced norms and expectations about what is "masculine" and what is "feminine." The difference between sex and gender, feminists have insisted, is a difference produced by complex and varied processes of socialization. These occur differently in different times and places but some can be easily recognized. Raising children is one place where dominant notions of gender are difficult to escape. From the moment of birth, children are often immediately "gendered." It starts with the division of pink for girls and blue for boys (a reversal of older color codings). As they grow up, boys were presented with cars and computer games of one kind or another. Daughters in affluent households gain enviable collections of dolls and princess-themed toys and clothing. Once, my wife and I gave our daughter a short haircut. She was mistakenly referred to as a boy on a number of occasions and insisted on growing her hair to princess-worthy length. Boys are called smart or brave. Girls are called pretty and kind. Sometimes these gendering activities break down and mostly they are repeated and (obviously) internalized. Gender is everywhere.

These binary ways of thinking about gender have become increasingly troubled within and beyond feminism (Cofield and Doan 2021; Todd 2021). One form of complication has been to add another level to the sex/gender distinction. Physical sex is defined by aspects of our biological bodies including, but not limited to, our genitals and secondary sex characteristics, and our X and Y chromosomes. Even with these, however, there is more of a continuum than a clear binary. How these characteristics are labeled as male or female is a social and cultural process that can and does change. Gender is not straightforward either. "Gender expression" refers to a set of practices and codes that either do or do not fit into dominant expectations concerning what men and women should, or should not, do. These expressions do not neatly map onto physical sex characteristics. In some societies, both in the present day and historically, there have been more than two genders. Judaism, for instance, has historically recognized six genders. "Gender identity" refers to people's sense of themselves and how that sense fits or does not fit with dominant gender binaries. People can identify as male and female or as genderfluid or genderqueer among other identities. Increasing numbers of people identify with the pronoun "they" instead of "he" or "she." Physical sex, gender expression, and gender identity combine in a multitude of ways that upset binary distinctions between men and women. This has caused a number of problems for feminism which has roots in a much clearer binary of men and women. One aspect of feminist politics in the 1970s and 1980s, in particular, was the creation of safe "women-only" spaces. The multiplication of identities and the variable mapping of physical sex, gender expression, and gender identity, and particularly the existence and increasing visibility of trans people (particularly trans women) makes the policing and definition of such spaces part of a fraught and complicated politics. Some feminists, known as "Gender Critical" feminists, have maintained a commitment to physical sex as the defining characteristic of what counts as a woman (Barefoot 2021). Most feminists however, and certainly those in geography, have recognized the importance of including trans women in the identity "woman."

Despite recent and rapid changes in how we think about gender, it remains the case that as we grow up, we are expected to be masculine (if we are cisgendered men) and feminine (if we are cisgendered women). Any transgression of these codes risks public disapproval. Cisgendered women who are assertive are castigated while cisgendered men who stay at home looking after families are treated as strange. The differences between masculine and feminine are not natural – they are produced socially. Expectations of masculinity and femininity have been different historically and vary geographically. The differences are also valued in hierarchical ways so that traits associated with masculinity are more likely to be positively valued than those associated with femininity. Gendering tends to advantage cisgendered men over cisgendered women (Royal Geographical Society (with The Institute of British Geographers). Women and Geography Study Group 1997).

Masculinism in Geography

Feminists argue that there is a gendered basis of knowledge itself. Ideas of rationality, objectivity, logic, and distance as well as the authority gained from the kind of language games employed by science – an objective disinterested tone along with the codes and practices of publication – are codes made up to serve the interests of masculine knowledge production. The idea that rational knowledge should be, independent from the social position of the knower, free from emotion and separate from the body, feminists argue, is a **masculinist** position produced historically by men. The idea that truth should be context free, value free, objective, and universal was produced at the time men were defined as separate from nature (in Europe since the sixteenth century) and women were defined as not separate but connected to nature, emotion, and the body. This historical context for the rise of science and rationality defined masculinity through its distance from the body and defined women as not able to produce rational (and therefore, important) knowledge because of their inescapable embodiment. As a result, knowledge produced by men was thought to be truthful, objective, and universal, yet it was dependent upon the exclusion of women and those areas defined as feminine.

This gendering of knowledge had two main effects: it excluded women from active participation in knowledge production and at the same time excluded those subjects which were not considered important or knowable through the methods of rational and positivist science. It structured both what counts as knowledge and who could be a knowing subject. Geography, as part of this rise of knowledge from the sixteenth century onwards, has been central to this production of masculinist knowledge, an argument made forcefully by feminist geographer Gillian Rose.

Rose's book *Feminism and Geography* switched the focus of feminist geography from "the geography of gender" to a thorough analysis of the role of gender in the thought and practice of geography itself. She focused on "the gender of geography." She made the case that geography is "masculinist":

> Masculinist work claims to be exhaustive and it therefore assumes that no-one else can add to its knowledge; it is therefore reluctant to listen to anyone else. Masculinist work, then, excludes women because it alienates us in its choice of research themes, because it feels that women should not really be interested in producing geography, and also because it assumes that it is itself comprehensive. (Rose 1993: 4)

Rose is not, primarily, talking about the long history of excluding women from the institutions of geography here. Rather she is arguing that geographic thought itself is a way of thinking that excludes women. As a masculinist way of thinking, geography assumes that it is complete and, in making this assumption, succeeds in reproducing masculinist patterns of thought.

Masculinist knowledge sees itself as the only knowledge. It is knowledge that does not acknowledge the fact that it is achieved from a particular position and in relation to other kinds of knowing that it submerges and pushes aside. The "master subject" of the White, male, heterosexual constructs all other subjects as different – women, non-White people, gay people, children, disabled people. Masculine knowledge, Rose argues, is most often formulated as "reason" or "rationality" – knowledge which is, by definition, detached from the social position of the knower. Femininity and the feminine are thus constructed as the non-rational or irrational other – outsider of the scope of reason. The feminine has often been constructed through recourse to notions of hysteria and emotions out of control – the very opposite of rationality.

We can see masculinism through all the episodes in the story of geographic thought we have explored so far. Varenius's "general geography" and Ptolemy's division of the world into a grid are both excellent examples of a desire to know completely. Or think of the world of geographers as explorers, as detached observers, producers of a seemingly whole knowledge of the world – knowledge that made the world into an ordered hierarchical system. More recent examples include the all-knowing explanations of Marxist historical materialism or the universal essences of phenomenology described by humanistic geographers.

A significant problem facing feminist geographers is how to relate the category of "woman" to the "feminine." One impulse is to build up these categories and to celebrate them in the face of masculinist marginalization. But there is also the opposite reaction – to deconstruct and tear apart the very gendered categories that have been produced in a masculinist way.

This kind of paradox is often central to the ways in which feminists think about and approach the world. Rose describes the seductions of being an academic and thinking about theory:

> My own desire as a student to be part of the academy was intense. I was first introduced to the powers and the pleasures of theory by tutors, lecturers and supervisors – almost all men – and listening to their arguments and conversation I desperately wanted to be able to join in, to be part of the debates among knowledgeable men, to *speak*. (Rose 1993: 15)

Rose saw the pursuit of theory (the kinds of ideas outlined in this book) as an exciting endeavor, but one that seemed to exclude her and other women. She describes the space of geographical theorizing as one which is produced within a White, bourgeois, heterosexual, masculine world and outside of which it is difficult to find a space from which to resist it. She wonders, for instance, how it is possible to represent women without recourse to the idea of "Woman" (an idea created within this dominant space).

A Feminist Epistemology

One strong claim that is often made by feminist geographers is that feminism necessitates new forms of epistemology. **Epistemology** is a theory of knowledge – an explanation for how we know what we know. There are a variety of approaches to epistemology perhaps the simplest of which (at least at first glance) is **empiricism** – the belief that we can know something when we have observed it (at second glance, of course, we realize that we cannot observe things like gravity, magnetism, racism, or the distance–decay effect straightforwardly, yet many of us believe in one or more of these). We have already seen how Monk and Hanson have suggested that knowledge is a social construct – it is humanly produced rather than simply lying around waiting to be discovered, counted, tabulated, and accounted for. We (feminists and many others argue) make up knowledge, we do not discover it. And given the diversity of human experience, it is also clear that there will be different knowledges produced by different people. As Meghan Cope has put it, "Knowledge is not something that we can passively or actively *acquire* because we are always involved in its production and interpretation. Similarly, knowledge production is never a 'value-free' or unbiased process" (Cope 2002: 43).

This view of knowledge is quite a challenge to traditional beliefs about scientific knowledge which is most often considered to be objective, value free, and unbiased. To many feminists the traditional view of science is "masculinist" (Harding 1986; Haraway 1988; Rose 1993). Science claims to be universal but is, in fact, knowledge produced from particular positions, where these positions are neither recognized nor acknowledged. The particular is made out to be universal. Feminists point out that in most realms of knowledge, and in geography in particular, this knowledge is produced by men who misrecognize their own knowledge as all knowledge.

Rose argues for what she calls a politics of location – a politics which recognizes the location of the knower and known in relation to multiple matrices of power (class, race, sexuality, etc. in addition to gender). She draws on the writings of Adrienne Rich:

> To write 'my body' plunges me into lived experience, particularly: I see scars, disfigurements, discolorations, damages, losses, as well as what pleases me. Bones, well nourished from the placenta; the teeth of a middle-class person seen by a dentist twice a year from childhood. White skin, marked and scarred by three pregnancies, an elected sterilization, progressive arthritis, four joint operations, calcium deposits, no rapes, no abortions, long hours at a typewriter – my own, not in a typing pool – and so forth. (Adrienne Rich cited in Rose 1993: 139)

Here, Rich acknowledges the particularity of her own body. This does two things which would be unusual in mainstream (masculinist) scientific discourse. First, it acknowledges that knowledge is produced within a body – is "embodied" – rather than being the product of a pure mind. Second, it acknowledges the particularity of that body as being both female and, in various ways, privileged. Rich also acknowledges being White and Jewish in the United States – both aware of persecution and being part of the power hierarchy.

Rose, along with many feminist geographers, insists on the complexity of being located in multiple ways. Not just as a woman but as someone with a particular set of positions based in class, sexuality, national origins, and race, for instance. For Rose, the space of location is always contradictory and paradoxical – pushed and pulled by a multitude of positions. This is a long way from the disembodied vision of masculinist objectivity. In this sense, as for Lise Nelson and Joni Seager, "[t]he body is the touchstone of feminist theory" (Nelson and Seager 2005: 2):

> This sense of space offered by these feminists dissolves the split between the mind and body by thinking through the body, their bodies. This way of thinking also seems to disregard any distinction between metaphorical and real space; spaces are made meaningful through experience and interpretation, which makes feminist spaces resonate with an extraordinary richness of emotion and analysis. Spaces are felt as part of patriarchal power. (Rose 1993: 146)

Feminist geographers have been inspired by the work of Sandra Harding (Harding 1986) and Donna Haraway (Haraway 1988) in their work on epistemological questions. These authors produced feminist philosophies of science and came to very similar conclusions about what counts as good knowledge, as truth, and as objectivity. Harding argues that the traditional view of objectivity as knowledge achieved from a "view from nowhere," untainted by the subjectivity of the scientist, is fatally flawed. She advocates instead a **standpoint theory** which posits that all knowledge is valid from the position or standpoint of the person producing the knowledge. The masculine position is one such standpoint but is hampered by the lack of recognition it gives to other standpoints. A better form of objectivity, she argues, is achieved when multiple standpoints (including those of normally excluded and marginalized people) are taken into consideration to produce a "strong objectivity" which "requires a wider array of questions, interpretations, different perspectives, and inclusion of researchers and subjects from marginalized groups to strengthen the claims of 'truth'" (Cope 2002: 48). In Harding's work, the standpoints of the marginalized and excluded are given particular weight. It is from these positions, she argues, that certain kinds of truth can be produced that are unavailable to those in positions of power. Women, therefore, have a particularly privileged position in respect to knowledge about women and gender inequality. Their knowledge of this, in other words, is likely to be better knowledge than that of the supposedly disinterested outside observer. This view of knowledge completely reverses the traditional view that values the knowledge of the disinterested, objective scientist (McDowell 1993; Falconer Al-Hindi 2002). Feminist standpoint theory in many ways reflects the kind of standpoint theory in Marxism where the members of the proletariat are the only ones likely to see the true nature of capitalism owing to their oppressed and exploited position in the relations of production.

A similar argument is made by Donna Haraway when she argues for **situated knowledges** (Haraway 1988). To Haraway, forms of knowledge that are recognized as context specific are more reliable than those that pretend to be universal and neutral. She suggests that we rid ourselves of any pretense to the view from nowhere and, instead, foreground our multiple perspectives, political biases, and cultural values and proceed from there. The resulting knowledge, she suggests, will be more reliable and accurate than knowledge that does not recognize its situatedness and pretends to be universal.

Feminist Geographies

We have seen how feminist geographers have made the case for a greater epistemological awareness in geographical enquiry and how they have championed versions of situated knowledge. The role of gender, in particular, has been foregrounded as a key factor in the construction of knowledge. These theoretical positions have opened up geographical enquiry to domains that had previously been largely unexplored. These include geographies of fear, approaches to nature, geographies of mobility, and the role of gender in development.

Geographies of fear

Crime has long been an object of enquiry in social geography (Pyle 1974; Herbert 1982; Pain 1998). So, what does a feminist perspective add to this? Susan Smith suggested in 1987 that women in the UK experienced crime, and the fear of crime, in very different ways from men. She revealed how women lived in constant fear of crime in urban areas in the west and how their everyday mobility was adapted to take account of the possibility of violent attack and rape. In addition, women were more likely not to use the spaces of the city at night and to lock themselves into their homes (Smith 1987). Crime against women by men has been a constant theme in feminist literature and has been theorized as playing a central role in the production of patriarchy (Brownmiller 1975). Smith's review of women's fear of crime, however, was not primarily a feminist intervention.

A feminist theorization of fear and crime can be found in the work of Rachel Pain (Pain 1991) who argued that women's fear of crime was very different from men's fear of crime. She linked fear of sexual violence to wider issues of social control. Pain also points out that methods of counting crime are not straightforward and that much sexual violence goes unreported. The relationship between geographies of fear and geographies of crime was taken up by Gill Valentine (Valentine 1989, 1992; Listerborn 2002). Valentine worked with a diverse group of women in the British town of Reading. She examined both the actual distribution of violent crime against women and women's perception of particular (often public) spaces as dangerous. In addition, she explored the sources of these perceptions (often in the media) and suggested practical steps to make urban space both less dangerous and less fearful. Central to her work was the common perception of public space as dangerous for women. Private space is often described as a safe sanctuary for women (Blunt and Dowling 2006), while public space is perceived as "masculine" and threatening. This is despite the fact that the vast majority of violence against women, and particularly rape, occurs in private space – at home. The reason for this, she argues, is that public acts of violence are disproportionately reported in the media, while private acts of violence are unlikely to be reported in the media. This leads to women excluding themselves from public space yet continuing to believe that the space of home will be safe. This mismatch between the geography of crime and the geography of fear leads to the reproduction of gendered geographies of violence and fear.

As with all accounts of general themes, feminist geographies of violence and fear look different when they come from different places. The Indian feminist geographer, Anindita Datta has explored what she calls "genderscapes of hate" in India (Datta 2016). Datta's paper was written in response to the horrific gang rape and murder of a young woman on a public bus in 2012 in addition to a series of other brutal rape and murder cases in subsequent years. "The genderscapes of hate" Datta writes "are the lived spaces over which women are constantly devalued, degraded, humiliated and subject to different forms of violence hinging from such discrimination and devaluation within what can be termed a culturally sanctioned misogyny" (Datta 2016: 179). Here, Datta defines an interconnected landscape over which gender is performed in both everyday and exceptional ways. By defining this landscape, she links seemingly mundane performances of gender with acts as horrific as the very public cases she starts out with. Recognizing this genderscape of hate leads to the conclusion that typical responses to seemingly exceptional cases – such as stiffer sentences for rapists, or the increased provision of public toilets for women – do not address the problem that is rooted in the unexceptional practice of everyday life where gender norms in Indian and other societies are constituted on a daily basis. "Dismantling these hatescapes," Datta argues, "necessitates going beyond legislation to include the strategies of resistance and search for gendered agency in the praxis of everyday life, media spaces and in popular culture" (Datta 2016: 180–181). Datta's concept of "genderscapes of hate" asks us to go beyond thinking of hate, violence, and fear in terms of public and private space and instead asks us to think about the interconnected gendered spaces that make it possible for gendered violence of all types to become normalized. A similar argument is made by Nalini Khurana in her discussion of the role of sexual harassment in the navigation of space on Delhi's new Metro system (Khurana 2020). Work on gender and mobility (see below) is another important realm of research in feminist geography and geographies of fear overlaps with it. The rape case that Datta highlights notably occurred on a public bus and is also noted by Khurana. Public transport allows access to services and experiences that may be otherwise unavailable and the fear of using public transport can lead to a lack of access to these services and experiences. This hampers women's participation in public life. Khurana explores women's use of space in the Delhi Metro, which includes the provision of women-only carriages as a measure intended to ensure safety while travelling. This, like Datta's observations on public toilets, is an example of policy makers attempts to solve the problem of gendered fear and violence through design while ignoring the social and cultural processes that lead to the fear and violence in the first place.

The findings, along with insights drawn from existing concepts and literature, highlight a central theme: though the creation of women-only coaches has provided women with a safe mode of travel and enhanced their confidence and ownership of space, it has also led to contestations over their right to occupy general spaces on the train, thus reinforcing patriarchal structures and keeping women 'boxed in' within both the figurative and literal space of the women only coaches. (Khurana 2020: 31)

So, while some of the women in Khurana's study stated that their experience of women-only coaches increased their general confidence, it also continued a general sense of women's segregation from a more general public space. Design-led solutions to women's fear and violence against women can only ever be part of a more general solution.

Recently, another aspect of geographies of fear has arisen in response to the increasing visibility of transgender identities. One of the main points of contention that has emerged around the existence of trans women is their use of public toilets for women. Social media has been full of expressions of fear about the possibility that men might identify as women in order to access women's restrooms and commit acts of sexual violence. This is despite a lack of evidence that this happens and the clear evidence that trans women are themselves often the target of harassment and violence in restrooms and elsewhere. As Rachael Cofield and Petra L. Doan have written, public toilets remain a space where the binary identity of man/woman is most clearly coded (Cofield and Doan 2021). This only became the case when the larger divisions of men's (public) and women's (private) spaces began to break down in the late nineteenth century and it was deemed necessary to identify separate spaces for women who were exploring their newly found freedom to be in public. Since then, public restrooms have been specifically designed to conform to expectations of binary gender identities, and are accompanied by signs that use gender stereotypes to announce their gender.

Cofield and Doan investigate ongoing moral panics around trans bodies and public restrooms in the United States. They trace the passage of a law in Houston known as HERO (Human Equal Rights Ordinance) in 2015 that forbade discrimination based on, among other things, gender identity, in the provision of restrooms. A petition led to the Supreme Court of Texas requiring that the ordinance be subject to a referendum. In the referendum, 61% of voters voted against it. This followed a campaign which raised fears of transgender individuals threatening the safety of women and children in public restrooms. Opponents of the bill would refer to trans women as "gender-confused men" thus delegitimizing or outright denying the existence of transgender identities and, at the same time, correlating trans identities with those of sexual predators. Cofield and Doan argue that such reactions might have other underlying reasons.

Often, the argument against gender desegregation revolves around protecting women from a perceived transgender threat. However, it becomes clear, considering the lack of data about such a threat, that this argument is only superficially about paternalistic concerns of protecting women and children. Instead it serves as a smokescreen for an unspoken fear that gender nonconformity might lead people to question the perceived naturalness of gendered bodies and destabilize patriarchy. We argue that gendered toilets are a means of controlling public space to maintain gender normativity and keep women and transgender individuals in "their place" within the supposed gender binary. (Cofield and Doan 2021: 72)

Cofield and Doan chart the production of a particular geography of fear based on the erroneous argument that trans women threaten the safety of ciswomen and children in restroom spaces. They show how this is related to another geography of fear – the fear that breaking down the binary spatial divide of restroom provision will lead people to ask further questions of gender binaries. Space, gender, and fear are interrelated in complicated ways that leave open the possibility of different and more emancipatory gendered geographies.

The idea that gender binaries are reified and perpetuated by the built environment is evocative, as it suggests that the built environment is malleable and could evolve to be indicative of another construction of gender. In other words, the matter is still unsettled, leaving room for further destabilizing of borders between bodies and genders. (Cofield and Doan 2021: 76)

Feminist geographies of nature

We saw in Chapter 3 how geography has often been configured as the study of human/environment interactions. This has been described, until quite recently, as "man and nature" or "man/environment." There has been a long tradition of associating women with nature in western thought since the Renaissance (Fitzsimmons 1989; Rose 1993; Castree 2005). Gillian Rose charts how dualistic thought has divided western thought in general and geography in particular into two, unequal halves:

Culture	Nature
Man	Woman
Masculine	Feminine
Reason	Emotion
Public	Private
Space	Place

This kind of list can be extended almost infinitely as we think of other aspects of human life – White and Black, for instance. For the purposes of this chapter, however, it is most important to note how the feminine and woman are mapped onto nature, and, second, to note that these things are always the weaker or inferior part of a dualism. It has been one of feminism's most consistent theoretical contributions to highlight and then challenge such dualistic thinking, and the culture/nature dualism has been firmly at the center of attention.

Broadly speaking, feminists have taken two approaches to the relationship between the feminine and nature. One has been to embrace it and to use it as a mirror to those who are out to damage both women and nature. Under the banner of **ecofeminism**, many have used the associations between women, and particularly women's bodies, and the natural world to critique seemingly cold, unnatural, rational masculinity (Merchant 1989; Daly 1992; Plumwood 1993; Griffin 2000; Emel and Urbanik 2005). Consider this example:

> Nothing links the human animal and nature so profoundly as women's reproductive system which enables her to share the experience of bringing forth and nourishing life with the rest of the living world. *Whether or not she personally experiences biological mothering*, it is in this that woman is truly a child of nature and in this . . . lies the well-spring of her strength. (Collard and Contrucci 1988: 106)

Caroline New suggests that this kind of argument is based on an essential and necessary connection between the biology of sexual difference and the destructiveness of a patriarchal society. It also gives women a privileged position from which to critique and protest ecological destruction (New 1997).

Other feminists, however, have distanced themselves from this kind of feminism and refused to be easily conflated with nature. This view is well summarized in Barbara Kruger's cover illustration for Rose's *Feminism and Geography* which shows a woman's face with her eyes covered with leaves and the words "We won't play nature to your culture." Being labeled as part of nature has, after all, had few positive consequences for those so labeled, a point made by the Australian ecofeminist philosopher Val Plumwood:

> To be defined as 'nature' . . . is to be defined as passive, as non-agent and non-subject, as the 'environment' or invisible background conditions against which the 'foreground' achievements of reason or culture

(provided typically by the white, western, male expert or entrepreneur) take place. It is to be defined as *terra nullius*, a resource empty of its own purposes or meanings, and hence available to be annexed for the purposes of those supposedly identified with reason or intellect, and to be conceived and moulded in relation to these purposes. It means being seen as part of a sharply separate, even alien lower realm, whose domination is simply 'natural', flowing from nature itself and the nature(s) of things. (Plumwood 1993: 4)

In her book *Earth Follies*, geographer Joni Seager provides a feminist analysis of both the abuse of nature and attempts to halt that abuse. She steers a course between the embrace of an essential nature in some forms of ecofeminism and the complete rejection of the natural. She highlights the way in which the institutions most responsible for the degradation of nature – the military, corporations, and bureaucrats – are "masculine" institutions. In turn, she shows how some of the institutions which attempt to rescue nature – particularly science and environmental organizations such as *Earth First* – are also masculine. The impacts of environmental degradation are also most often experienced first and hardest by women who attempt to construct daily lives in a poisoned and polluted world. Pollution often becomes apparent in the processes of pregnancy and childbirth. Environmental disasters such as droughts and hurricanes often hit women harder as they have been comparatively less able to escape. These are not arguments which are often found in the daily news, where environmental problems are most often described in technical/scientific terms or simply blamed on pollution or deforestation (for instance), with little idea of who is polluting and who is deforesting. When blame is attached, it is often attached to arenas of life most often associated with women. Issues such as the use of green cleaning products or recycling are seen as part of the spheres of the domestic and consumption – spheres which are frequently thought of as feminine.

Seager is careful to resist essentialism. As we have seen, there are some feminists, broadly described as "ecofeminists," who suggest that women's bodies are uniquely connected to "mother earth" and that women, therefore, have a natural (or essential) connection to the environment (see Figure 8.1). She rejects this argument in favor of one that points out that women *do* have a special connection to nature but one which is produced within a patriarchal society that has forced such a connection. Consider the way she interprets the metaphor of "mother earth":

> The earth is *not* our mother. There is no warm, nurturing, anthropomorphized earth that will take care of us only if we treat her nicely. The complex, emotion-laden, conflict-laden, quasi-sexualized, quasi-dependent mother relationship (and especially the relationship between *men* and their mothers) is not an effective metaphor for environmental action. It suggests a benign distribution of power and responsibility, one that establishes an erroneous and dangerous assumption of the relations between us and the environment. It obfuscates the power relations that are really involved when we try to sort out who's controlling what, and who's responsible for what, in the environmental crisis. It is not an effective political organizing tool: if the earth is really our mother, then we are children, and cannot be held fully accountable for our actions. (Seager 1993: 219)

Seager also points out instances in which the "mother earth" metaphor has been used to disempower the environmental movement by suggesting that "nature" can look after itself – that the earth is the kind of mother who will sort out your mess. In 1989, for instance, the vice-president of the oil company Exxon made the following argument when challenged about his company's clean-up operation following the *Exxon Valdez* oil spill in waters off Alaska:

> I want to point out that water in the Sound replaces itself every 20 days. The Sound flushes itself out every 20 days. Mother Nature cleans up and does *quite* a cleaning job. (Charles Sitter, vice-president of Exxon, quoted in Seager 1993: 221)

Figure 8.1   Is the earth your mother? Source: Astronaut photograph AS17-148-22727, courtesy NASA Johnson Space Center Gateway to Astronaut Photography of Earth (http://eol.jsc.nasa.gov).

Seager offers a kind of situated theory – suggesting that the actual positions of women in the world and the relation of that position to nature allows them access to a special kind of truth that does not result from an essential characteristic but from the fact of women's position in an unequal world – a world in which women are frequently cleaning up after men.

Feminist geographies of mobility

Feminist geographers have been at the forefront of what has become known as the "mobility turn" in recent social science (Cresswell 2006; Sheller and Urry 2006; Urry 2007). Well before there was anything with the label "mobility turn" or "new mobilities paradigm," there were feminist analyses of forms of mobility ranging from the movements of the body (Young 1990) to daily travel patterns (Hanson and Pratt 1995) to travel and exploration (Blunt 1994). Traditional explanations of mobility patterns, often within the field of spatial science (see Chapter 5), created the sense of a genderless, unmarked "rational mobile man" who made mobility decisions based on rational decisions about the relative merits of "here" and "there." This led one feminist geographer to label the subject of early work on daily travel "the neuter commuter" (Law 1999).

Some of the most important contributions to our theorizations of mobility come from feminists (Wolff 1992; Braidotti 1994; Kaplan 1996). While mobility has often been equated with freedom and agency, feminists are not so sure. On the one hand, some recognize the (relative) freedoms that some women have attained through their insistence on the ability to move, whether walking through the city or conducting transnational mobilities (Wilson 1991; Domosh 1998). Rosi Braidotti has written of her love of airports in this vein:

But I do have special affection for the places of transit that go with traveling: stations and airport lounges, trams, shuttle buses, and check in areas, in between zones where all ties are suspended and time stretched to a sort of continuous present, oases of nonbelonging, spaces of detachment. No-(wo)man's land. (Braidotti 1994: 18)

Here there is a familiar sense of detachment from roots and belonging that is often celebrated by men. For many feminist scholars, this sense of detachment is illusory.

At the level of the body, the work on the gendered geography of fear (above) indicates that simply walking through the city can be a very different experience for men and women. These differences are further marked by differences in skin color or physical ability (for instance). Daily travel patterns for men and women in family units can be gendered too, with women more frequently having to mix the journey to work with school runs, shopping, and visits to medical facilities. Some of the earliest work in feminist geography considered these differences and how they challenged the assumptions behind spatial science's rational "man" (Hanson and Johnston 1985; Law 1999). Again, when mixed with other social markers such as race, this becomes even more complicated. Some of this complexity is revealed in an account of activism among bus riders in Los Angeles. A radical organization known as the Bus Riders Union has long protested city investment in expensive light rail systems that serve mainly White commuters moving between affluent suburbs and downtown areas. The money spent on these rail corridors directly impacts the provision of less glamorous buses that serve predominantly poor, minority, and female populations in Los Angeles. Burgos and Pulido recount an incident that shines light on the gendered (and classed and raced) politics of urban mobility:

The geography of work and travel reflects the spatiality of patriarchy, structural racism, and the division of labor. Domestic workers are a case in point. Often when organizing on buses traveling to affluent suburbs such as San Fernando Valley or Pacific Palisades, organizers have encountered entire busloads of immigrant women. Once while negotiating with the MTA [Metropolitan Transit Authority], a staff person expressed surprise at the overcrowding of buses going from downtown to the Valley. She said that "the rush hour traffic should be going downtown." Another person joked, "that's your maid going to your house." (Burgos and Pulido 1998: 80–81)

Here we see a complete misunderstanding of the ways in which everyday forms of urban mobility are marked by gender along with race and class. The buses are very different forms of transport from either cars or trains. They have different people on board, traveling different routes for different reasons. The kinds of sensibility raised by an awareness of the play of gender in the world allow scholars and others to notice these things – unlike the representative of the MTA in this account.

Feminist geographies of development

Development geography was one of the first of the discipline's subfields to engage with feminist theory and, often, feminist theory originating in the Global South. Up to the 1970s, development theorists of all theoretical persuasions tended to talk about development in a general way, paying no attention to the different ways men and women might experience the processes of development. This was challenged by Ester Boserup in her book *Women's Role in Economic Development* in 1970 (Boserup 1970). In this book, Boserup used United Nations statistics to show how women played a much more significant role in developing economies than had hitherto been recognized. Economic development, she argued, led to an increased specialization of labor moving away from the family as the main unit of production and consumption to a specialized economy based on a market. This has very serious consequences for women, as she summarizes in the preface to the 1986 edition:

> The process of increasing specialization of labour is accompanied both by an increasing hierarchization of the labour force and a gradual adaptation of the sex distribution of work, both in the family and in the labour force, to the new conditions. Since men are decisionmakers both in the family and in the labour market, are better educated and trained than women, and are less burdened with family obligations, they are much more likely to draw benefit from these changes than women, who tend to end up at the bottom of the labour market hierarchy. (Boserup 1986: npn)

In the years that followed the publication of Boserup's book, a number of issues became clear. Most official data was collected under the heading of "head of household," which was usually considered to be a man despite the fact that in large parts of the global south households are effectively headed by women. Similar issues arose with definitions of what counts as "work," when women appeared to do most of the work yet it did not count as it was often unpaid. By 1980, it was acknowledged at the United Nations Conference on Women in Copenhagen that the female population undertake over 60% of the world's labor and provide 44% of the world's food (Women and Geography Study Group of the Institute of British Geographers 1984).

Soon after the publication of *Geography and Gender* in 1984, the Women and Geography Study Group of the IBG compiled *Geography of Gender in the Third World* (Momsen and Townsend 1987). In this book, the editors suggested that theories of development (whether pro- or anti-capitalist) tended to focus only on economic factors and, therefore, fail to take the particular position of women into account. Development theory, they argued, was patriarchal.

An example of a writer who takes a wider view of development theory is Vandana Shiva, a feminist development theorist from India. Shiva has argued that development theory comes out of a western tradition that has little sway in her own context. "Contemporary western views of nature," she writes, "are fraught with the duality between man and woman, and person and nature. In Indian cosmology, by contrast, person and nature (Purusha-Pakriti) are a duality in unity. They are inseparable complements of one another in nature, in woman, in man" (Shiva 1997: 175):

> The dichotomized ontology of man dominating woman and nature generates maldevelopment because it makes the colonizing male the agent and model of "development." Women, the Third World and nature become underdeveloped, first, by definition, and then, through the process of colonization, in reality. (Shiva 1997: 176)

Here, Shiva is taking a much wider view than simply seeing development as an economic issue as she reminds us of the different kinds of theorization that might arise from a different position – in her case out of India.

More recent work on gender and development has moved on from a critique of genderblindness in development discourse and looked towards debates on **transnationalism** and **postcolonialism** that do not focus entirely on development but on wider issues that connect the Global North and the Global South (Radcliffe 2006). There is now a rich literature on the complicated role of gender in global relations. One example that illustrates some of the complexities in this project is Juanita Sundberg's work on environmental conservation in Guatemala. While recognizing the centrality of gender to feminist research, she notes how "gender intersects with other systems of power to produce multi-faceted, complex, and potentially contradictory identities" (Sundberg 2006: 46).

Sundberg's research takes place in the Maya Biosphere Reserve in Guatemala. She shows how identities are produced in relation to others through an exploration of an Indigenous Women's Group for the Rescue of Itza' Medicinal Plants (Agrupación) and the way in which they interacted with Indigenous men (BioItzá), the members of an NGO (CI/ProPetén), and herself as a "White" researcher. She argues that gender and race are always being produced in relation to each other in an ongoing, iterative process. She outlines how the male leaders of an historically marginalized Indigenous group formed an organization (BioItzá) to channel NGO funds into local projects – such as the creation of a forest reserve – with no involvement from women at management level. She then traces the formation of a subgroup of women to promote medicinal knowledge of local plants. The paper proceeds through a series of encounters in which race and gender are continuously intertwined. Here I focus on the first encounter.

In her first encounter, Sundberg describes a meeting called to prepare the women for their first meeting with CI/ProPetén. In this meeting, led by Rosalia, one of the women, a leader of BioItzá, Don Jaime, instructed the women on how to talk to the NGO and made sure that they agreed on their role of rescuing the medicinal uses of native plants – a form of knowledge traditionally associated with women. He also advised the women to avoid mentioning western medicine in order to confirm to the NGO that they were properly Indigenous and in touch with nature – a narrative that sells well to fund givers:

> Although a number of interesting dynamics emerge, I want to highlight how the narratives and performances enacted in the meeting (re)configure gender and race in contradictory ways. One of the most striking elements of this encounter is the (re)assertion of male superiority. Although Rosalia directs the meeting, Don Jaime interrupts at will. He advises the women on how to perform their gendered and racialized identities and he mediates the Agrupación's relationship with me, the researcher (to be discussed in the third encounter). When Rosalia asserts that the women wished to change their gendered behavior through participation in the group, Don Jaime responded by subsuming the Agrupación within the BioItzá and its wider goals for cultural revitalization. Indeed, he is suggesting that women are important to the revitalization movement because of their gender specific activities. Ultimately, his narratives act to discipline women's gendered behavior and reproduce patriarchy in the name of preserving cultural traditions. (Sundberg 2006: 50)

Despite these attempts to direct the purpose of the meeting, however, it is also clear that some women in the room had a second aim, to produce a space for women to transform established gender relations. In this encounter, then, "interlocking systems of power, shaped by histories of sexism and racism, produce gendered and racialized subject identities as unequal" (Sundberg 2006: 52). Don Jaime establishes a role of male authority figure; he instructs the women on how to play on their Indigenous identity in a way that underlines their inferior status. He also insists on the women reproducing their gendered roles as "healers." At the same time, however, "the medicinal plants group creates a space wherein women have the opportunity to disrupt male dominance and (re)configure gender relations" (Sundberg 2006: 52). Sundberg goes on to explore a number of further encounters, including her own as a White woman who conformed to the role of an incomer who was able to help otherwise helpless Indigenous women. The research, she concludes, led her to consider her own role as a White woman in the research process as well as more general interactions of gender and race in the development and conservation processes.

This kind of feminist research is a long way from the original insistence of taking women, and then gender, seriously in the process of development. Contemporary work by feminists in the Global South is no longer necessarily focused on development as such but includes an array of interactions between north and south, gender, and other forms of group identity.

Conclusions: Feminist Geography, Difference, and Intersectionality

All of the above illustrations of feminist geographic theory in practice illustrate how a focus on gender as an analytical category is more than just an added ingredient. This focus fundamentally challenges the way we think about the geographical world and the ways in which we study it. It is also clearly a focus that is grounded in, and frequently highlights, the politics of real-world struggle.

There are important senses in which feminist geography (rather than simply geographies of gender) had been perceived as a Western concern within the discipline of geography. Anindita Datta has argued that, in India, feminism, along with postmodernism and poststructuralism, were seen by many as imported Western ideas inappropriate for an Indian context. In India, Datta notes, geographers tended to eschew qualitative research and, even when studying gender differences, tended to favor quantitative accounts that could be used to influence policy makers (Datta 2013). This contrasts with feminist theory that starts with a recognition and theorization of the differences between genders, but rarely stops there.

Since the early 1990s, feminist geographers have grappled, more forthrightly and often more agonizingly than others, with the problems produced by the recognition of difference. Women, as Sundberg's research clearly shows, are not just women. They are also classed, sexed, and raced in different ways. Rose has outlined the necessity of avoiding complicity with racism and heteronormativity in the process of constructing feminist geography. She insists on the importance of thinking beyond the dualism of gender but across multiple identities, embracing, rather than denying, difference. This embrace of difference, she suggests, can serve to displace the dualistic thinking of masculinism. Recently, for instance, Katherine McKittrick has considered the absence of Black voices and Black subjects in geography, not just in mainstream geography but in feminist geography too. Being Black makes as much difference as being a woman and the two can be compounded (McKittrick 2006). The idea of **intersectionality** was introduced by Kimberlé Crenshaw in 1989 as a way of giving a name to how interconnected forms of power combine to effect marginalized people in society in distinctive ways that are irreducible to any single axis of identity. Thus, axes of gender intersect with axes of race, class, sexuality, age, and other forms of identity which are shaped by and shape power relations. As feminist geography has progressed its concerns have become notably more intersectional as evidenced by the essays in *Feminist Geography Unbound* that are both by and about a range of intersectional positions. It is important to note that the idea of intersectionality arose out of Black feminist thought. Sometimes the word intersectionality has been used to refer to any combination of identities – an intersectional analysis of disability and sexuality for instance – while, in actuality, it has arisen out of a specific recognition of the entanglement of race and gender within Black feminism (Hopkins 2019). As it is taken up without due recognition of its origins, there is a danger that intersectionality may become a diluted form of critique that, once again, washes over the importance of race within critical thought and re-whitens the project of feminist geography. Having said that, it is important to note that the race/gender intersection has moved beyond its origins in North American thought and been fruitfully applied in other, particularly postcolonial, settings. This has been compellingly argued, for instance, by Sharlene Mollett and Caroline Faria.

> In part, its application is globally relevant as racial violence, segregation and white supremacy operate in place-specific ways *everywhere*. Set in the context of international development, both colonial histories and contemporary global relations demand that we pay attention to race and by extension caste and ethnicity too. (Mollett and Faria 2018: 571)

Resisting the "feel good" feminism that Gökariksel and colleagues (Gökariksel *et al.* 2021) write about means paying attention to the difficult levels of tension that arise from heeding the differences within feminism and heeding the complexities of intersectionality.

> Rather, working in solidarity, *across and through the interrogation of difference*, with agreement and discord, we are better positioned to understand this particular manifestation of racial-gendered-sexual violence, to trace its genealogies in colonial-pasts, and to spatialize our resistance to such violence in the present, on behalf of our feminist geographic futures. It is in these spaces that, not only our contradictions, but our complex commonalities as human beings, are indelibly laid bare. (Mollett and Faria 2018: 574)

References

* Barefoot, A. (2021) Women-only spaces as a method of policing the category of women, in *Feminist Geography Unbound: Discomfort, Bodies, and Prefigured Futures* (eds B. Gökariksel, M. Hawkins, C. Neubert, and S. Smith), West Virginia University Press, Morgantown, pp. 158–179.

Blunt, A. (1994) *Travel, Gender and Imperialism: Mary Kingsley and West Africa*, Guilford, New York.

Blunt, A. and Dowling, R. M. (2006) *Home*, Routledge, London.

Boserup, E. (1970) *Woman's Role in Economic Development*, Earthscan, London.

Boserup, E. (1986) *Woman's Role in Economic Development*, Gower, Aldershot.

Braidotti, R. (1994) *Nomadic Subjects: Embodiment and Sexual Difference in Contemporary Feminist Theory*, Columbia University Press, New York.

Brownmiller, S. (1975) *Against Our Will: Men, Women and Rape*, Secker and Warburg, London.

Burgos, R. and Pulido, L. (1998) The politics of gender in the Los Angeles Bus Riders' Union/Sindicato De Pasajaros. *Capitalism Nature Society*, 9, 75–82.

Castree, N. (2005) *Nature*, Routledge, London.

* Cofield, R. and Doan, P. L. (2021) Toilets and the public imagination: planning for safe and inclusive spaces, in *Feminist Geography Unbound: Discomfort, Bodies, and Prefigured Futures* (eds B. Gökariksel, M. Hawkins, C. Neubert, and S. Smith), West Virginia University Press, Morgantown, pp. 69–87.

Collard, A. F. E. E. and Contrucci, J. (1988) *Rape of the Wild: Man's Violence against Animals and the Earth*, Women's Press, London.

Cope, M. (2002) Feminist epistemology in geography, in *Feminist Geography in Practice* (ed. P. J. Moss), Blackwell, Oxford.

Cresswell, T. (2006) *On the Move: Mobility in the Modern Western World*, Routledge, New York.

Daly, M. (1992) *Pure Lust: Elemental Feminist Philosophy*, HarperSanFrancisco, San Francisco, CA.

Datta, A. (2013) Wildflowers on the margins of the field: on the geography of gender in India, in *Paradigm Shift in Geography* (ed. M. H. Qureshi), Manak Publications, Delhi, pp. 234–249.

* Datta, A. (2016) Genderscapes of hate: on violence against women in India. *Dialogues in Human Geography*, 6(2), 178–181.

* Domosh, M. (1991) Toward a feminist historiography of geography. *Transactions of the Institute of British Geographers*, 16, 95–104.

Domosh, M. (1998) Those "gorgeous incongruities": polite politics and public space on the streets of nineteenth-century New York City. *Annals of the Association of American Geographers*, 88, 209–226.

Emel, J. and Urbanik, J. (2005) The new species of capitalism: an ecofeminist comment on animal biotechnology, in *A Companion to Feminist Geography* (eds L. Nelson and J. Seager), Blackwell, Oxford, pp. 445–457.

Falconer Al-Hindi, K. (2002) Toward a more fully reflexive feminist geography, in *Feminist Geography in Practice* (ed. P. J. Moss), Blackwell, Oxford, pp. 103–115.

Fitzsimmons, M. (1989) The matter of nature. *Antipode*, 21, 106–120.

* Foord, J. and Gregson, N. (1986) Patriarchy: towards a reconceptualisation. *Antipode*, 18, 186–211.

* Gökarıksel, B., Hawkins, M., Neubert, C., and Smith, S. (2021) *Feminist Geography Unbound: Discomfort, Bodies, and Prefigured Futures*, West Virginia University Press, Morgantown.

Griffin, S. (2000) *Woman and Nature: The Roaring Inside Her*, Sierra Club Books, San Francisco, CA.

Hanson, S. and Johnston, I. (1985) Gender differences in work trip length: explanations and implications. *Urban Geography*, 6, 293–219.

Hanson, S. and Pratt, G. J. (eds) (1995) *Gender, Work, and Space*, Routledge, New York.

Haraway, D. (1988) Situated knowledges: the science question in feminism and the privilege of partial perspective. *Feminist Studies*, 14, 575–599.

Harding, S. G. (1986) *The Science Question in Feminism*, Open University Press, Milton Keynes.

Herbert, D. T. (1982) *The Geography of Urban Crime*, Longman, New York.

Hopkins, P. (2019) Social geography I: intersectionality. *Progress in Human Geography*, 43, 937–947.

Hubbard, P., Kitchin, R., and Valentine, G. (eds) (2004) *Key Thinkers on Space and Place*, Sage, Thousand Oaks, CA.

Johnson, L. C., Huggins, J., and Jacobs, J. M. (eds) (2000) *Placebound: Australian Feminist Geographies*, Oxford University Press, Oxford.

Kaplan, C. (1996) *Questions of Travel: Postmodern Discourses of Displacement*, Duke University Press, Durham, NC.

Khurana, N. (2020) Geographies of fear: sexual harassment and women's navigation of space on the Delhi Metro. *South Asian Journal of Law, Policy, and Social Research*, 1(1), 18–39.

Kitchin, R. (ed.) (2008) *International Encyclopedia of Human Geography*, Elsevier, Boston, MA.

Law, R. (1999) Beyond "women and transport": towards new geographies of gender and daily mobility. *Progress in Human Geography*, 23, 567–588.

Listerborn, C. (2002) Understanding the geography of women's fear: toward a reconceptualization of fear and space, in *Subjectivities, Knowledges, and Feminist Geographies: The Subjects and Ethics of Social Research* (ed. L. Bondi), Rowman & Littlefield, Lanham, MD.

Livingstone, D. N. (1993) *The Geographical Tradition: Episodes in the History of a Contested Enterprise*, Blackwell, Oxford.

Maddrell, A. (2009) *Complex Locations: Women's Geographical Work in the UK 1850-1970*, WileyBlackwell, Oxford.

McDowell, L. (1993) Space, place and gender relations: part II. Identity, difference, feminist geometries and geographies. *Progress in Human Geography*, 17, 305–318.

* McKittrick, K. (2006) *Demonic Grounds: Black Women and the Cartographies of Struggle*, University of Minnesota Press, Minneapolis, MN.

Merchant, C. (1989) *The Death of Nature: Women, Ecology, and the Scientific Revolution*, Harper & Row, New York.

Millett, K. (1971) *Sexual Politics*, Hart-Davis, London.

* Mollett, S. and Faria, C. (2018) The spatialities of intersectional thinking: fashioning feminist geographic futures. *Gender, Place & Culture*, 25, 565–577.

* Momsen, J. H. and Townsend, J. (1987) *Geography of Gender in the Third World*, State University of New York Press, Albany, NY.

Monk, J. and Hanson, S. (1982) On not excluding half of the human in human geography. *Professional Geographer*, 34, 11–23.

Nelson, L. and Seager, J. (2005) Introduction, in *A Companion to Feminist Geography* (eds L. Nelson and J. Seager), Blackwell, Oxford, pp. 1–11.

New, C. (1997) Man bad, woman good? Essentialisms and ecofeminisms, in *Space, Gender, Knowledge: Feminist Readings* (eds L. McDowell and J. Sharp), Arnold, London, pp. 177–192.

Pain, R. (1991) Space, sexual violence and social control; integrating geographical and feminist analysis of women's fear of crime. *Progress in Human Geography*, 15, 415–431.

Pain, R. (1998) *Geography and the Fear of Crime: A Review*, University of Northumbria at Newcastle, Division of Geography and Environmental Management, Newcastle.

Plumwood, V. (1993) *Feminism and the Mastery of Nature*, Routledge, New York.

Pyle, G. F. (1974) *The Spatial Dynamics of Crime*, University of Chicago, Chicago, IL.

* Radcliffe, S. A. (2006) Development and geography: gendered subjects in development processes and interventions. *Progress in Human Geography*, 30, 524–532.

* Rose, G. (1993) *Feminism and Geography: The Limits of Geographical Knowledge*, Polity, Cambridge.

Rose, G. (1995) Tradition and paternity: same difference? *Transactions of the Institute of British Geographers*, 20, 414–416.

* Royal Geographical Society (with the Institute of British Geographers). Women and Geography Study Group. (1997) *Feminist Geographies: Explorations in Diversity and Difference*, Longman, Harlow.

* Seager, J. (1993) *Earth Follies: Coming to Feminist Terms with the Global Environmental Crisis*, Routledge, New York.

Sheller, M. and Urry, J. (2006) The new mobilities paradigm. *Environment and Planning A*, 38, 207–226.

Shiva, V. (1997) Women in nature, in *Space, Gender, Knowledge: Feminist Readings* (eds L. McDowell and J. Sharp), Arnold, London, pp. 174–192.

Smith, S. (1987) Fear of crime: beyond a geography of deviance. *Progress in Human Geography*, 11, 1–23.

Stoddart, D. (1986) *On Geography and Its History*, Blackwell, Oxford.

Sundberg, J. (2006) Identities in the making: conservation, gender and race in the Maya Biosphere Reserve, Guatemala. *Gender, Place and Culture*, 11, 43–66.

Todd, J. D. (2021) Exploring trans people's lives in Britain, trans studies, geography and beyond: a review of research progress. *Geography Compass*, 15, e12556.

Urry, J. (2007) *Mobilities*, Polity, Cambridge.

Valentine, G. (1989) The geography of women's fear. *Area*, 21, 385–390.

Valentine, G. (1992) Images of danger: women's sources of information about the spatial distribution of male violence. *Area*, 24, 22–29.

Wilson, E. (1991) *The Sphinx in the City*, University of California Press, Berkeley, CA.

Wolff, J. (1992) On the road again: metaphors of travel in cultural criticism. *Cultural Studies*, 6, 224–239.

*Women and Geography Study Group of the Institute of British Geographers. (1984) *Geography and Gender: An Introduction to Feminist Geography*, Hutchinson, London, in association with The Explorers in Feminism Collective.

Young, I. M. (1990) *Throwing Like a Girl and Other Essays in Feminist Philosophy and Social Theory*, Indiana University Press, Bloomington.

Zelinsky, W. (1973) The strange case of the missing female geographer. *Professional Geographer*, 25, 101–106.

Zelinsky, W., Hanson, S., and Monk, J. (1982) Women and geography: a review and prospectus. *Progress in Human Geography*, 6, 317–366.

Chapter 9

Postmodernism and Beyond

In the late 1980s, when I was studying for my doctorate, a new word started to pop up in conversation and at conferences. The word was "postmodernism" (and its variants: "postmodern" and "postmodernist"). The first conference I ever went to was the Annual Meeting of the Association of American Geographers in Phoenix in 1988. It seemed that almost every session was in some way about postmodernism. This was picked up on by others too – sometimes less than approvingly. Referring to the 1988 meeting, Linda McDowell wrote:

> Reference to a totally new set of literature is now essential: everyone (or at least those in the know) is now a deconstructionist. Difference, diversity, the 'Other', situatedness, positionality, polyvocality – these are the new words that rise in the hum of intellectual exchange in the rooms and corridors of geography meetings. (McDowell 1992: 58)

The reading groups that postgraduate students formed at the University of Wisconsin, Madison, while I was there gradually shifted from groups reading Marx and Harvey to groups reading Baudrillard, Foucault, and Soja. Following the Phoenix conference, I attended an American Cultural Studies conference at New Orleans and it seemed once more that the whole meeting was dedicated to aspects of postmodernism. Some rooms were full of excited and largely unintelligible (to me, then) discussions of postmodern panic and the writings of Jean Francois Lyotard or Jean Baudrillard, while others featured huddles of enthusiasts expounding on the postmodern trickery of the TV series *Moonlighting* (in which a knowing young Bruce Willis would often turn to the camera and say things like "now it's time for the car chase scene"), or the postmodern simulation of national identity in Disney's Epcot Center. It was an exciting (and confusing) time to be a doctoral student.

Geographic Thought: A Critical Introduction, Second Edition. Tim Cresswell.
© 2024 John Wiley & Sons Ltd. Published 2024 by John Wiley & Sons Ltd.

The world of the late 1980s was one in which many of the things we now see around us were coming into being. I think I became aware of the internet around then, MTV was wildly popular, and music videos with their rapid-fire cut up techniques were all the rage. Movies like *Bladerunner* and *Robocop* reflected on the increasingly unclear boundaries between "reality" and some manufactured version of it. In literature, scholarly types (and I wanted so much to be one of them) were reading novels like *If on a Winter's Night a Traveler* by Italo Calvino, a novel in which we are presented with the problem of reading a book called "*If on a Winter's Night a Traveler*" interspersed with bits of the book that is being referred to. It is a novel about reading a novel (Calvino 1981). In John Fowles' novel *The French Lieutenant's Woman*, we are presented with two endings, one historical and one contemporary, which, in the movie, turned into an ending in the movie and then an ending for the actors acting in the movie (Fowles 1971). It was all cool, clever, and very aware of the process of production and reception, a process that had traditionally been kept well hidden.

Think about the world we now inhabit. We are thoroughly used to a multi-channel existence of on-demand news and entertainment. Where I live I am bombarded with forms of advertising everywhere. In shops I can buy food from almost anywhere I want, while the word "local" attached to anything is a sure sign of both cultural prestige and a premium price. Words like "hybrid" and "fusion" are used in everything from restaurant menus to names for cars. The music I listen to freely mixes forms from around the world such that the term "world music" that arose in the postmodern 1980s is almost meaningless. Words like "collage" and "pastiche" are freely used to describe the landscape. These kinds of observations were beginning to emerge across the social sciences and humanities in the 1980s. One example is the work of the British cultural theorist Iain Chambers, who described the world of the late 1980s, in metropolitan Britain, as follows:

> Everywhere, there is a sense of a new cultural economy in which everything is a little less precise, a little more complex. Where there are boys *and* girls, black *and* British, youth *and* Asian, hetero- *and* homosexual, leisure *and* work (or unemployment), public *and* domestic cultures, older distinctions begin to collapse and give way to less traditional, multiple, less historicized, hence 'lighter' (not to be confused with less serious) and more open prospects, the allegory of class revealed in youthful style statements and decentred by the plural coordinates of gender, sexuality and ethnicity. In youth and not so youthful cultures it has been collage dressing and musical eclecticism, where, for example, it has been not only the confidently declaimed 'roots' authenticity of contemporary black music (dub, toasting, rap, scratch), but equally their (post)modernist aesthetics of chance construction, deft bricolaging, sonorial archaeology and recycling, that has dominated the 1980s. Where subcultures once offered a 'strong' sense of stylistic opposition to the status quo and the world of 'them', … this has been extended and then gradually reworked into a wider sense of detailed differences; the 'Other' becomes more simply, but no less significantly, the 'others'. (Chambers 1990: 69–70)

This new world was not easy to map. Everything was mixed up.

This was one definition of postmodernism that emerged – as a new time period or style. It was a description of the observable changes in the world of the developed west. In this sense, it was perfectly possible to react to it as a Marxist, a humanist, or a feminist, and many did. At around the same time, however, another version of postmodernism emerged in the form of postmodern theory which, in many ways, reflected the eclecticism of the world described by Chambers and others. Chambers appeared willing to embrace the possibility of what he called "weak thought" that drew back from making grand explanations:

> Social and cultural sense, then, becomes not a goal but a discourse, not a closure but a trace in an endless passage that can only aspire to a temporary arrest, to a self-conscious drawing of a limit across the diverse possibilities of the world. As Gilles Deleuze puts it, sense is a surface-effect, an event, and not the sign or symptom of an absent origin, a lost totality, or a pure consciousness. It is precisely the lack of fixed

referent or stable foundation that produces meaning. For to produce it does not mean to touch a sacred stone or turn the right key that will reveal the nature of things, but involves tracing out a recognizable shape on the extensive complexity of the possible. (Chambers 1990: 11)

To Chambers and others, the new world demanded new ways of thinking about the world informed by French philosophers such as Gilles Deleuze. To others, the postmodern world could be adequately interpreted, or even dismissed, from already established perspectives such as humanism (Ley 1989), Marxism (Harvey 1989), or feminism (McDowell 1992). This chapter necessarily flits between accounts of the postmodern and thought, which is, itself, postmodern.

Two Buildings

The difference between **modernism** (and modernist) and postmodernism (and postmodernist) can be illustrated through two contrasting buildings: the Pruitt-Igoe housing development in St. Louis, Missouri and the Bonaventure Hotel in Los Angeles. The origins of postmodernism, in its late-twentieth-century guise, comes from the world of architecture. Charles Jencks, an architectural theorist, wanted to describe a new kind of architecture emerging after 1972 when a major modernist architectural project, the Pruitt-Igoe Project in St. Louis, was demolished (Jencks 1977). It is worth spending some time on the issue of modern and postmodern architecture because it illustrates some of the central themes in the emergence of postmodernism in theory too.

Modern architecture, although a contested term, has normally been applied to architectural movements of the early twentieth century and is most closely associated with the work and ideas of Le Corbusier. At the heart of modern architecture was and is a belief that architecture can serve a social purpose – that it can improve the lives of the mass of people. Aesthetically it is marked by quite strict geometries – often quite grid-like, with straight lines and right angles predominating. It paid little attention to the local context. A good modern building, it was believed, could work wherever it was and therefore would look pretty much the same whether it was in New York, Tokyo, or Chandigarh. Such architecture would form a "machine for living in" and would be marked by rationality and efficiency. These buildings would improve people's lives and produce a more just society. Rational order was linked to freedom. There was great belief in the idea of progress. Le Corbusier believed, for instance, that the historical fabric of Paris was a relic of a bygone age that got in the way of a utopian future of modernist tower blocks and rational life. Becoming modern was tied to universal notions of truth and reason. Through rationality it would be possible to find the one best way to do something and then to do it everywhere. The rigid geometries of a building such as Pruitt-Igoe are linked to a kind of thought that puts faith in foundations (literally and figuratively).

We have already seen (in Chapter 5) how reason and rationality are supposed to exist out of place – in a space above the world – a view from nowhere. This is how science in general and spatial science in particular had attempted to come to law-like generalizations about geography. Spatial science was actually symptomatic of high modernism and can be linked more or less directly to the ambitions of modernist architecture. The people behind the construction of the Pruitt-Igoe Project had sought to improve people's lives through the application of rationality. Its demolition in 1972 (Figure 9.1) has been seen as a mark of not only the failure of that building but of the kind of theory that went behind it.

As postmodernism started to circulate in geographical circles, one building in particular was used to exemplify it – the Bonaventure Hotel in Los Angeles. The hotel has appeared in a number of accounts (both sympathetic and hostile) of postmodernism. The hotel was a product of John C. Portman, who designed a number of giant "urban renaissance" projects across the United States including the Renaissance Center in Detroit (surely ironic given the current state of Detroit following

Figure 9.1 The demolition of Pruitt-Igoe, St. Louis, 1972 – the end of modernism? Source: US Department of Housing and Urban Development. http://commons.wikimedia.org/wiki/File:Pruittigoe_collapse-series.jpg (accessed May 29, 2012).

the ravages of late capitalism!) and the Peachtree Center in Atlanta. All of these are spaces within neglected downtowns that attempt to produce "cities within cities" and turn their backs (actually mirrored façades) on the city beyond. His works are marked by a sense of disconnection from the street-level cityscapes in which they are located. They often feature skywalks connecting the buildings to nearby office blocks and shopping areas – pathways free from the possible intrusion of the city's disturbing "others" – the poor, the racially other, and the homeless. Buildings such as this appear to have no social project apart from one which separates the wealthy from the poor. More recently, he has been partly responsible for the awe-inspiring transformation of Beijing and Shanghai – surely the archetypal cities of the new century. It was the Bonaventure Hotel, however, that caught the attention of the theorist Frederic Jameson as he sought to describe postmodernism. Jameson describes the mirror-glass-clad hotel as a popular building that forgoes any attempt to "improve society" as a modernist Pruitt-Igoe might have done. Rather the building plays along with the "tawdry and commercial sign system of the surrounding city" (Jameson 1991: 39). Similarly, there appears to be no overarching spatial logic to its interior design, which Jameson described in some detail:

> The entryways of the Bonaventure are, as it were, lateral and rather backdoor affairs: the gardens in the back admit you to the sixth floor of the towers, and even there you must walk down one flight to find the elevator by which you gain access to the lobby. Meanwhile, what one is still tempted to think of as the front entry, on Figueroa, admits you, baggage and all, onto the second-story shopping balcony, from which you must take an escalator down to the main registration desk. What I first want to suggest about these curiously unmarked ways in is that they seem to have been imposed by some new category of closure governing the inner space of the hotel itself (and this over and above the material constraints under which Portman had to work) … (Jameson 1991: 39)

The Bonaventure, to Jameson, is attempting to cut itself off from the city. In some ways this was also a criticism made of modern(ist) architecture that sought to work in a manner that would be possible with little regard to the place it was built (there was little that was specifically "St. Louis" to the Pruitt-Igoe Project). To Jameson, though:

> … this disjunction from the surrounding city is different from that of the monuments of the International Style, in which the act of disjunction was violent, visible, and had a very real symbolic significance …. [T]he Bonaventure, however, is content to 'let the fallen city fabric continue to be in its being' (to parody Heidegger); no further effects, no larger protopolitical Utopian transformation is either expected or desired. (Jameson 1991: 41–42)

This spatial confusion was picked up by geographer Edward Soja, in his account of Los Angeles as a proto-typical postmodern city. Like Jameson, Soja was baffled by the spatial layout of a hotel that seemed designed to make you lost and contained shops that nobody could find:

> Everything imaginable appears to be available in this micro-urb but real places are difficult to find, its spaces confuse an effective cognitive mapping, its pastiche of superficial reflections bewilder co-ordination and encourage submission instead. Entry by land is forbidding to those who carelessly walk but entrance is nevertheless encouraged at many different levels, from the truly pedestrian skyways above to the bunkered inlets below. Once inside, however, it becomes daunting to get out again without bureaucratic assistance. In so many ways, its architecture recapitulates and reflects the sprawling manufactured spaces of Los Angeles. (Soja 1989: 243–244)

Jameson terms this dizzying experience of space – "postmodern hyperspace" – a form of space that "has finally succeeded in transcending the capacities of the individual human body to locate itself, or organize its immediate surroundings perceptually, and cognitively to map its position in a mappable external world" (Jameson 1991: 44).

Figure 9.2 The Westin Bonaventura Hotel, Los Angeles. Source: Photo by Geographer, CC BY-SA 3.0. http://commons.wikimedia.org/wiki/File:Westin_Bonaventure_Hotel.jpg (accessed May 29, 2012).

It has been over 30 years since the Bonaventure Hotel made such a splash in the world of spatial theory. I have been there, ironically enough, for a geography conference and can confirm that the spatial experience is baffling. But it does not seem remarkable and I was left wondering what the fuss was all about. It is pretty much like high-end atrium hotels all over the world that I have visited for conferences. I have also felt it in modern hospitals. Perhaps this feeling of being underwhelmed was a sign of the success and spread of postmodern hyperspace (see Figure 9.2).

Key Points in Postmodern Theory

Reading texts about postmodernism, or texts which are themselves postmodern, can be a dizzying experience not unlike attempting to navigate the Bonaventure Hotel. This is often the point in the "difficult" theory course where things become too complicated for many. It is also the point at which imaginations are stretched to breaking point. One reason for this is that geographers suddenly find themselves reading French theory in translation. This kind of writing is often convoluted and tortuous. Sometimes this is because the writing is either bad or badly translated, or it is because the writer is deliberately attempting to challenge the conventions of "transparent" and clear writing to make a suitably postmodern point. My task here is to filter out a few key points in postmodern theory. Such a task is not, itself, very postmodern.

Against metanarratives

One of the key texts in the emergence of postmodernism in academic life was Jean Francoise Lyotard's *The Postmodern Condition* in which he argues that contemporary life and thought are marked by a deep suspicion of big stories that purport to explain everything (or close to it). Lyotard describes this as an "incredulity toward metanarratives" (Lyotard 1984: xxiv).

Metanarratives are accounts which provide the foundations for various kinds of theoretical, practical, and political schemes in the modern world. There are all kinds of metanarratives, ranging from religious ones (particularly fundamentalist ones) to forms of science to political projects such as liberalism or Marxism. These forms of knowledge, which attempt to be generally applicable and to include as much within their sphere as possible, are called "totalitarian" by Lyotard. He suggests replacing such "big theories" with a multitude of little, local forms of knowledge. While metanarratives attempt totality, Lyotard favors diversity. Local knowledge, to Lyotard, is inherently critical – especially of the kinds of projects inherent in metanarratives. As with much of postmodern theory, there is an irony here, of course. In his attempt to undercut the power of metanarratives, Lyotard begins to construct another one – a metanarrative that holds the local and the critical as a model for thought – an anti-totality metanarrative if you like.

Such a suspicion of metanarratives was evident in the destruction of the Pruitt-Igoe Project which had promised much, based on modernist stories of emancipation, progress, and reason, and had failed. At a grander level it can be seen in many post–World War II responses to the failures of science (the atom bomb, the green revolution), Marxist political movements (the Soviet Union, China), or capitalism (the free market, development). Grand claims for "reason" and "rationality" had often led to horrendous consequences. You do not have to look toward complicated theory to see Lyotard's incredulity. It is evident in everyday life and popular culture as people question the efficacy of medicine and seek out "alternative therapies" or as they refuse to believe the scientists who assure them of the reality of global warming. We can see how political parties in the west have moved away from proclamations based on grand ideologies (usually referred to as "dogma") and have instead focused in on small issues that leave little difference between them. Political protest groups are less likely to be driven by a need for global revolution and more often by "single issues" (animal welfare, anti-roads, fair-trade, etc.) around which social movements are formed. This is all a far cry from a widely held belief in scientific progress or the mortal combat of capital and labor.

Against "foundations" and "essences"

Despite their many differences, the approaches to geography we have examined so far also have important things in common. Foremost among these is the belief in some bedrock truth that theory can be built on. For some, this is the evident truth of the world that is observable; this reached its height in some forms of positivist spatial science. For some Marxists there is the "last instance" of the forces and relations of production which can be used to explain much of social life. For some feminists it is the relations between men and women. For humanists it is experience. Theorists refer to such beliefs as "foundational" and "essentialist." These are all forms of *modernist* thought because they share a view, which has been dominant since the Renaissance, that somewhere there are foundations which can be used to explain things and to determine what is, in some sense, "true." All of these were challenged in the 1980s by the arrival of a diffuse body of ideas that emerged under the heading of "postmodernism."

Before postmodernism, most, if not all, theoretical statements about the world included some notion of essence. Essentialism is the belief that things are made up from certain properties that simply have to be there in order for that thing to be what it is. The thing, say a book, is constituted by its essence. For an essence to be an essence, we only have to consider what would happen if it were removed. Would your chosen thing still be what it was? Could we have a book without words? Without pages? Without a cover? Even e-books replicate these in virtual form. This gets more complicated the more complicated the thing is – a human being for instance. Or what about non-material things such as love or equality? In one sense things seem to have essences – this seems like common sense. But on another level it is actually quite hard to arrive at aspects of things it is impossible to remove.

The question of whether or not things have essences gets more controversial when we are considering things such as race and sex. For most of geography's history, for instance, it was taken for granted that such a thing as "race" existed and that race, as well as each individual race, had some kind of essence (behavioral, genetic, etc.). This meant, for instance, that races could be mapped and counted. It is only recently that race has been seen as an invention – a social and cultural construct. To think of race as a social construct is to take an "anti-essentialist" approach to it – an approach that became popular with the rise of postmodernism in the 1980s. When we take an anti-essentialist approach to race it is much harder, if not impossible, to map and count it.

Postmodernism is **anti-foundational**. All of the historical metanarratives that seek to explain wide swathes of human life rely on some foundation which gives their theory power. Perhaps the clearest example of this is classical Marxism where the simplest form of the base–superstructure model (see Chapter 7) declared the forces and relations of production and their historical development the basis of human history. In the "last instance" everything (that mattered) could be explained by recourse to the economy. Another obvious example is the belief in God for Christian and other faiths. Postmodern theorists claim that there is no such foundation to appeal to. Not only is God dead, but all His replacements too.

Problems with representation

Postmodernists do not believe in Truth (with a big T) nor in any consistent and knowable "reality" that can be known and represented. Most scientific and social scientific research and writing, including that in human geography, has assumed that there is some "out there" which is being researched and then written about in as clear a way as possible. There is a sense that our words, equations, pictures, maps, and diagrams bear some relation to an external real world and that the idea is to make the text as close to possible to the world the text represents. Language (in its widest sense) is supposed to be transparent. This is a belief shared by spatial scientists, Marxists, and humanists. Postmodernists do not believe in the external incontrovertible reality and so cannot believe in transparent acts of representation. A range of theorists including Foucault, Derrida, and Rorty insist on the active role of representation in the constitution of the world – even to the point where there is no world outside of representation.

Once in Los Angeles

So-called "schools" of theory often emerge in particular places where a critical mass of scholars are able to generate the sense of something greater than the efforts of single writers. A good deal of the writing about postmodernism in geography emerged in Los Angeles and frequently took Los Angeles as its subject-matter (Soja 1989, 1996; Davis 1992; Dear 2000). To geographers such as Soja and Dear, the Bonaventure Hotel was simply one facet of a new kind of urbanism that they labeled postmodern. Soja writes that the Bonaventure "simulates the restructured landscape of Los Angeles and is simultaneously simulated by it. From this interpretive interplay of micro and macro-simulations there emerges an alternative way of looking critically at the human geography of contemporary Los Angeles, of seeing it as a mesocosm of postmodernity" (Soja 1989: 244):

> With exquisite irony, contemporary Los Angeles has come to resemble more than ever before a gigantic agglomeration of theme parks, a lifespace comprised of Disneyworlds. It is a realm divided into showcases of global village cultures and mimetic American landscapes, all embracing shopping malls and crafty Main Streets, corporation-sponsored magic kingdoms, hightechnology-based experimental prototype communities of tomorrow, attractively packaged places for rest and recreation all cleverly hiding the buzzing workstations and labour processes which help to keep it together. (Soja 1989: 246)

Soja describes a Los Angeles of splintered and fractured spaces arranged in a way that makes little sense for those who have thought of urban spatial arrangement as a series of rings around a central business district (CBD) – a model derived from the archetypal modern city: Chicago. Los Angeles is split into enclaves with their own city government structures – places like Santa Monica and Culver City. Industries cluster in unlikely places. Residents cut themselves off in gated communities with their own private police forces. "Villages" attempt to connect into deep (but fabricated) histories all called "Rancho something-orother" (Soja 1989: 245). The city goes on for miles and miles with zones of governance and industry dotted around, seemingly randomly. Business is unlikely to be central and more likely to occur in what were once called suburbs. This fracturing was apparent in the local forms of street signs:

> A recent clipping from the *Los Angeles Times* (…) tells of the 433 signs which bestow identity within the hyperspace of the City of Los Angeles, described as 'A City Divided and Proud of It'. Hollywood, Wilshire Boulevard's Miracle Mile, and the Central City were among the first to get these community signs as part of a 'city identification programme' organized by the Transportation Department. (Soja 1989: 245)

In Los Angeles, Soja tells us, we have a new kind of city that undoes our understandings of centers and peripheries. Los Angeles, he insists, deconstructs the urban and produces something new that may become a model for cities elsewhere – sprawling and multicentered, hidden under a mass of marketing and signs that sells an image of the city.

Michael Dear also makes the move from the modernity of Chicago, based as it is on mass production, heavy industry, the slaughterhouses, and the centralization of transport, to the postmodernity of Los Angeles. Dear's Los Angeles is made up of a number of intermingled forms. One of these is "edge city" – a new kind of urbanity based on automobility and freeway development; even in Chicago it is possible to see this developing in the corridor around O'Hare airport. Then there are "privatopias" – private housing developments with strict planning regulations and elaborate forms of defense against the city outside. In addition, there is the "city as theme park" – simulations in the style of Disneyworld that sell images to residents. The notion of "simulation" is key to theories of postmodernism. It suggests the perfect copy of something that never existed – even better than the real thing. Orange County is given by Soja as an example of this – a place that pretends to be a city. And then there are all the dystopian elements. The fortress of downtown Los Angeles with its stark rejection of the homeless and others, the highly surveiled spaces littered with CCTV cameras and security signs reading "warning – armed response." The list goes on. It is a long way from the ideas associated with the Chicago School of Sociology with their CBD, surrounding zone of transition, and affluent suburbs. Los Angeles, as a postmodern city, is infinitely more complicated. Dear attempts an outline of a synthesis of postmodern urbanism:

> It suggests a 'proto-postmodern' urban process, driven by a global restructuring that is permeated and balkanized by a series of interdictory networks; whose populations are socially and culturally heteroge-nous, but politically and economically polarized; whose residents are educated and persuaded to the consumption of dreamscapes even as the poorest are consigned to carceral cities; whose built environ-ment, reflective of these processes, consists of edge cities, privatopias, and the like; and whose natural environment, also reflective of these processes, is being erased to the point of unlivability while at the same time providing a focus for political action. (Dear 2000: 149)

A Postmodern Geography?

The two most well-known books in geography which concern postmodernity are David Harvey's *The Condition of Postmodernity* and Edward Soja's *Postmodern Geographies* (Harvey 1989; Soja 1989). Harvey's book is quite clearly an account of postmodernity (in the world and in theory) but not in itself a call for postmodern theory. Quite the opposite – it is a Marxist account of the condition of postmo-dernity in the world. Here, postmodernity is a reflection (superstructure even) of new forms of econ-omy described as "flexible accumulation." Although not identical, this is similar to another important book – Frederick Jameson's *Postmodernism: or, the Cultural Logic of Late Capitalism* (Jameson 1991). Both Harvey and Jameson describe postmodernism in the cultural sphere as a product of features of late capitalism. In this argument, capitalism has indeed changed, from a form of accumulation based on mass production (Fordism) to a much more flexible, footloose, and just-in-time mode (post-Fordism). This, the argument goes, was accompanied by a fragmented cultural sphere.

To other geographers engaging with postmodern theory, there seemed to be a clear connection between the nature of geography as a discipline and the emerging lessons of postmodernism. The challenges of postmodernism to the certainties of more traditional forms of academic knowledge opened up opportunities for a reinvigorated geography. Dear, for instance, argued that we should take advantage of internal incoherence in the discipline and the rise of social theory to make human geog-raphy central to debates in the humanities and social sciences. Looking over the plethora of "specialty groups" within the Association of American Geographers, he noted astonishing topical diversity, some of which he referred to "as a symptom of acute intellectual obsolescence" (Dear 1988: 262):

> The first thing to notice is the extreme eclecticism of the Association's view of geography. It endorses proficiencies in librarianship as well as audio-visual techniques. But there is apparently no need for pro-ficiency in theoretical matters, philosophy or method as they relate to geography. These are presumably of less import than military geography *inter alia*. (Dear 1988: 263)

Dear contrasts geography's apparent interest in, say, "military geography" with nineteenth-century thought where geography was at the center of important developments in intellectual life. The prob-lem was, he argued, that geography had become disengaged from the mainstream of philosophy and social thought. It was threatened with becoming redundant. Postmodernism offered a chance for geography to place itself back at the center of wider debates:

> There is no more important task for contemporary human geography than to confront the ontological and epistemological challenges posed by the postmodern movement. Most of our current dialogue in geography pales to insignificance by comparison. Moreover, if we choose to ignore this challenge, then outsiders will be justified in labeling geography as moribund and irrelevant. (Dear 1988: 267)

At the heart of Dear's argument is that geography's engagement with space gave it a special position in the new debates. Just as Kant had argued that space and time were the two most privileged categories of thought (and thus geography and history are privileged disciplines), so a focus on space, in the context of social theory, could place geography back at the center of human enquiry. In this sense, Dear was foreshadowing the reemergence of space in the wider world of social theory.

It was this cause that animated the work of Edward Soja in his book *Postmodern Geographies* (Soja 1989). In this book and others that followed it, Soja made the argument that space had been neglected in social theory and that postmodernism provided an opportunity to place it back at the center of debates (Soja 1999, 2000). "Space," he argued, "still tends to be treated as fixed, dead, undialectical; time as richness, life, dialectic, the revealing context for critical social theorization" (Soja 1989: 11). In other words, Soja and others argued, space had been relegated in social theory to a dead stage for history. Time was where the action was. In postmodernism, Soja sensed an opportunity:

> A distinctively postmodern and critical human geography is taking shape, brashly reasserting the interpretive significance of space in the historically privileged confines of contemporary critical thought. Geography may not yet have displaced history at the heart of contemporary theory and criticism, but there is a new animating polemic on the theoretical and political agenda, one which rings with significantly different ways of seeing time and space together, the interplay of history and geography, the 'vertical' and 'horizontal' dimensions of being in the world freed from the imposition of inherent categorical privilege. (Soja 1989: 11)

Similar arguments were made by Frederick Jameson, from a position outside of geography, as he argued for the importance of "cognitive mapping" for understanding the postmodern world: "a model of political culture appropriate to our own situation," he wrote, "will necessarily have to raise spatial issues as its fundamental organizing concern" (Jameson 1991: 89). Perhaps most famously, the French theorist Michel Foucault had suggested in his essay "Of Other Spaces" that:

> The great obsession of the nineteenth century was, as we know, history: with its themes of development and of suspension, of crisis and cycle, themes of the ever-accumulating past, with its great preponderance of dead men and the menacing glaciation of the world. The nineteenth century found its essential mythological resources in the second principle of thermodynamics. The present epoch will perhaps be above all the epoch of space. We are in the epoch of simultaneity: we are in the epoch of juxtaposition, the epoch of the near and far, of the side-by-side, of the dispersed. We are at a moment, I believe, when our experience of the world is less that of a long life developing through time than that of a network that connects points and intersects with its own skein. (Foucault 1986: 22)

Needless to say, this statement has proven to be very popular with geographers and others who are keen to promote the reinsertion of space into social theory and to show how space is not dead, but an active and dynamic component in the constitution of society.

Derek Gregory, too, found some reasons to embrace elements of the postmodern challenge to geography. This challenge, he argued, differed from previous arguments in human geography between competing paradigms of thought (Marxists vs. spatial scientists, etc.) but was rather post-paradigmatic. Postmodernism refused to have any foundation from which to construct some notion of Truth:

> All of these systems of thought are – of necessity – incomplete, and if there is then no alternative but to pluck different elements from different systems for different purposes this is not a licence for an uncritical eclecticism: patching them together must, rather, display a sensitivity towards the differences and disjunctures between them .... The certainties which were once offered by epistemology – by theories of knowledge which assumed that it was possible to 'put a floor under' or 'ground' intellectual inquiry in some safe and secure way – are no longer credible in a post-modern world. (Gregory 1996: 213–214)

The hostility to totalizing discourses, or incredulity toward metanarratives in Lyotard's terms, militated against structurally coherent accounts that sought to explain in a grand theoretical way the totality of our everyday lives. Here, postmodernism was equally opposed to the "last instance" of structural Marxism and the coherent knowing subject of humanism. Postmodernism was happy to proclaim that there was no superordinate logic to which we could appeal in our explanations.

These kinds of theoretical standpoints led Gregory and others to a familiar place for geographers – the importance of particularity and uniqueness. In its opposition to metanarratives, foundations, and totalizing theory, postmodernism put the accent on difference. Spatial difference or "areal differentiation" was once more at the center of theoretical attention. Various kinds of postmodern practice pointed toward the necessity of taking place and context seriously when providing explanations which were always only local. In anthropology, Clifford Geertz's notion of "deep mapping" and "thick description" – of working with local knowledge to provide rich and always particular accounts of a particular place and its inhabitants – proved to be one inspiration (Geertz 1975). Geographers, Gregory argued, could utilize lessons from postmodernism to return to an interest in the particular "but armed with a new theoretical sensitivity towards the world in which we live and to the ways in which we represent it" (Gregory 1996: 229).

There were two clear ways in which postmodernism represented an opportunity for geography. The first was the reemergence of the importance of space in social theory. This project was already ongoing when postmodernism arose, as Marxist geographers engaging with Lefebvre and others were already outlining accounts of the "production of space" and the active role of space in the production of social reality (Harvey 1982; Smith 1991). David Harvey or Neil Smith hardly needed to be told that space was important. It remains the case, however, that there had been a close connection between a focus on time and theoretical metanarratives that sought to explain history. A focus on space, meanwhile, emphasized difference and particularity in a way that could be connected to the deep history of geography. If there are no universal truths and no foundations, then we are left only with particular truths. Explanation, rather than coming from overarching theories, might once again emerge from a deep involvement with the local.

And this is the second postmodern opportunity for geography – the possibility of a re-enlivened encounter with place, region, and locality. This is Gregory's suggestion – for regional geography *plus*. An approach to the particular with a more theoretical orientation than the descriptive regional accounts of the early part of the twentieth century. There is of course a paradox in the cautiously positive engagements of Gregory and Dear with postmodernism. The thrust of postmodernism suggests the collapse of centers and foundations through the erosion of faith in big stories or big theories. Yet here we have geographers claiming that this is an opportunity for geographers to play a central role. Perhaps, even, that geographers have been natural postmodernists all along.

New Geographies of Difference

One of the key terms in the postmodern lexicon is "difference" – a term emerging from the work of Derrida, Deleuze, and Guattari in particular (Derrida 1978; Deleuze 1994). A suspicion of metanarratives and denial of essences almost automatically necessitates an interest in difference. While metanarratives seek to explain many disparate things through one logic, a postmodern approach seeks both to dwell in the specificity of things and to take apparent totalities and show the differences within them. Difference of some kinds, of course, has always been a concern of theorists. At the heart of a Marxist account of the world is the difference between capital and labor. The differences between men and women similarly undergird much of feminist theory. The problem for postmodernists is that these are very simple either/or kinds of differences. Feminists who have embraced aspects of postmodernism, for instance, raise awareness of the differences between women – black women, bourgeois women, lesbian women, women from the Global South, etc. Postmodernist approaches seek to rescue concealed differences from invisibility and explain how such differences have come to be submerged under one kind of difference deemed the most important.

One, influential, way in which difference was brought to the fore was in the realm of political theory through the work of Iris Marion Young and her conception of a "politics of difference" (Young 1990). Politics, in a postmodern framework, is a politics of constant multiplication and critique – a politics which seeks to consistently pull apart forms of certainty as they emerge. This is very different from a modernist politics that is based on a foundational notion of what kind of world would be a just world. This is the kind of politics that informs most theories of justice:

> A theory of justice typically derives fundamental principles of justice that apply to all or most societies, whatever their concrete configuration and social relations, from a few general premises about the nature of human beings, the nature of societies, and the nature of reason .... It assumes a point of view outside the social context where issues of justice arise, in order to gain a comprehensive view. The theory of justice is intended to be self-standing since it exhibits its own foundations. As a discourse it aims to be whole and show justice in its unity. It is detemporalized, in that nothing comes before it and future events will not affect its truth or relevance to social life. (Young 1990: 4)

So here, again, we have an outline of a modernist theory that seeks to incorporate an apparently neutral "view from nowhere." As in architecture, so in political theory. One of the key universalizing ideas in most theories of justice is what Young calls "the ideal of impartiality." Think of some of the most significant social and political movements of the past 100 years. The women's rights movement from the late nineteenth century onward was directed at women achieving equality with men – being allowed to vote, to earn the same as men, to participate in government or the military. The civil rights movements of the 1960s onward in the United States sought to give Black people an equal standing with White people. More recently, gay rights and disabled rights activists (to name but two) have sought similar forms of equality. At the heart of these movements have been notions of impartiality and equality – notions which seek to erase difference – to make differences inconsequential. Implicit in this is what Young labels an "assimilationist ideal" – an idea of increasing sameness. A postmodern politics of difference, however, argues that "the participation and inclusion of all groups sometimes requires different treatment for oppressed or disadvantaged groups" (Young 1990: 158).

One example of a concrete argument given by Young surrounds American approaches to law and rights around pregnancy and childbirth. In an effort to appear neutral, women who are pregnant or have just given birth in the United States take leave under the heading of "disability" leave. The basis for this is that it would be unfair to treat women differently from men. Men and women can both be disabled but they cannot both be pregnant or give birth. Thus the ideal of impartiality is fulfilled. Young argues forcefully that law and rights, in this case and many others, need to recognize difference:

> In my view an equal treatment approach to pregnancy and childbirth is inadequate because it either implies that women do not have any right to leave and job security when having babies, or assimilates such guarantees under the supposedly gender-neutral category of "disability." Such assimilation is unacceptable because pregnancy and childbirth are usually normal conditions of normal women, because pregnancy and childbirth themselves count as socially necessary work, and because they have unique and variable characteristics and needs (…). Assimilating pregnancy and childbirth to disability tends to stigmatize these processes as "unhealthy." It suggests, moreover, that the primary or only reason that a woman has a right to leave and job security is that she is physically unable to work at her job, or that doing so would be more difficult than when she is not pregnant and recovering from childbirth. (Young 1990: 175–176)

Young's account is an articulate and convincing case for theories of justice based on difference rather than notions of "equality" or "impartiality." It is one instance of a focus on difference inspired by postmodern theory.

In geography, the late 1980s and 1990s saw an explosion of suddenly visible difference as geographers became interested in forms of difference such as age, sexuality, and disability that had previously been close to invisible (Philo 1992; Bell and Valentine 1995; Kitchin 1998). As an example, consider the idea of a postmodern rural geography. In 1992, the British geographer Chris Philo published a review essay on a book about childhood in the countryside by the anarchist author Colin Ward (Ward 1988). The review essay is a sympathetic account of a book which seeks to raise awareness about the lives of children in rural areas and the ways that these have changed historically. Toward the end of the essay, Philo links this to the postmodern suspicion of metanarratives and concern for multiple "others" that was emerging in social and cultural geography in the early 1990s:

> Those who relate the 'metanarratives' tend to be white, middle-class, middle-aged, able-bodied, sound-minded, heterosexual men living in the major urban centres of the West; those who tell the other and often more local 'stories' tend not to be such things, and as a result their voices are rarely heard and are even more rarely allowed to qualify (let alone to dismantle) the grand moves of the grand masters. (Philo 1992: 199)

The rural is probably a conceptual and actual space most often associated with degrees of homogeneity – especially in Britain. It is almost a given that the city is characterized by heterogeneity. Philo's essay asks those involved in rural studies to be alive to, not only children, but all kinds of "others." An approach inspired by postmodernism, he argues, prompts "a very genuine attempt to craft a new form of human geographical inquiry open to the circumstances *and* to the voices of 'other' peoples in 'other' places: a new geography determined to overcome the neglect of 'others' which has characterized much geographical endeavour to date" (Philo 1992: 199). Philo asks geographers, and others, to pay attention to the lack of "difference" that appears in the work of the subdiscipline:

> … the social life of rural areas is indeed fractured along numerous lines of difference constitutive of overlapping and 'multiple forms of otherness', all of which are surely deserving of careful study by rural geographers (who therefore need to look more closely at the contents of their catch-all categories such as the 'locals', 'newcomers', 'deprived' and 'well-off'). (Philo 1992: 201)

In a thoughtful response to Philo's essay, Murdoch and Pratt go on to suggest the existence of the "post-rural" (Murdoch and Pratt 1993). Following this exchange, there was a notable flowering of research and writing on multiple forms of difference in the rural, including issues around homelessness, sexuality, and gypsies/travelers (Halfacree 1996; Cloke *et al.* 2000; Little 2002; Woods 2005).

The turn toward difference was not without its critics. At the heart of some unease among Marxists and feminists in particular was the issue of whether there was any way to differentiate differences that matter from differences that do not. To put it crudely, why do we take the difference between classes or genders more seriously than, say, the differences between people who prefer Coke and people who prefer Pepsi? If we have no foundational bedrocks to allow us to judge between differences, why not take all differences equally seriously? David Harvey, for one, did not want to take all differences equally seriously. In a chilling account of a fire at a chicken-packing factory in North Carolina, Harvey described how the chicken-processing industry was owned by a few large companies, how salmonella contamination was rife, and how most employees were African-American, female, and paid the minimum wage of $4.25 an hour (in 1991). On Tuesday, September 3, 1991, the plant in Hamlet caught fire, many of the exit doors were locked, and 25 workers died. Harvey focuses on the lack of political response to the fire, which contrasted with responses to the beating of a black man, Rodney King, in Los Angeles (which led to urban riots as well as concerted political reaction). Harvey suggests that the reason for the lack of concerted response to the fire at Hamlet was the lack of an organized form of class politics that could get beyond what he refers to as the rise of politics focused on special issues and "so-called new social movements focusing on gender, race, ethnicity, ecology, sexuality, multiculturalism, community and the like" (Harvey 1996: 341). Harvey blames postmodernism's suspicion of any universal sense of social justice for apathy:

> According to most common-sense meanings of the word, many of us would accept that the conditions under which men, women and minorities work in the Hamlet plant are unjust. Yet to make such a state-ment presupposes that there are some universally agreed upon norms as to what we do or ought to mean by the concept of social justice and that no barrier exists … to applying the full force of such a powerful principle to the circumstances of North Carolina. But 'universality' is a word which conjures up doubt and suspicion, downright hostility even, in these 'postmodern' times; the belief that universal truths are both discoverable and applicable as guidelines for political-economic action is nowadays often held to be the chief sin of 'the Enlightenment project' and of the 'totalizing' and 'homogenizing' modernism it supposedly generated. (Harvey 1996: 341–342)

Clearly Harvey believes that the postmodern focus on difference disables the ability to work in a unified way for social justice. It is equally clear that he believes that a focus on the workings of capitalism and class identity does provide a way forward. To Harvey, the difference which we call "class" is more impor-tant, most of the time, than the multitude of differences introduced by postmodernists such as Young.

For others, of course, there were other differences that were more important. Feminist geogra-phers, for instance, were also ambivalent about the embrace of difference because the dizzying array of difference within the category "woman" might potentially dilute the impact of a feminist political agenda that needs an ongoing commitment to the existence of commonalities between women (McDowell 1991).

Geography and the Crisis of Representation

Postmodernism was not only a challenge to epistemologies founded on the existence of forms of Truth that exist out there in the world. It was simultaneously a challenge to the ways we write or, more broadly, represent, the world. Before postmodernism there was a general shared faith in a rep-resentational view of the world – a faith shared by most Marxists, spatial scientists, humanists, and regional geographers. In this view, the dominant view in the west since the Renaissance (Foucault 1971), the meaning of our forms of **representation** (writing, pictures, equations, charts, etc.), is based on the things (objects, emotions, ideas, actions, etc.) to which they refer and represent. Meaning, in other words, existed in the world, and representation was simply a way of relaying this meaning. With the rise of postmodernism, this simple equation became problematic. Anti-representational views of representation suggest that representations do not refer to some aspect of reality but simply to other forms of representation – to other signs. For this reason, things such as "truth" and "reality" have no reliable existence outside of and prior to representation. Rather they are products or effects of representation. Our representations produce reality. Everything is a form of "text" (Derrida 1976). This view, for many the crux of postmodern theory, is clearly stated by Ulf Strohmayer and Matt Hannah in an early, rigorous account of the postmodern challenge:

> The idea(l) that shared language (e.g., a definition) reflects a shared reality is an article of faith which unfortunately can only be 'substantiated' in more language and thus not really substantiated at all. The truth of any statement, scientific or otherwise, which ultimately must rely on some anchoring in order to avoid being completely arbitrary, is *undecideable. This in turn does not imply that there is no truth, but rather that if there is, we are incapable of pinning it down.* (Strohmayer and Hannah 1992: 36)

Such a view (or one very close to it) reached its most dramatic form in the writings of the French philosopher Jean Baudrillard, who suggested that we are living in a world of "simulations" – exact copies of worlds which never existed (Clarke *et al.* 1984; Baudrillard 1994). Baudrillard's argument is that the world of signs (texts, popular culture, advertising, television, etc.) has long since ceased to refer to any reality. Signs refer to more signs. In postmodernity, the simulacra has replaced the original and there is no distinction between representation and reality. They are the same thing. It is in this spirit that Baudrillard suggested that the first Gulf War was in fact a war over the image of war – fought on TV screens rather than territory. A similar argument was made by the novelist Umberto Eco when he suggested that a visit to Disneyworld was a visit to a perfect copy of a nonexistent world where the alligators can be guaranteed to appear in a way that cannot be assured in a visit to the Florida Everglades. Why bother with the original? He referred to this as "hyper-reality" (Eco 1986).

Postmodernist geography involved looking at the process of representation not as simply an issue of technique but as a deeply problematic element in the construction of knowledge. Representation was not primarily "about" some reality, but an important element in the production of what would come to be thought about as reality. One form of representation that lies at the heart of the discipline is mapping. This came under particular scrutiny by the historian of cartography, Brian Harley (Harley 1989). Maps, in their modern form, are a form of representation that is supposed to be scientific or neutral. They are supposed to convey information about the world that is accurate and useful. Old maps, such as the **Mappa Mundi**, seem strange or even amusing to modern eyes because they are clearly inaccurate. Our own maps, say in a city A to Z or in a road atlas, appear as simply functional. It was this idea that Harley took to task – the idea that cartographers' work is a scientific and value-free form of knowledge creation, with their maps serving as mirrors to the world:

> The first set of cartographic rules can thus be defined in terms of a scientific epistemology. From at least the seventeenth century onward, European map-makers and map users have increasingly promoted a standard scientific model of knowledge and cognition. The object of mapping is to produce a 'correct' relational model of the terrain. Its assumptions are that the objects in the world to be mapped are real and objective, and that they enjoy an existence independent of the cartographer; that their reality can be expressed in mathematical terms; that systematic observation and measurement offer the only route to cartographic truth; and that this truth can be independently verified. (Harley 1989: 4)

All of these assumptions, you will recall, are the assumptions of philosophical positivism. Harley sought to break the mimetic link between the real world and the cartographic representation – a link which he claimed had "dominated cartographic thinking" and "led it in the pathway of 'normal science' since the Enlightenment, and has also provided a ready-made and 'taken for granted' epistemology for the history of cartography" (Harley 1989: 2). Instead, Harley draws on the social theory and philosophy of Derrida and Foucault to suggest an epistemology which insists that even the most apparently neutral maps have their own rules of order which reflect and reinforce norms and values that exist outside of the map and are far from neutral or transparent.

This emphasis on the mimetic properties of good and proper maps, Harley argues, was partly based on a process of "othering" that marginalized and looked down upon maps that did not fit the scientific standard (non-western or historical, for instance). Other forms of map were seen as inferior. A proper map is an accurate scientific map and all others in some sense fail:

> Its central bastions were measurement and standardization and beyond there was a 'not cartography' land where lurked an army of inaccurate, heretical, subjective, valuative, and ideologically distorted images. Cartographers developed a 'sense of the other' in relation to nonconforming maps. (Harley 1989: 4–5)

One effect of this othering is to deflect attention from the ways in which the seemingly accurate maps are themselves brimming over with metaphorical power that is far removed from the ideal of neutrality.

The sheen of science hides the ways in which maps serve to mask the social and political worlds that are simultaneously legitimated by the appearance of neutrality. Modern maps, Harley insists, are just as much an "image of the social order as they are a measurement of the phenomenal world of objects" (Harley 1989: 7). Harley considers the kinds of road atlases most of us have in our cars:

> What sort of an image of America do these atlases promote? On the one hand, there is a patina of gross simplicity. Once off the interstate highways the landscape dissolves into a generic world of bare essentials that invites no exploration. Context is stripped away and place is no longer important. On the other hand, the maps reveal the ambivalence of all stereotypes. Their silences are also inscribed on the page: where, on the page, is the variety of nature, where is the history of the landscape, and where is the space-time of human experience in such anonymized maps? (Harley 1989: 13)

It follows from these observations that we have to consider our own forms of representation, our maps but also our writing, very carefully. One model for academic representation (and writing in particular) is that it should be a transparent window on the world telling us what we need to know in the clearest and most straightforward way. The writing should be invisible – otherwise it is an obstacle to understanding. This model relies on representational modes of understanding and is symptomatic of a faith in the good match between representation and reality. If we have no such faith, then how should we write, map, or diagram? To some geographers this challenge was one of the most important postmodernism had to offer and one way out is to pay more attention to the craft of writing, recognizing its power to produce realities. In his account of postmodernism, Derek Gregory makes the following argument:

> It may well be true that we are not trained to be painters or poets but, like Pierce Lewis, I don't think we should boast about it. For if we cannot evoke landscapes, if we cannot provide descriptions of the relations between people and places 'so vivid that they move our emotions', then – to adapt a phrase from Geertz – we radically 'thin' our geographies. (Gregory 1996: 227)

Since the late 1970s, some geographers have experimented with creative forms of writing to raise awareness of the role of writing in the creation of the world. Perhaps the best known of these experiments is the writing of the Swedish geographer Gunnar Olsson who appeared to be postmodern before his time. The first book of his I encountered as an undergraduate student could be read front to back and back to front and was called *Birds in Egg/Eggs in Bird* (Olsson 1980). Before that, I had read his not particularly "humanistic" essay in the "Humanistic Geography" collection edited by David Ley and Marwyn Samuels (Olsson 1978). This essay appeared a good 10 years before the sudden explosion of postmodernism in geography but it certainly prefigures postmodernism in important ways and Olsson remains an inspiration for geographers informed by postmodernism and poststructuralism.

In this essay, Olsson reflects upon the "(impossible) problem of telling the truth, of creating, and of communicating all at the same time. Triple bind!" (Olsson 1978: 109). So far, so clear. "How," he continues, "can we distill meaning from the marks on a map? What do the signifying symbols signify? What are the relations between the geographic words of spatial coordinates and the objects of human action?" (Olsson 1978: 110). The issue of representations and its impossibility was central to Olsson's enterprise and still is (Olsson 1991, 2007). His writing examines the limits of language and the kinds of reason that seek to rid us of ambiguity. Olsson wants us to embrace the ambiguous and the multiple from within the process of communicating:

> One of the most crucial fronts in this constant war between social simplicity and individual complexity is in the communication process itself. Thus, it is in the interest of social cohesion to impoverish language. It is my conviction that these forces must be fought, or our very survival as a species is at stake. As a consequence, we should continue to read modern writers such as Baudelaire, Mallarmé, Joyce, Kafka, and Beckett. For even though they never managed to escape from the prison house of language, they nevertheless bent its walls and thereby expanded our common universe. (Olsson 1978: 110)

In the bulk of the essay, Olsson considers the interplay of certainty and ambiguity, reflecting on the apparent certainties of the language of spatial science (a mathematical language he was trained in) and the apparent ambiguities of poetry. The language Olsson looks to as inspiration and example is that of James Joyce, particularly in *Finnegans Wake*. At the end of the essay, he asks us to "heed Beckett's call for unifying form and content so that the text is what it expresses and expresses what it is" (Olsson 1978: 118). Olsson is not afraid to try this himself. Words are crossed out to express the process of erasure: "Through sensual games of deconstruction, new sets and relations grow out of well-delineated categories; in change, equivalence is both asserted and questioned, for whatever we erase always leaves a trace" (Olsson 1978: 111). Sure enough, a trace is left visible. At some points the text turns into something close of experimental poetics:

> The trumpet sounds: Silence in thought!
> SPREACH?
> Break and preach, search and reach
> Through the speech of each.
> And Babble's walls come mumbling down.
>
> (Olsson 1978: 117)

Another great experimenter with writing was Allan Pred. He was inspired by the writing of the German urban theorist/philosopher Walter Benjamin, who had sought to mirror the confusing modernity of early-twentieth-century Paris with a montage form that interlocked his own words with direct passages taken from a dizzying array of other sources (Benjamin 1999). Pred used this form and combined it with a kind of free verse structure, working with repetition to make his points about aspects of urban (post)modernity. Here is a passage from *Recognizing European Modernities*:

> In short,
> these were encounters
> with plural complexities,
> with shifting heterogeneities, and
> with shockingly new meanings
>   that repeatedly dislocated and displaced,
>   that confirmed rupture,
>   that invited or demanded cultural contestation
> by calling the meanings
>   of locally preexisting practices and social relations
>   into question,
> and thereby called individual and collective identities
>   into question.
>
> (Pred 1995: 14–15)

You will recall that it was Allan Pred that Patricia Limerick took to task for his obfuscating writing style (see Chapter 1). It is equally clear that Pred is at least trying to break out of what Olsson, borrowing from Wittgenstein, calls the "prison house of language." If the world is complicated and ambiguous, then maybe this world should be reflected in our attempts to represent it. If we have given up on the possibility of a transparent language that truly represents the world (truth, reality), then maybe we do need to draw attention to the process and form of the way we represent it. This is the gamble being made by Olsson and Pred. Only the readers can decide to what extent the gamble has paid off.

Such attempts at experimental writing have certainly been criticized as being overly playful and obscurantist. Bondi and Domosh, for instance, agree with the general postmodern critique of language as transparent but take the experiments of Olsson and others to task for writing that is "marred by ... general incomprehensibility." These strategies, they argue,

> only add to the mystery of authorship, sometimes producing cult-like groups of cognoscenti able to converse in a language inaccessible to all but themselves and able to increase their status accordingly .... Indeed, the elitism of a good deal of postmodern discourse is clear. It may break with dominant assumptions about the centred speaking subject but it does not divest itself of a specifically masculine authority: it is decentred but not desexed. (Bondi and Domosh 1992: 208)

It is hard to know exactly how to perform the trick of divesting the author of masculine authority apart from by not writing. Perhaps one answer lies in forms of collaborative writing. Indeed, it is not only in the form and texture of writing that geographers have sought to ask postmodern questions arising from the crisis of representation. Another strategy has been to attempt to undermine the idea of the singular knowing author – an idea that has been prevalent in human geography if not in physical geography where team writing is commonplace. Some human geographers have sought to go beyond jointauthorship to suggest hybrid forms of authorship. Perhaps the best known is the author J. K. Gibson-Graham (see Chapter 7) – a combination of Julie Graham and Katherine Gibson – who have playfully combined postmodernism with Marxism and feminism in their important work on alternative economies (Gibson-Graham 2006a,b). Other examples include Ian Cook *et al.* Ian Cook, who researches and writes about the travels of things (and particularly food things) from production to consumption, always adds "*et al*" to his name to indicate the collaborative nature of all his work and writing (Cook *et al.* 2008). Finally there has been the hybrid name mrs kinpaisby (Mike Kesby, Rachel Pain, and Sara Kindon), used to underscore the nature of collective and participatory research and writing (kinpaisby 2008). While these hybrid authors do not directly engage postmodernism substantively, their authorial strategies would have been hard to imagine before the postmodern questioning of the knowing subject as author.

Conclusions: Feminism and Postmodernism

One of the key points of postmodernism is the refusal to believe in the ability to perform the "god-trick" of being able to see the world objectively, from on high. This belief, however, is not unique to postmodernism. Humanists critiqued spatial science by pointing out that humans are subjective beings who are not always, or even ever, entirely rational (although they tended to posit a universal "man" unmarked by difference). Feminists too have argued for forms of positioned standpoints that produce a different kind of truth than the truth of science. The sudden emergence of postmodernism in the late 1980s in geography looked like yet another attempt by men to colonize ground already staked out by feminism. While some feminists welcomed the insights of postmodern theory – and even referred to themselves as postmodern feminists – others were more skeptical (McDowell 1991). Liz Bondi, for instance, argued that feminism could even be seen as an "instigator" for postmodernism:

> For example, feminism has been highly critical of claims to universality in philosophy and political theory, on the grounds that they are rooted in a culturally specific conception of the individual as a masculine subject (...). Feminism can, therefore, be viewed as a particular version, and perhaps an instigator, of the postmodern attack on discourses that claim privileged access to truth. (Bondi 1990: 161)

Reading feminist engagements with postmodernism from the end of the 1980s and beginning of the 1990s, there is a strong sense that the critique being performed by postmodernists not only stole ideas from feminism without paying due respect but made it quite difficult to pursue theoretical and political agendas. One argument made by postmodern theorists, for instance, is that our identity is never stable and always fractured. Labels like "woman" or "man" have no foundations but are continuously made up over and over again. This is an anti-essentialist argument and one not unfamiliar to feminist theorists. Bondi and others wondered, though, whether the radical critique of a knowing subject might not undermine the possibilities of a feminist politics based on the identity of "woman." Postmodernism's anti-essentialism is so radical, Bondi suggests, that the subject becomes so fragmented that it cannot easily be distinguished (in its effects) from standard liberal formulations of coherent stable selves. Both undercut any ability to refer to group identities such as class and gender which, in postmodern terms, are social constructs and therefore (the argument goes) incapable of supporting a radical insistence on justice and truth. "Gender relations and gender hierarchies," Bondi insists, "cannot just, playfully, be wished away" (Bondi 1990: 162). Linda McDowell makes a similar argument:

> [Postmodernists] in the main white men, have been the most forceful in their rejection of humanism's whole, integrated ideal of humanity. Other groups cannot afford to reject the subject wholesale, largely because they have never been allowed access to it. Fragmentation has less appeal to those of us who have neither been whole nor at the centre. (McDowell 1992: 61)

Another aspect of feminist geographers' critical engagement with postmodernism was the suggestion that postmodernism allows men to dabble playfully in feminist themes without really committing to the full consequences of a feminist critique. Bondi cites the feminist writer Suzanne Moore who had characterized postmodernism as "gender tourism, whereby male theorists are able to take package trips into the world of femininity, in which they 'get a bit of the other' in the knowledge that they have return tickets to the safe, familiar and, above all, empowering terrain of masculinity" (Suzanne Moore in Bondi 1990: 163).

Despite reservations, some feminist geographers were happy to identify affinities with postmodern theory. Linda McDowell, for instance, identified the joint challenge that postmodernism and feminism posed to the idea of a universal human subject:

> Both reject the doctrine of a unity of reason, of a universal subject striving towards common moral and intellectual aims, and hence the notion of humanity as a unitary subject. Thus 'universal rational man' is dislodged from the centre of both discourse and politics and space is opened up for the voices of the previously marginalized others to speak. (McDowell 1992: 61)

Liz Bondi and Mona Domosh identify five shared approaches. First, they note the shared critique of an objective, scientific form of knowledge which had its roots in the Enlightenment and, feminists argue, relied on masculinist ways of knowing the world. Second, they chart a shared critique of an epistemology based on the idea of "detached observers" who are able to know the world by performing the God-trick of detaching themselves from their embeddedness in the world. Their third and related point of agreement is that it is impossible and undesirable for researchers to attempt to separate themselves from either their own context or the context of whatever is being researched. Fourth, they identify a shared suspicion in the process of representation that occurs through academic writing. Here they are critiquing the notion that writing can be "transparent" and not involved in the construction of whatever is being written about. Postmodernists and feminists, they argue, share an interest in problematizing the process of writing and experimenting with diverse forms of writing. Finally, they oppose the prevalence of dualistic thinking both in geography and in modern thought in general – the kind of thought that is structured around opposing binaries such as man/woman, where the first part of the binary is "better" than the second, which is, somehow, "other." In short, postmodernism and feminism, they argue, share a skepticism about the claims to truth that legitimate much of western culture since the Enlightenment.

Bondi and Domosh also see points of disagreement between feminism and postmodernism. One suggestion is that postmodernism is largely a men's game. Certainly the big figures in debates around postmodernism – Lyotard, Foucault, Baudrillard, Derrida, Rorty, etc., tend to be men (and mostly French!). The absence of women here is not due to lack of interest, they assert, but due to the fact that groundbreaking work by women often gets usurped by men with little or no recognition. Regardless of why, it can be argued that postmodernism reproduces familiar hierarchies. In addition, some of the main points of postmodernist theory dramatically undercut the emerging power of feminist theory – a point made by Nancy Hartstock:

> Somehow it seems highly suspicious that it is at this moment in history, when so many groups are engaged in 'nationalisms' which involve redefinitions of the marginalized Others, that doubt arises in the academy about the nature of the 'subject', about the possibilities for a general theory which can describe the world, about historical 'progress'. Why is it, exactly at the moment when so many of us who have been silenced begin to demand the right to name ourselves, to act as subjects rather than objects of history, that just then the concept of subject-hood becomes 'problematic'? Just when we are forming our own theories about the world, uncertainty emerges about whether the world can be adequately theorized? Just when we are talking about the changes we want, ideas of progress and the possibility of 'meaningfully' organizing human society becomes suspect? And why is it only now that critiques are made of the will to power inherent in the effort to create theory? (Hartstock 1987: 196)

In reference to Hartstock's argument, Bondi and Domosh argue that feminists should be wary of the process by which postmodernism assimilates feminism. Just as it becomes possible (barely, just about) to become a singular knowing feminist author, the author is declared dead. Singularity and knowing are also deemed suddenly problematic by men. You can see the problem!

The postmodern critique of a singular knowing subject who is able to detach himself from the world and both understand it and represent it unproblematically came as no surprise to feminists who had never had access to such a position. Women had always had to locate knowledge in some way and often engaged in more collective forms of knowing and writing through necessity rather than for any essentially "feminine" reasons. The starting point for feminists dealing with postmodernism was simply different. Postmodernists (i.e. postmodernist men), Bondi and Domosh argue, engage in "gender tourism," visiting some carefully selected aspects of feminist theory but without submitting to the full implications of what that might mean as a full critique of their own practice as theorists. In this sense, they argue, postmodern theorists are like the explorers that are so central to the heritage of the discipline. Both are engaged in the exploration and appropriation of "exotic" other worlds and both have the possibility to be transformed in the process – a possibility that offers some hope to Bondi and Domosh. "Our intellectual journeys, too, are uncertain and ambivalent," they write, "although the inequalities of existing social relations are often intensified, there are always possibilities for reorderings of a different character. If a postmodern geography can assist and participate in that type of change, it is to be welcomed" (Bondi and Domosh 1992: 211).

References

Baudrillard, J. (1994) *Simulacra and Simulation*, University of Michigan Press, Ann Arbor.

Bell, D. and Valentine, G. (eds) (1995) *Mapping Desire: Geographies of Sexualities*, Routledge, New York.

Benjamin, W. (1999) *The Arcades Project*, Belknap Press of Harvard University Press, Cambridge, MA.

Bondi, L. (1990) Feminism, postmodernism, and geography: space for women? *Antipode*, 22, 156–167.

* Bondi, L. and Domosh, M. (1992) Other figures in other places: on feminism, postmodernism and geography. *Environment an Planning D: Society and Space*, 10, 199–213.

Calvino, I. (1981) *If on a Winter's Night a Traveler*, Harcourt Brace Jovanovich, New York.

Chambers, I. (1990) *Border Dialogues: Journeys in Postmodernity*, Routledge, New York.

Clarke, D., Doel, M., and Gane, M. (1984) The perfection of geography as an aesthetic of disappearance: Baudrillard's America. *Cultural Geographies*, 1, 317–323.

Cloke, P., Milbourne, P., and Widdowfield, R. (2000) Homelessness and rurality: 'out-of-place' in purified space? *Environment and Planning D: Society and Space*, 18, 715–735.

Cook I. *et al*, (2008) Geographies of food: mixing. *Progress in Human Geography*, 32, 821–833.

Davis, M. (1992) *City of Quartz: Excavating the Future in Los Angeles*, Vintage, New York.

* Dear, M. (1988) The postmodern challenge: reconstructing human geography. *Transactions of the Institute of British Geographers*, 13, 262–274.

Dear, M. J. (2000) *The Postmodern Urban Condition*, Blackwell, Oxford.

Deleuze, G. (1994) *Difference and Repetition*, Columbia University Press, New York.

Derrida, J. (1976) *Of Grammatology*, Johns Hopkins University Press, Baltimore.

Derrida, J. (1978) *Writing and Difference*, Routledge and Kegan Paul, London.

Eco, U. (1986) *Travels in Hyperreality: Essays*, Harcourt Brace Jovanovich, San Diego, CA.

Foucault, M. (1971) *The Order of Things: An Archaeology of the Human Sciences*, Pantheon Books, New York.

Foucault, M. (1986) Of other spaces. *Diacritics*, 16, 22–27.

Fowles, J. (1971) *The French Lieutenant's Woman*, World Books, London.

Geertz, C. (1975) *The Interpretation of Cultures: Selected Essays*, Hutchinson, London.

Gibson-Graham, J. K. (2006a) *The End of Capitalism (as We Knew It): A Feminist Critique of Political Economy*, University of Minnesota Press, Minneapolis.

Gibson-Graham, J. K. (2006b) *A Postcapitalist Politics*, University of Minnesota Press, Minneapolis.

* Gregory, D. (1996) Areal differentiation and post-modern human geography, in *Human Geography: An Essential Anthology* (eds J. A. Agnew, D. Livingstone, and A. Rogers), Blackwell, Oxford, pp. 211–232.

Halfacree, K. (1996) Out of place in the country: travellers and the "rural idyll." *Antipode*, 28, 42–72.

* Harley, J. B. (1989) Deconstructing the map. *Cartographica*, 26, 1–20.

Hartstock, N. (1987) Rethinking modernism: minority vs. majority theories. *Cultural Critique*, 7, 187–206.

Harvey, D. (1982) *The Limits to Capital*, Blackwell, Oxford.

* Harvey, D. (1989) *The Condition of Postmodernity*, Blackwell, Oxford.

Harvey, D. (1996) *Justice, Nature and the Geography of Difference*, Blackwell, Oxford.

Jameson, F. (1991) *Postmodernism, or, the Cultural Logic of Late Capitalism*, Verso, London.

Jencks, C. (1977) *The Language of Post-Modern Architecture*, Academy Editions, London.

Kinpaisby, M. (2008) Taking stock of participatory geographies: envisioning the communiversity. *Transactions of the Institute of British Geographers*, 33, 292–299.

Kitchin, R. (1998) 'Out of place', 'knowing one's place': space, power and the exclusion of disabled people. *Disability and Society*, 13, 343–356.

Ley, D. (1989) Modernism, postmodernism and the struggle for place, in *The Power of Place* (eds J. A. Agnew and S. Duncan James), Unwin Hyman, London, pp. 44–65.

Little, J. (2002) *Gender and Rural Geography: Identity, Sexuality and Power in the Countryside*, Prentice Hall, Harlow.

Lyotard, J.-F. (1984) *The Postmodern Condition: A Report on Knowledge*, University of Minnesota Press, Minneapolis.

McDowell, L. (1991) The baby and the bathwater – diversity, deconstruction and feminist theory in geography. *Geoforum*, 22, 123–133.

McDowell, L. (1992) Multiple voices: speaking from inside and outside 'the project'. *Antipode*, 24, 56–72.

Murdoch, J. and Pratt, A. (1993) Rural studies: modernism, postmodernism and the 'post-rural'. *Journal of Rural Studies*, 9, 411–427.

Olsson, G. (1978) Of ambiguity or far cries from a memorializing mamasta, in *Humanistic Geography: Prospects and Problems* (eds D. Ley and M. Samuels), Croom Helm, London, pp. 109–120.

Olsson, G. (1980) *Birds in Egg; [and], Eggs in Bird*, Pion, London.

Olsson, G. (1991) *Lines of Power/Limits of Language*, University of Minnesota Press, Minneapolis.

Olsson, G. (2007) *Abysmal: A Critique of Cartographic Reason*, University of Chicago Press, Chicago.

Philo, C. (1992) Neglected rural geographies: a review. *Journal of Rural Studies*, 8, 193–207.

Pred, A. R. (1995) *Recognizing European Modernities: A Montage of the Present*, Routledge, New York.

Smith, N. (1991) *Uneven Development: Nature, Capital, and the Production of Space*, Blackwell, Oxford.

* Soja, E. W. (1989) *Postmodern Geographies: The Reassertion of Space in Critical Social Theory*, Verso, New York.

Soja, E. W. (1996) *Thirdspace: Journeys to Los Angeles and Other Real-and-Imagined Places*, Blackwell, Oxford.

Soja, E. W. (1999) Thirdspace: expanding the scope of the geographical imagination, in *Human Geography Today* (eds D. Massey, J. Allen, and P. Sarre), Polity, Cambridge, pp. 260–278.

Soja, E. W. (2000) *Postmetropolis: Critical Studies of Cities and Regions*, Blackwell, Oxford.

* Strohmayer, U. and Hannah, M. (1992) Domesticating postmodernism. *Antipode*, 24, 29–55.

Ward, C. (1988) *The Child in the Country*, Hale, London.

Woods, M. (2005) *Rural Geography: Processes, Responses, and Experiences in Rural Restructuring*, Sage, Thousand Oaks, CA.

Young, I. M. (1990) *Justice and the Politics of Difference*, Princeton University Press, Princeton, NJ.

Chapter 10

Toward Poststructuralist Geographies

It's up to you, who are directly involved with what goes on in geography, faced with all the conflicts of power which traverse it, to confront them and construct the instruments which will enable you to fight on that terrain. And what you should basically be saying to me is, 'You haven't occupied yourself with this matter which isn't particularly your affair anyway and which you don't know much about'. And I would say in reply, 'If one or two of these "gadgets" of approach or method that I've tried to employ with psychiatry, the penal system or natural history can be of service to you, then I shall be delighted. If you find the need to transform my tools or use others then show me what they are, because it may be of benefit to me. (Foucault 1980: 64)

Since the height of discussions about postmodernism in the late 1980s and early 1990s, human geography has witnessed a proliferation of theoretical perspectives. While human geography has always been much more messy than a linear development of theoretical perspectives might suggest, the current scene is particularly fractured, with little sense of the big paradigmatic arguments of the past. Postmodernism, as a way of thinking characterized by a rejection of metanarratives, foundations, and essences, helped to put an end to a search for big theoretical shifts. We appear to live in an intellectual world characterized by groups of people plowing their own theoretical furrows, with little outright objection to others doing their own thing. In some sense, this is to be welcomed. The posturing of grand movements was often less than collegial and sometimes pointlessly self-aggrandizing. There is also a general acceptance of hyphenated theoretical positions such as poststructuralist-feminism.

One way of thinking about theory is as a serious and overarching commitment to a body of ideas that is superior to others. Often this commitment is linked to firmly held political, ethical, and moral beliefs. Another way of thinking about theory is as a kind of toolbox where particular sets of ideas allow us to understand particular bits of living in the world. With notable exceptions, this latter, more pragmatic, approach appears to be the main way of doing theory in twenty-first-century geography. This is the approach to theory put forward by Michel Foucault (above) who was suggesting, in an interview by French geographers, that they should not look to his work for an overarching approach but, instead, make tactical use of whatever "gadgets" he might provide.

Having said this, many human geographers would be happy to call themselves poststructuralist. Poststructuralism names an incredibly diverse array of ideas which are united only by the fact that they are not structuralist and yet contain some reference to structuralism within them – just as postmodernism necessarily contains something of modernism within it. Post-1980s human geography has seen a proliferation of "post" positions. Other key terms have been "postcolonialism" and "post-humanism." Poststructuralism contains within its name a particular approach to the problem of structure (and therefore agency). For this reason, this chapter begins with an account of this tricky problem as it has played out in geography since the 1980s, leading up to a properly poststructuralist position.

Structure and Agency in Geographic Thought

You are watching a DVD on a Saturday evening. The film is halfway through. You notice that you are hungry. Pressing pause, you make your way into the kitchen and open the freezer. In front of you there are tubs of ice-cream – vanilla, strawberry, and chocolate. You pick one. Why? This is a small individual action, the kind of action that all of life consists of, accumulating year after year, endlessly. A lot depends on how we think about that moment when you chose vanilla. Or was it strawberry? There are many social scientists, philosophers, and human biologists who still do not believe anyone has a suitable explanation of what happens in the split second when you made the choice. Human geographers would argue over it. Some, including some humanistic geographers, behavioral geographers, and nonrepresentational theorists (who we will encounter later), would pay close attention to a moment like this – a moment when an individual makes some kind of choice – a choice which is either conscious and intentional or largely unconscious. Others, such as most Marxists, would consider the whole thing unimportant. They might ask why these were the only flavors available – why no pistachio? They might point to the economic structures of capitalism that made this choice possible in some places, but not in others. They might even examine the packaging that makes vanilla seem "plain" and strawberry seem exciting. They would point out that your choice was only really the illusion of choice and that focusing on what happened when you chose vanilla is really not the important question. These are questions of structure and agency.

Human geographers, and social scientists in general, have tended to fall into camps that emphasize individual agency and camps that emphasize the ways in which our choices are determined or constrained by various kinds of structure. We have consistently seen how the themes of humanity's relations to the natural world and the relative importance of the general and the specific have run through the history of geographic thought. Another key theme, particularly since the 1970s, has been this endless battle between structure and agency. Are people more or less controlled, in their activities, by deep and pervasive structures, or are they more or less in charge of their own destinies? Is the appropriate scale for a good understanding of human geography a large and systematic scale or the scale of the individual human being? These are questions that run through all the social sciences and much of the history of philosophy as well as human geography. At the outset, we should note that very few people believe in the extreme ends of this spectrum, but it is clear that many geographers prefer positions closer to one end or another. This is perhaps most often portrayed, with some justification, as an intellectual battle between some Marxists (as structuralists) and some humanists (as idealists). Marxists believe the correct level at which to diagnose society is at the level of economic structures and the relations of production they embody. Humanists tend to focus on the individual subject, the mind, and the senses.

Varieties of structuralism

Structuralism refers to a diverse body of thought that assumes that there is some structure, at a deep level, that undergirds the world of appearances. This structure, once discovered, can be used to explain the surface events of life. The most clearly structuralist body of thought in geography is Marxism – particularly the kind of Marxism that invests in the notions of base and superstructure outlined in Chapter 7. The economic base made up of the productive forces and relations of production is the structure that explains the seemingly chaotic nature of life as lived. But there are other forms of structure in geography. Environmental determinism can be seen as a form of structuralism where the deep, objective structures of ecological reality determine human life. Even in humanism, often thought of as the opposite of structuralism, it is possible to see the influence of forms of structuralism. Yi-Fu Tuan, for instance, often roots his explanations in the structural imperatives of the human body with its clearly demarcated "structure" of "front and back," "up and down," and "left and right" (Tuan 1977).

Structuralism also works at the level of culture and language where people are supposed to think and behave in ways that are pre-programmed linguistically or neurologically. The term "structuralism" probably arose in the "general linguistics" of Ferdinand de Saussure in which he asserted that language had "deep" structures (regular connections between parts of grammar) that made language able to make sense in all the particular ways in which it is used (Saussure and Baskin 1959). Think about the way we talk in English. Any instance of talking is likely to be close to unique, given the affectively infinite range of combinations that can be made from the millions of words. And people change language all the time, using slang or inverting the meaning of words. And yet we can still, most of the time, understand each other. This is because there is a structure to language that we can tap into to make the particular instances understandable – a structure of grammar. It is also important to note that we may not even be able to articulate what this structure is exactly – we can leave that to experts in grammar or linguistics. This even works across all languages, which makes it possible for us to learn other languages with reference to a basic set of rules about tenses, the need to have a verb in a sentence, etc.

Forms of structuralism can be seen across disciplines. One of the main advocates of structuralist thinking was the French anthropologist Claude Lévi-Strauss, who proposed that there were underlying patterns of human thought that pervade cultures and explain the surface appearance of different cultures. These patterns follow rule-like behavior and are generally invisible or unarticulated by people who perform a particular culture. The role of the structural anthropologist is to discover, describe, and explain these patterns and rules (Lévi-Strauss 1963). A structure, then, lies beneath any particular instance in the world and can be used to explain all the variety that we experience as we lead our lives. It is like a code that, once known, makes things make sense. In other words, structuralists may disagree with each other over which structures are the most important ones. Marx, Lévi-Strauss, and Saussure were unlikely to agree on this. What they share, however, is a sense that there is something bigger than the individual, that is often invisible at first glance, that explains most of the features of human life that they consider important.

In the 1980s, geographers began to think up ways to overcome what appeared to be a dead-end struggle between structure and agency in the construction of theory. This problem was at the heart of approaches such as time-geography, critical realism, the "new cultural geography," and, most of all, structuration theory.

Time-geography

Time-geography concerns the trajectories in time and space of individuals and groups. It was the brainchild of the Swedish geographer Torsten Hägerstrand (Hägerstrand 1967; Hägerstrand and Pred 1981). Hägerstrand was concerned that the modelers of spatial science had neglected the micro scale of what we as individuals do, geographically speaking, in our everyday lives. Rather than turning to phenomenology, like humanistic geographers, Hägerstrand continued to make models:

> Time-geography rests on the premise that each of the actions and events consecutively occurring between the birth and death of an individual has both temporal and spatial attributes. Thus, the biography of a person is ever on the move with her and can be conceptualized and diagrammed at daily or lengthier scales of observation as an unbroken, continuous path through time-space (...). While unwinding her path in the course of sleeping, contemplating, and participating in everyday practices, the individual is constantly in physical touch with, (or close proximity to), other individuals, other living organisms belonging to the animal and vegetable worlds, or man-made or natural objects, each of which also traces out an uninterrupted path in time-space between the point of its inception or creation and the point of its death or destruction. (Pred 1981: 9)

Hägerstrand's time-geography models map the existence of individual humans as they make their way through a day or longer. They account for both space (where they are at a given point in time) and time (when they are at a given point in space). Such an accounting can take place over a day, a year or even a lifetime (though the illustration of a whole lifetime gets tricky). These models show how every human action has both spatial and temporal aspects. Additionally, each action is part of a longer trajectory – "a weaving dance through time-space" (Pred 1977: 208). Pred describes this well:

> The most significant elementary steps, or events, in such a 'choreographic' depiction occur at physically fixed buildings or territorial units of observation – referred to as 'stations' or 'domains' – either when two or more individuals meet to form a group (or 'activity bundle') or when such groups dissolve, with one or other 'activity bundles' at other 'stations'. The choreographic point of view – by which the structure and process of outward physical existence become one – focuses on the constraints which in both obvious and subtle ways limit the individual's freedom to move from station to station and to choose activity bundles. (Pred 1977: 208)

In one sense these paths through time and space can be read as diagrams of choices made by individuals. In another sense they depict constraints on the movements of people. These constraints include what we might think of as natural "capability constraints" such as the need to sleep and eat and the impossibility of getting from a to b in a given time span. Then there are "coupling constraints" which refer to the need for individuals to be with others in a specific place and time period (such as being at work). Finally there are "authority constraints" which include all the prohibitions that prevent people from entering places and forbid access for political, economic, or socio-cultural reasons. These basic building blocks of time-geography are most often applied to humans but can be applied to all "nonhuman" populations too, meaning that, in Pred's terms, "time-geography can: (1) specify the necessary (but not the sufficient) conditions for virtually all interaction between men and elements of the natural environment, and man and man-made objects; and (2) provide a bridge to cover the gap between human ecology and biological ecology" (Pred 1977: 209). Time-geography thus has the capacity to include everything from the individual path through the day to the reunification of the physical and human components of geography. Hägerstrand's work balances notions of structure (constraints) and agency (movements) and makes space more dynamic by mixing it with time, getting around the tendency to think of space as a static cross-section of a more dynamic axis of time. Time-geography is one way of thinking about a hybrid **space–time** (Merriman 2012).

Time-geography is another way in which geographers can provide **contextual explanations** which account for processes within a particular space or region. Pred suggests that it can provide the basis for a reinvigorated regional geography. Regional geography, he argues, had sought to provide endless inventories of physical and human components of a region "in the vein hope of somehow being able to create an integrated picture of the bounded region as a whole" (Pred 1977: 213).

Hägerstrand, Pred suggests, creates the possibility of thinking of a region as consisting of a set of human, nonhuman, and technological actors involved in an intricate ballet that is constantly in process. Not an achieved thing but a state of becoming. This state of becoming, this dance of social life, necessarily involves the constant balancing of choice and constraint, structure and agency.

Time-geography has not been without its critics. It was taken to task for appearing to reduce human life to dots and lines in diagrams. One aspect of human life for which Hägerstrand's diagrams cannot account is human intentions and all the accompanying imaginations, dreams, and desires. Time-geography shares with spatial science the danger of reducing human life to a kind of social physics where humans are like atoms in Brownian motion. Gillian Rose provides a feminist critique of time-geography, suggesting that the diagrammed actions of individuals fails to account for the very specific forms of feminine sociality and subjectivity that result from the "routine work of mothering and domesticity" (Rose 1993: 27). Time-geography, like spatial science, presents, in Rose's view, a masculinist, disembodied geography where bodies appear as abstract things rather than socially marked individuals. Despite these criticisms, however, time-geography remains one of the most influential bodies of ideas informing subsequent work on the role of human action in social life, as well as theorizations of time and space, from a number of perspectives (Parkes and Thrift 1980; Schwanen *et al.* 2008; Merriman 2012).

Toward structuration theory

In the 1980s, several attempts were made to solve the structure and agency debates once and for all by collapsing them into each other. One approach was evidenced in the "new cultural geography" which sought to connect the seemingly opposite instincts of humanism and Marxism in a way that reflected the revisionist Marxism of cultural theorists such as Raymond Williams (see Chapter 7). Williams consistently used terms that sought to integrate base and superstructure, the social and the individual. Consider just one of these – "*structure of feeling*" (Williams 1977). The words "structure" and "feeling" are seeming opposites. One refers to the permanent, foundational, and shared while the other signifies the personal and fleeting. So what can such a phrase mean? Williams did not like the distinction between a definition of the "social" as something complete, structural, and finished that is presented in the past tense on the one hand, and a "present" that is always other than social:

> We are talking about characteristic elements of impulse, restraint, and tone: specifically affective elements of consciousness and relationships: not feeling against thought, but thought as felt and feeling as thought: practical consciousness of a present kind, in a living and interrelating continuity. (Williams 1977: 132)

In the notion of "structure of feeling," Williams sought to make the idea of structure less fixed and feeling more structured. This kind of "cultural Marxism" shared much with structuration theory and heavily influenced geography through the "new cultural geography" (Cosgrove and Jackson 1987; Jackson 1989).

Critical realism is an approach that attempts to disentangle conditions that are "necessary" (and therefore "structural") for a particular event to occur and those that are "contingent" (local, accidental, replaceable) (Bhaskar 1986; Sayer 2000). Something that is necessary has to be present for an event to occur. This does not mean the event will occur – this might need something "contingent" to also be present. A common example is provided by gunpowder:

> Gunpowder has the (necessary) causal power to explode, but it does not explode at all times in all places. Whether or not it does so depends on the right contingencies being co-present, in this case the presence

of a spark or other form of detonation. As with gunpowder, so in society certain causal powers are argued to exist necessarily by dint of the characteristics and form of the objects that possess them, but it is *contingent* whether these causal powers are released or activated. (Cloke *et al.* 1991: 147)

Derek Gregory described critical realism as follows: "Realism is a science based on the use of abstraction to identify the (necessary) causal power and liabilities of specific structures which are realized under specific (contingent) conditions" (Gregory quoted in Cloke *et al.* 1991: 134–135). As these ideas developed under realism, "theory" became the search for the "necessary" (often hidden, structural) causes of surface phenomena while the world of the "contingent" (on the surface, often visible) was left to empirical research. The relationship between critical realism and arguments about structure and agency is not clear. This is because while there is a fairly strong fit between the realist notion of the "necessary" and a wider notion of structure, the "contingent," on the other hand, is not necessarily about agency, but often is.

The most systematic approach to integrating structure and agency in human geography and beyond has been given the catch-all term **structuration theory** – a term associated with the British sociologist Anthony Giddens. The work of Giddens came into geography through the work of Derek Gregory, Alan Pred, Benno Werlen, and others working around the idea of the region in the 1980s (Thrift 1983; Pred 1984; Werlen 1993; Gregory 1996). Giddens had used elements of "time-geography" in his account of the structuration of social life. The basic insight at the heart of structuration theory is that "structuration" is not a solid and complete thing (like a structure tends to imply), but a process in which the structural properties within a particular society are constantly reiterated and expressed in everyday actions at the level of the individual. Simultaneously these everyday actions are reproducing the structural properties of that society.

Although he did not use the term, structuration is also associated with the work of the French sociologist Pierre Bourdieu. In what follows I focus less on Giddens's work and more on the work of Bourdieu, whose thinking has proved to be influential across a range of disciplines and in my own work. The broad outlines of the approach are very similar to that of Giddens.

So what does structuration refer to? One, often cited, passage from cultural theory that illustrates the essentials of structuration theory nicely is by Michel de Certeau. In his book *The Practice of Everyday Life*, de Certeau considers the everyday act of walking as a moment of creativity that is never fully encompassed or determined by the **spatiality** of the city. In the following passage, he describes walking as similar to talking. At first, talking would seem to be heavily determined by the structure of language with all its preexisting structures of grammar. Similarly, walking in the city would appear to be reliant on the structure of the spaces that we walk through. We cannot walk through walls and it is hard to cross an eight-lane highway safely. But just as talkers play with language, so walkers can play with space:

> First it is true that the spatial order organises an ensemble of possibilities (e.g. by a place in which one can move) and interdictions (e.g. by a wall that prevents one from going further), then the walker actualizes some of these possibilities.... But he also moves them about and he invents others, since the crossing, drifting away, or improvisation of walking privilege, transform or abandon spatial elements.... And if on the one hand he actualizes only a few of the possibilities fixed by the constructed order (he goes only there and not here), on the other he increases the number of possibilities (for example, by creating shortcuts and detours) and prohibitions (... he forbids himself to take paths generally considered accessible or even obligatory). He thus makes a selection. (de Certeau 1984: 98)

In this passage there is clearly a structure at work – the structure of urban space. At the same time the walker is equally clearly not determined by this structure – within certain limits he can play with this structure and use it to create new possibilities – short cuts. We might apply a similar logic to skateboarders who are enabled by the presence of park benches or concrete ramps to perform new and unexpected tricks.

I used another commonplace example in my book *In Place/Out of Place*, which was heavily indebted to Bourdieu's social theory (Cresswell 1996). In that book I used the example of a library as a place that is supposed to be quiet. The quietness of a library depends on the people in it being quiet. In general, people enter libraries and remain silent or whisper without anyone telling them to be quiet. The actions of all the people before them reaffirm the necessity of silence and by remaining silent they reproduce the silence of the library. In this way structures are continually reproduced through reiteration of individual practices. This is the process at the heart of structuration theory. Social structures cannot exist without the individuals in them acting in ways that produce and reproduce the structures. If everyone suddenly acted otherwise, the structure would (after a suitable time lag) be transformed. That is what revolutions aspire to. At the same time individual actions would have little meaning or efficacy without structures to hang on to. Imagine I was writing this sentence without the English language, or any language, to make these marks on a page meaningful. Imagine I tried to buy a loaf of bread with some coins when there was no such thing as money to make it make sense. My individual actions, as well as yours, are constantly authorized by structures. At the core of structuration theory, then, is a cyclical sense of causality in which structure and agency constantly reproduce each other, with neither having logical priority.

Allan Pred used structuration theory (as well as time-geography) in a careful outline of the idea of place. The idea of a constant, mutually constitutive interplay between structure and agency, he argues, should be at the heart of thinking about places as never finished or complete but always in process:

> The assemblage of buildings, land-use patterns, and arteries of communication that constitute place as a visible scene cannot emerge fully formed out of nothingness and stop, grow rigid, indelibly etched in the once-natural landscape. Whether place refers to a village or a metropolis, an agricultural area or an urban-industrial complex, it always represents a human product. Place, in other words, always involves an appropriation and transformation of space and nature that is inseparable from the reproduction and transformation of society in time and space. As such, place is not only what is fleetingly observed on the landscape, a **locale**, or setting for activity and social interaction (…). It also is what takes place ceaselessly, what contributes to history in a specific context through the creation and utilization of a physical setting. (Pred 1984: 279)

Pred located structuration in place by combining it with time-geography's insights about choice and restraint in life-paths. He sought to be more precise about how "any given time-space specific practice can simultaneously be rooted in past time-space situations and serve as the potential roots of future time-space situations" (Pred 1984: 281). In order to do this, he focused on the allocation of time and space in place, contrasting individual paths with what he called "dominant institutional projects." This latter term was used to add a conceptualization of power into the often quite descriptive accounts of time-geography. In the becoming of place, Pred argued, some projects take precedence in the allocation of time and space and these can be referred to as "dominant." The requirement that you be at work at a particular time supersedes your desire to go for a sudden walk in the park on a bright spring day. Work, in this sense, is part of a dominant institutional project. Pred's account of place as a process of structuration is similar to de Certeau's account of walking in the city in that it constantly balances the structures of "power" with the paths of individuals.

In the work of Pierre Bourdieu, there is a constant sense of him grappling with what he calls "subjectivism" and "objectivism." Subjectivism refers to the work of phenomenologists and ethnomethodologists. The closest bodies of work in geography to this are humanistic geography and nonrepresentational theory. Objectivism refers to those who believe there are objective structures that provide explanatory frameworks for everyday actions. The most obvious examples of this in geography are some forms of Marxist geography. Bourdieu believed both are inadequate:

> Of all the oppositions that artificially divide social science, the most fundamental, and the most ruinous, is the one that is set up between subjectivism and objectivism. The very fact that the division constantly reappears in virtually the same form would suffice to indicate that the modes of knowledge which it distinguishes are equally indispensable to a science of the social world that cannot be reduced either to a social phenomenology or to a social physics. (Bourdieu 1990: 25)

The problem with phenomenology, for Bourdieu, is its inability to account for the possibilities of individual experiences. It does not ask the question of what makes specific experiences (such as the choice of ice cream we considered earlier) possible, or particular acts creative:

> But it [phenomenolology] cannot go beyond a description of what specifically characterizes 'lived' experience of the social world, that is, apprehension of the world as self-evident, 'taken-for-granted'. This is because it excludes the question of the conditions of possibility of this experience, namely the coincidence of the objective structures and the internalized structures which provides the illusion of immediate understanding, characteristic of practical experience of the familiar universe, and which at the same time excludes from that experience any enquiry of its own conditions of possibility. (Bourdieu 1990: 26)

Structuralism, on the other hand, fails to account for the very everyday experiences and practices that make structures work – they just appear to be there without any process that results in their production:

> ... because it [objectivism, structuralism] ignores the relationship between the experiential meaning which social phenomenology makes explicit and the objective meaning that is constructed by social physics or objectivist semiology, it is unable to analyse the conditions of the production and functioning of the feel for the social game that makes it possible to take for granted the meaning objectified in institutions. (Bourdieu 1990: 26–27)

Bourdieu's way around these two ways of thinking is to construct a number of concepts that bring structure and agency into constant interaction, with each producing the other. The most important of these concepts is **habitus**:

> The conditionings associated with a particular class of conditions of existence produce *habitus*, systems of durable, transposable dispositions, structured structures predisposed to function as structuring structures, that is, as principles which generate and organize practices and representations that can be objectively adapted to their outcomes without presupposing a conscious aiming at ends or an express mastery of the operations necessary to attain them. Objectively 'regulated' and 'regular' without being in any way the product of obedience to rules, they can be collectively orchestrated without being the product of the organizing action of a conductor. (Bourdieu 1990: 53)

Here we can see Bourdieu, in a baroque way, dancing around the interplay of the objective and the subjective, structure and agency. The habitus describes the way that the external (structure) is internalized and the internal (subjectivity) produces structures. These structures are not in some way externally "orchestrated" but are produced by individuals acting together in concert. In this way some form of coherence emerges and history repeats itself: "The *habitus*, a product of history, produces individual and collective practices – more history – in accordance with the schemes generated by history" (Bourdieu 1990: 54). Replace "history" with "geography" in this sentence and you get a sense of what a structurationist geography looks like. Particular spatial arrangements – places, territories, regions – at different scales are constantly enacted, reproduced, and transformed in contexts of already existing spatial arrangements. Spatiality is thus before, after, and within specific practices and experiences.

Bourdieu frequently illustrated his point with the notion of a game – such as a game of basketball. A game has rules that are written and a kind of pre-given structure, but knowledge of this will not make you a good player. Watching an accomplished player catch a ball in mid-flight it is possible to see the coming together of both structure (the rules of the game) and agency (the player's ability to play). It is an image of simultaneous freedom and constraint. The key here is the difference between the *rules of the game* and a *sense of the game*:

> Nothing is freer or more constrained at the same time than the action of the good player. He manages quite naturally to be at the place where the ball will come down, as if the ball controlled him. Yet at the same time, he controls the ball. Habitus, as the social inscribed in the body of the biological individual, makes it possible to produce the infinite acts that are inscribed in the game...(Bourdieu 1986: 113)

Structuration theory found its way into geography in several ways. Although I did not refer to structuration theory directly, my book *In Place/Out of Place* was an attempt to grapple with some of the issues emerging at the time around poststructuralism and cultural studies in a geographical way. It is a book that is heavily influenced by reading Bourdieu. In it, I attempt an account of place that sees places as both providing an ideological frame for human actions and as themselves constantly being produced by our action – place as preexistent structure, as tool in human action, and as product. I did this through an examination of moments when the commonsense and taken-for-granted attributes of place were disrupted by people acting in ways that were deliberately or accidently "out-of-place." These moments (graffiti, traveling communities, peace protests) were moments where the expected relations between place and behavior were broken and had to be defended and reinstated by those with an interest in existing notions of common sense (Cresswell 1996). Such moments of transgression, I argued, had the potential to lead to societal transformation. To continue Bourdieu's sports analogy, though, it is possible to think of moments when the game is changed by a player doing something outside of our normal expectations. A clear example of this is when someone picked up the ball when playing football (soccer) and thus invented the game of rugby – a new "structure" was born.

Despite these attempts to balance or integrate structure and agency in human geography, many geographers, increasingly influenced by French social theory and philosophy, found the whole enterprise far too structuralist. My work on transgression was taken to task for just such a shortcoming (Rose 2002):

> Although Cresswell acknowledges the constructed nature of structuring forces, he remains oddly committed to conceptualizing them as stable and, thus, causal. Like Bourdieu, he sees structures as both socially constructed and effective. Thus, they are capable of constituting a structured subject. In response, I argue that there is no reason to believe that forces designed to structure thought actually succeed in sedimenting or stabilizing meaning – that is, there is no reason to believe that structuring acts actually result in a structured being: *all we know is that they are represented as such.* (Rose 2002: 392)

By the turn of the last century, influenced by the turmoil of debates about postmodernism, a poststructuralist geography was born.

Poststructuralism and Geography

In at least a small sense, all of the approaches we have considered in this chapter so far continue to hold on to a sense of a "real" structure (or structures) that relates in some way to the everyday practice of leading our lives. At the same time, they all attempt to escape the sense of a structure which *determines* social life in a straightforward way. Poststructuralism finally attempts to do away with the notion of a coherent structure which precedes everyday life entirely. If other approaches hold on to some sense of "reality" or "truth" that preexists thought and action, then poststructuralism is a body of thought that insists that "reality" and "truth" are an effect of what we think and do, rather than a cause.

Writing in 2023 it is hard to think of any recent work in geography that has mobilized the term "postmodernism" or made claims to being "postmodernist." The terms "poststructuralism" and "poststructuralist," however, are everywhere. To many, postmodernism and poststructuralism are more or less the same thing. This is not a view shared by Jonathan Murdoch in his book *Post-Structuralist Geography*. He only mentions postmodernism once – in the following way:

> The prefix 'post' often leads to the assumption that post-structuralism has much in common with 'post-modernism' – that is, is yet another attempt to delineate specific features of contemporary society of the basis of some assumed historical shift from one distinct social condition to another. (Murdoch 2006: 4)

If you read many of the papers and books referred to in the last chapter, you will see the two terms promiscuously intermingled. Many saw them as versions of the same thing. Many of the chapters in Benko and Strohmayer's edited collection *Space and Social Theory: Interpreting Modernity and Postmodernity* use the term poststructuralism more than postmodernism (Benko and Strohmayer 1997).

Some of the differences are nicely illustrated in a smart and amusing paper by Deborah Dixon and John Paul Jones III which imagines a conversation between a poststructuralist (i.e. either of the authors) and a spatial analyst (i.e. Jones in a past life):

> SA: So, did you simply tire of the label 'postmodernist'?
>
> PS: For many, yes, postmodernism has gotten tiresome; for me, it's all too redundant. Postmodernism is a vessel that contains too much ... has been (mis)taken far too often to retain its usefulness....
>
> SA: What makes poststructuralism any different?
>
> PS: Poststructuralism is a form of analysis that relies upon the critical scrutiny of foundational moments in social thought. In particular, poststructuralism rejects the essentialism and fixity assumed to inhere in the knowing subject, both as an object of inquiry and as a producer of meanings. As such, it refuses the stability of all social relations that are productive of the 'individual' on the one hand, while problematizing the presumption of the bounded concept, and the certainty of knowledge, on the other. (Dixon and Jones 1998: 248)

The confusion between postmodernism and poststructuralism is hardly surprising. A structure, after all, is very like a "foundation." It is a systematic set of generative principles. Those who can access these structures are special people – scientists or theorists or experts in grammar, who have learned the code and are able to examine life dispassionately and objectively. From this description it should be clear why poststructuralism shares much with theory that goes under the label of "post-modernism." If we return to the failed Pruitt-Igoe Project in St. Louis, we can see that it was built on the belief that deep structures had been identified and that humanity could be improved by reengineering these structures. A structure is "deep" (i.e. invisible to those who do not know how to look for it), foundational, and universal. While postmodernism made many things possible, you are unlikely ever to encounter a postmodern structuralist.

Poststructuralists do not believe that it is possible to identify deep, generative structures beneath the infinite variety of the surface of life. Deborah Dixon and John Paul Jones III (Dixon and Jones 1996, 1998) argue that geography, along with most other sciences, has utilized a view of the world based on dividing elements of life into neat categories and drawing clear boundaries between them. This allows some kind of illusory clarity when engaging in critical thought but does not do justice to the messiness of the world. These categories (such as the "natural" and the "social" for instance) are the product of boundary-drawing processes that are historical artifacts rather than matters of fact:

> From a poststructuralist perspective, however, ontological assumptions put the cart before the horse, for any ontology is itself grounded in an epistemology about how we *know* 'what the world is like'; in other words, the analysis of ontology invariably shows it rests upon epistemological priors that enable claims about the structure of the real world. For example, the ontological divisions between physical and social phenomena, or between individual agency and sociospatial structure, to mention just two that are prevalent in geographic thinking, is the result of an epistemology that segments reality and experience in order to comprehend both. (Dixon and Jones 1998: 250)

They name this way of approaching things the "epistemology of the grid" – a way of thinking and acting that carves up reality into segments so that it can be made legible to the social scientist. This is, once again, the view from nowhere or God-trick that insists on the necessity of seeing the world from above and laying a grid over reality in order to make it measurable and intelligible. The grid is not necessarily literal (as in a map) but metaphorical – denoting a way of knowing – an epistemology. The poststructuralist does not deny the utility of categories but asks us to take on board the processes through which they are produced:

> A major component of poststructuralist research, therefore, involves posing questions to categories. For example, who has the power to designate difference, and, in the process to draw the boundaries of the category? How do categories function in social life? Likewise, who has the power to draw the grid, that is, to implement categorization in social space? What we will invariably find is that the process of categorization is never neutral. (Dixon and Jones 1998: 254)

A structuralist identifies a set of things that exist in the world (however invisible they may be at first glance) and that generate actions and beliefs. Often these are presented as dualisms – pairs of opposites that, together, define the way the world is. Common examples include the social and the natural, structure and agency, or the local and the global. Analysis starts from these points. A poststructuralist starts by questioning the way of knowing that led us to think in these terms in the first place and insists on the ways in which the terms rely on each other for their meaning. They are, in other words, relational:

> This relational perspective asserts that no *A* exists without a *not A*, and thus the former is maintained to be constitutively dependent on the latter. In this view there is no outside to any epistemological category, and hence to any system of geographic thought (including poststructuralism). (Dixon and Jones 1996: 768)

Poststructuralism does not look for structures below the surface of reality – rather it assumes that reality is always being produced *on the surface* by various sets of relations – it is more of a horizontal model than a vertical one. Rather than relatively fixed, stable, structures, poststructuralism fixes on process and the act of "becoming."

The entry points for poststructuralism into human geography were the "new cultural geography" of the late 1980s and early 1990s alongside strands of feminist geography that shared some of the preoccupations with poststructuralist theory. The "new cultural geography" emerged in response to an apparent lack of interest in culture among human geographers in the UK, where there had not really been a tradition of cultural geography, and a questioning of both traditional North American cultural geography associated with Carl Sauer and the (largely) idealist traditions of humanistic geography. Geographers such as Peter Jackson, Denis Cosgrove, Stephen Daniels, and James Duncan were informed by a range of theories, some of which were clearly poststructuralist (especially in Duncan's case) and some of which were variations of cultural Marxism (Cosgrove 1984; Cosgrove and Jackson 1987; Jackson 1989; Duncan 1990; Daniels and Cosgrove 1993). Although each of these authors had slightly different projects, what united them was a desire to move away from both humanist idealism (which focused too much on a self-contained individual knowing subject and lacked any sense of politics) and from structural Marxism (which posited a generative economic structure). Writers like Antonio Gramsci and Raymond Williams were important as they attempted to hold on to insights from Marx while abandoning a structuralist perspective (Gramsci 1971; Williams 1980). They were, if you like, proto-poststructuralist theorists. At the same time French writers such as Roland Barthes (who moved from structuralist to poststructuralist positions), Pierre Bourdieu, and Michel Foucault were becoming influential in geography.

One important strand of work within new cultural geography, following on from the important writings of Peter Jackson (Jackson 1989), was an interest in all manner of human activity that might be considered as resistance (Pile and Keith 1997; Sharp *et al.* 1999). This, in turn, was informed by the Centre for Contemporary Cultural Studies in Birmingham under the leadership of Stuart Hall (Hall and Jefferson 1976). One reason for writing about resistance was to show how structures, if they did exist, were never complete and always contested. The very subject of resistance probed the limits of the possibility that social life was determined by structure. If all kinds of resistance could be identified, then clearly social life was not, in the last instance, determined by structure. Thinking such as this in cultural studies and cultural geography led to an array of work on the various ways in which people, individually or in groups, behaved in ways that did not blindly follow some structure and often appeared to oppose "the way things are." Power and resistance within cultural worlds became a key theme in human geography.

These themes of power and resistance are evident in the work of Soja as he developed the insights of Henri Lefebvre. Soja outlined a theoretical "trialectic" of first-, second-, and thirdspace. While firstspace, the space of the planner, neatly maps on to abstract notions of space from spatial science, secondspace is more imaginative and representational, more like "place." *Thirdspace*, however,

> ... is portrayed as multi-sided and contradictory, oppressive and liberating, passionate and routine, know-able and unknowable. It is a space of radical openness, a site of resistance and struggle, a space of multi-plicitous representations, investigable through its binarized oppositions but also where *il y a toujours l'Autre*, where there are always 'other' spaces, heterotopologies, paradoxical geographies waiting to be explored. It is a meeting ground, a site of hybridity and *mestizaje* and moving beyond entrenched bound-aries, a margin or edge where ties can be severed and also where new ties can be forged. It can be mapped but never fully captured in conventional cartographies; it can be creatively imagined but obtains meaning only when practiced and fully lived. (Soja 1999: 279)

Here, the emphasis is on radical openness, livedness, resistance, and mixing. This is not a clear bounded space with a singular identity that can be mapped on to it. It is a way of thinking about space where things come together rather than one where they are held apart. The sense of a lack of determinacy and of creative becoming that characterizes this thirdspace is clearly poststructuralist.

Soja was particularly influenced by the writings of the French urban theorist Henri Lefebvre. Another key influence on his work has been the French theorist Michel Foucault. While there are many philosophers and social theorists who have contributed to poststructuralist geography, the influence of Foucault has probably been the most significant across the various subdisciplines of human geography. It is to his influence that I now turn.

Foucault's Geographies

Michel Foucault was among the most influential of thinkers across the social sciences and humanities throughout the 1980s and 1990s. Geography was no exception (Driver 1985; Soja 1989; Philo 1992; Driver 1993; Gregory 1994). The original inspiration for many geographers was Foucault's book

Discipline and Punish (Foucault 1979). Here, Foucault describes how prisons (and particularly the ideal prison of Jeremy Bentham's panopticon) instill in their inmates the sense that they are constantly being watched. This in turn leads to a process of self-regulation which Foucault calls "discipline." This, he argued, replaced the direct infliction of punishment on the body through various means of torture as punishment. Geographers became excited about this as it put a particular spatial arrangement at the center of the process of disciplining. Foucault appeared to give space an active role. Foucault focused on particular forms of space and their relationships to the bodies that they housed in a series of books which included reflections on both the asylum and the clinic (Foucault 1965, 1975). Strangely, perhaps, it was a short interview with some French geographers that has continued to ignite the interest of geographers. In this interview, published as "Questions on Geography," Foucault asserts the importance of space for his thinking on power:

> A critique could be carried out of this devaluation of space that has prevailed for generations. Did it start with Bergson, or before? Space was treated as the dead, the fixed, the undialectical, the immobile. Time, on the contrary, was richness, fecundity, life, dialectic. For all those who confuse history with the old schemas of evolution, living continuity, organic development, the progress of consciousness or the project of existence, the use of spatial terms seems to have the air of an anti-history. If one started to talk in terms of space that meant one was hostile to time. It meant, as the fools say, that one 'denied history', that one was a 'technocrat'. They didn't understand that to trace the forms of implantation, delimitation and demarcation of objects, the modes of tabulation, the organization of domains meant the throwing into relief of processes – historical ones, needless to say – of power. The spatializing decription [sic] of discursive realities gives on to the analysis of related effects of power. (Foucault 1980: 77)

In the interview, Foucault initially resisted attempts by the geographers to get him to grasp the centrality of space to his work. By the end, however, he was convinced and recognized that geography was, indeed, central to his concerns.

Foucault's recognition of the centrality of space for the workings of power motivated those geographers who were insisting on space's centrality in the constitution of spatial power – geographers who saw space as an active agent rather than a passive product of other forces. He was particularly important in Soja's analysis of postmodern geographies and, later, "thirdspace" (Soja 1989, 1999). In a more precise way, he inspired work by geographers on the history of the asylum and madness as well as more metaphorical forms of disciplining through surveillance and ordering (Philo 1992; Hannah 2000).

While there are hints of structuralism in Foucault's trademark accounts of historical transformations such as the transformation from punishment to discipline, he is widely recognized as a key poststructuralist. This is because Foucault insists that there is no hidden deep truth which might explain what happens on the surface of life. Rather he insists that "truth" is an effect of the surface – something that is produced in particular times and spaces. While there are many ideas of Foucault's that have been used by geographers – such as **governmentality**, *heterotopias*, and **biopolitics** – perhaps the key term that reached across the various subdisciplines of human geography is **discourse** (Hetherington 1997; Hannah 2000; Braun 2007; Elden 2007).

Foucault's theory of discourse approaches epistemological questions of how we know what we know, where such knowledge comes from, who is authorized to profess it and how, ultimately, "truth" is established.

Throughout his writing, Foucault revealed how linguistic and other elements are brought together to form new kinds of objects of knowledge (madness, populations, the pathological, etc.), new sets of concepts and categories, new claims about what can be said and what must remain unsaid, and a new set of subject positions (the mad, the criminal, the deviant as well as the expert, the psychologist, etc.). He referred to this mixing of words, actions, institutions, and infrastructures that are more or less logically coherent and produce new forms of truth as *discourse*.

Discourse is at the center of Foucault's best-known book, *Discipline and Punish*. In this book he charts the eighteenth-century growth of a disciplinary society based on a set of legal and penal discourses which produced a new kind of criminal subject. This new kind of person had their own sets of characteristics which would have made no sense before the invention of these discourses. New forms of punishment and surveillance arose hand in hand with these subjects, as did new kinds of practice (in science, in law, in politics, etc.). These discourses were more than sets of spoken or written words (as suggested by general definitions of the meaning of discourse), but constellations of words, practices, institutions (such as the prison), and things. Neither are these discourses clearly referring to something that existed in the world before they did. Foucault's notion of discourse suggests the opposite – that new "truths," new possibilities of being, emerge out of particular discursive formations. Discourses can thus be seen as establishing new networks of meaning and practice which delineate, produce, and reinforce relations between what it is possible to think, say, and do. Objects and subjects are not external to this process but are thoroughly constituted through it.

The Foucauldian use of the term discourse thus has two primary differences from its more everyday use. First, discourse is considerably more than spoken and written words (it includes all forms of representation and, in some readings, institutions, objects, and practices) and, second, discourse is not simply "about" something (politics, illness, sexuality, etc.) but brings these things into being. In the sense that discourses produce reality they are performative. In this sense, it is more than a reflection of "reality" that can be true or false but is in fact performing "reality" and producing "truth effects."

One way to specify the Foucauldian meaning of discourse (and to illustrate the difference between structuralism and poststructuralism) is to compare it to the concept of ideology. Ideology, like discourse, considers the relations between meaning, power, and truth. Most theorists of ideology define it as sets of meanings in the service of the relatively powerful – meanings which reinforce already existing power relations. Discourse differs from this in important ways. While ideology relies on some notion of truth and objectivity ("real interests," for instance), that is external to it, discourse produces truth. Foucauldian discourse analysis is not concerned with separating out the truthful and the objective from the rest but with showing how historically "effects of truth" are produced discursively. In addition, ideology (at least in its Marxist formulations) is always part of a superstructure which is caused by some other realm of human interaction – usually the world of political economy. Discourse has no such "base" to appeal to or that can be used to explain it. Truth, in other words, is not in some other place but is within discourse itself. Finally the idea of ideology rests on the notion of a knowing and intentional subject – it is fundamentally humanist in its conception. Foucault's notion of discourse rejects humanism and, in particular, the humanist notion of the subject at the center of the universe. The subject in Foucault is not unitary and knowing but fragmented and decentered – produced discursively through the process of subjectivization. An example would be the ways that various medical texts, beliefs, and practices, as well as the construction of the physical environment, produce the "disabled" subject who is both marginalized and endowed with particular forms of competence.

Discourse analysis is thus very different from the analysis of **ideology**. While the analysis of ideology (and "critical theory" in general) seeks to peel away obfuscation to reveal the "truth" of a situation, the analysis of discourse tries to define not what is concealed by a discourse but what is produced by it – the rules of the discourse itself and the new regimes of truth that it enables.

The influence of Foucault's notion of discourse was certainly lurking in my own work on the tramp in the United States in the period following the Civil War (Cresswell 2001). One way of writing such an account would be to find out as much as possible about this phenomenon in the world and report it to the reader. My approach was different as I chose, instead, to look at how the sources I was using made the tramp possible – brought the tramp into being.

A careful observer would have noticed increasingly frequent references to tramps in the pages of the *New York Times* from the 1870s onward. In fact they would not have seen any references to tramps before February 1875. It would have been as though they did not exist. A year later the first "tramp law" came into being in New Jersey. A picture was beginning to emerge both in the press and in the definitions provided by law of what a tramp was. Apparently idle, without work, and frighteningly mobile – able to be in New York one day and in California two weeks later thanks to the recently completed transcontinental railroad. In the early 1890s, the New England reformer John McCook was writing sympathetic columns about tramp life which resulted from extensive questionnaires and interviews with tramps. By 1899, all kinds of books were appearing either about or by these tramps. Indeed, a new academic discipline, sociology, was dedicating a large amount of time to recording the lives of these people and further classifying them into different kinds of tramps, as well as distinguishing them from hobos and bums. One such "sociologist," Josiah Flynt, spent a year dressed as a tramp and traveling with them. He reported his adventures in the popular magazine *Atlantic Monthly*. All kinds of theories about these disorganized people began to emerge from the new Chicago School of Sociology. In the early twentieth century, the sociologists were joined by the medical profession and, particularly, by eugenicists, who suggested that the tramps suffered from an inherited predisposition to wander endlessly.

Tramps also became central to other new forms of knowledge. Cartoonists frequently featured tramps in cartoon strips in the daily paper as figures of fun. Vaudeville performers made fun of the tramp's hygiene and work aversion. The new silent movies were to put the tramp at center stage as Charlie Chaplin used his "little tramp" to poke fun at conventional society. The practice of documentary photography was producing picturesque portraits of tramps in New York's back alleys. Given that our careful observer would never have seen the use of the word tramp as a noun until 1875, it is surely remarkable that such an array of words, images, charts, movies, diagnoses, treatment, and punishments should have been brought to bear on this previously invisible and un-named character. It should be noted that sociology, newspaper cartoons, documentary photography, vaudeville, silent movies, and eugenics were also new to this period. It was as if they had arisen hand in hand with the tramp. There are clearly at least two ways to think about this. The conventional way would be to say that a new phenomenon has arisen – a new thing in the world – the tramp – and that, once noticed, knowledge had to account for it. The second way would suggest that with the construction of new discourses (sociology of deviance, vaudeville humor, etc.) it became possible to be a tramp. In this second sense the tramp was "made-up" discursively. Importantly, this is not simply to say that new ideas and images of the tramp were produced culturally (this would hardly be controversial) but that tramps were literally brought into being – it became possible to be a tramp. And with the arrival of the tramp came new possibilities for practice, new punishments, new medical diagnoses, new spaces and places.

So what is geographical about discourse? There are four principal ways in which geographers either have engaged, or potentially could engage, with discourse. First, the idea of discourse places a great deal of importance on context. Knowledge, texts, truths, practices, and realities are all products of particular times and spaces. They are not universal but contextual. Clearly, then, discourses have a distinctive spatial character. The work of Foucault, for instance, is often grounded in particular times and places and the knowledge that arises there is specific to that context. His studies of madness and discipline, for instance, were clearly set in France and the setting impacted upon the discourse. Discursive knowledge is not universal but particular. Its truths are not general but specific.

Second, and more important, discourses have very specific sites where they arise. Laboratories, clinics, universities, and asylums are all micro-geographies which are both the context for and part of discourse. Most clearly, perhaps, modern discipline is unthinkable without the modern prison as famously represented by Bentham's panopticon. However, all kinds of discourse require authority for someone to be able to use it and produce truth effects. Part of this authority comes from the place from which discourse was produced by the author.

Third, geographical knowledge itself can be subjected to discourse analysis. Geography departments and societies are sites from which discourse is produced, as well as material parts of discourses themselves. Somewhere like the Royal Geographical Society, for instance, is a site where geographical discourse constituted particular kinds of colonial and imperial subjects and produced particular kinds of colonial and imperial truths. Geography has its own discourses and some of them have been historically powerful (Driver 1999).

Fourth, discourse is implicated in the production of places and, in particular, in the judgment of people's practices within places. What counts as acceptable, appropriate behavior, for instance, is often determined by a nexus of place and discourse. Subjects are not simply constituted anywhere but on a particular terrain. The discourse of urban planning, for instance, is implicated in the production of the "ghetto." The ghetto then historically became a site associated with particular people (often racially coded) and particular kinds of practice.

These different aspects of geographies of discourse have, in one form or another, circulated through geography's subdisciplines since the mid-1980s. Cultural geographers, with their interests in textuality and representation, began to invoke the notion of discourse quite loosely alongside literary concepts such as metaphor and rhetoric. The general "cultural turn" in human geography saw the migration of discourse analysis, and particularly the Foucauldian version of it, into other subdisciplines such as political geography and economic geography. Critical geopolitics, for instance, often accounts for political power on the international stage through analysis of a range of discourses from the popular media to official governmental policy (Toal 1996). Rather than considering geopolitics as an objective form of knowledge about the world, critical geopolitics considers them as discursive constructions of the political world constructed from particular places by particular people with the authority to speak. Critical geopolitics thus considers all manner of representational forms from official documents to newspaper cartoons as elements in the formation of the political landscape.

The historical geographer Matthew Hannah conducted a particularly intensive discursive analysis of the development of the census in the United States alongside a new statistical discourse which brought the new category of the "population" into being. This analysis involved an understanding of the places from which statistics emerge, as well as how particular people are given the authority to speak the discourse of statistics about populations (Hannah 2000). Census statistics, it turns out, are not simply about populations – they create them. Once populations are invented, all kinds of statistical practice (and other kinds of practice) can be exercised on and through these populations. All of this is intensely geographical.

Conclusions

The influence of poststructuralism on geography is now so vast that it has been impossible to do it justice in this chapter. Some of poststructuralism's influence is evident in the chapters on Marxism (particularly the work of Gibson-Graham), feminism, and postmodernism and there is much in the next two chapters that is directly attributable to a poststructuralist orientation. I am aware that I have said nothing about the influence of Derrida or Delueze and Guatarri whose work has been used by many geographers over the past several decades (Doel 1999). What I hope to have achieved in this chapter is to provide a basis for thinking about geography in poststructural terms.

Poststructuralism has not been without its critics. To some, it is just too foundationless – too relativistic – leaving no grounds for certainty or political critique. An important ontological question, for instance, is "what lies beyond discourse or the text?" A strong reading of poststructuralist writing would seem to suggest that the answer is "nothing." A skeptic might point to the hardness of rocks or the laws of gravity and say that such things are clearly not created discursively. Indeed, a supporter of

Foucault, Ian Hacking, has gone to great pains to suggest that there is something importantly different about a "tramp," for instance, and a "waterfall" (Hacking 1999). While it is true that a waterfall would exist with or without our apprehending it, a tramp is literally made up through our knowledge about it and our representations of it. However, even a waterfall has to be apprehended and, a Foucauldian might argue, our understanding of waterfalls only happens through discourse. So here we have a weaker discourse theory that acknowledges things beyond discourse but says there is no way to know such things except discursively.

Among the more thought-provoking critiques of discourse theory is the observation that if objects are internal to discourses (i.e. are constituted by them) then how are we to judge that discourse has constructed its object appropriately, validly, or justly? Or to put it more straightforwardly, how is it possible to be right or wrong? This question is both an ontological one and a moral–political one. Unlike the theory of ideology, for instance, there does not seem to be a place from which to judge a discourse and the truths it constructs. How, for instance, could we say that Nazi discourse, or racist discourse, or homophobic discourse is wrong? On what grounds could we seek to change it?

Feminists have had a complicated relationship with poststructuralism. This chapter is notable for the dominance of male voices and there is some truth in the suggestion that poststructuralism, with its sometimes overly assertive theoretical language, is **masculinist**. The suggestion that "gender," for instance, is constituted through discourse can be seen as a deeply dis-empowering act for people who have promoted intellectual and political resistance under the banner of "women." These worries have been clearly articulated by Geraldine Pratt:

> Many feminists feel a deep ambivalence towards these views of theory and truth. On the one hand, many accept the social constitution of knowledge; on the other hand, most would also claim to have a hold on "the" truth about gender oppression. The antifoundationalism of poststructuralism seems to open the door to relativism, to the view that "might makes right," to the position that the view heard is of necessity the one whose supporters have the most clout; this is an extremely threatening world for a group that traditionally has not enjoyed much power and a profoundly unsatisfying one for those who remain committed to a shared normative framework for positive social change. (Pratt 2008: 55)

In the end, however, Pratt embraces what she sees as the emancipatory possibilities of a theoretical approach which insists on the production of gender (along with "truth" and "justice") from within discourse. Such an approach, Pratt argues, does not need to lead to thoughtless relativism where anything is possible, and it is impossible to make any kind of moral or political judgment:

> There is a real and obvious difference between jettisoning the category woman as a prelude to a plunge into individualism and a careful process of situating differences among women so as to understand how gender is constructed in a myriad of ways…. (Pratt 2008: 56)

Poststructuralist geography is one which is fundamentally opposed to apparently rigid and determining structures. For this reason, the first half of this chapter was devoted to charting a course through the ways in which geographers have negotiated structure and agency. By the time we arrived at poststructuralism, it was clear that any remaining sense of structures as determining but invisible entities had been abolished. Poststructuralist geography is a strongly anti-essentialist geography that argues that the meaning and nature of things are produced historically and geographically rather than "naturally." Poststructuralist geography is happy to stay at the surface of life rather than digging for deeper "truth." It embraces the particular as a basis for explanation which is contextual rather than universal. It tends to think in terms of flows, networks, and folds rather than clearly defined spaces. It constantly looks to moments in which difference is important (rather than sameness). Often it seeks to replicate these differences. Perhaps most importantly for geographers, poststructuralism insists on the lively and central presence of space in our understanding of society. This theme is continued in the next chapter.

References

* Benko, G. and Strohmayer, U. (1997) *Space and Social Theory: Interpreting Modernity and Postmodernity*, Blackwell, Oxford.

Bhaskar, R. (1986) *Scientific Realism and Human Emancipation*, Verso, London.

Bourdieu, P. (1990) *The Logic of Practice*, Stanford University Press, Stanford, CA.

Bourdieu, P. (1986) From rules to strategies: an interview with Pierre Bourdieu. *Cultural Anthropology*, 1, 110–120.

Braun, B. (2007) Biopolitics and the molecularization of life. *Cultural Geographies*, 14, 6–28.

Cloke, P. J., Philo, C., and Sadler, D. (1991) *Approaching Human Geography: An Introduction to Contemporary Theoretical Debates*, Chapman, London.

Cosgrove, D. E. (1984) *Social Formation and Symbolic Landscape*, Croom Helm, London.

Cosgrove, D. E. and Jackson, P. (1987) New directions in cultural geography. *Area*, 19, 95–101.

Cresswell, T. (1996) *In Place/Out of Place: Geography, Ideology and Transgression*, University of Minnesota Press, Minneapolis.

Cresswell, T. (2001) *The Tramp in America*, Reaktion, London.

Daniels, S. and Cosgrove, D. (1993) Spectacle and text: landscape metaphors in cultural geography, in *Place/Culture/Representation* (eds J. Duncan and D. Ley), Routledge, London, pp. 57–77.

De Certeau, M. (1984) *The Practice of Everyday Life*, University of California Press, Berkeley, CA.

* Dixon, D. and Jones III, J. P. (1996) Editorial: for a supercalifragilisticexpialidocious scientific geography. *Annals for the Association of American Geographers*, 86, 767–779.

* Dixon, D. and Jones III, J. P. (1998) My dinner with Derrida: or spatial analysis and post-structuralism do lunch. *Environment and Planning A*, 30, 247–260.

* Doel, M. A. (1999) *Poststructuralist Geographies: The Diabolical Art of Spatial Science*, Rowman & Littlefield, Lanham, MD.

Driver, F. (1985) Power, space and the body: a critical assessment of Foucault's *Discipline and Punish*. *Environment and Planning D: Society and Space*, 3, 425–446.

Driver, F. (1993) *Power and Pauperism: The Workhouse System, 1834–1884*, Cambridge University Press, Cambridge.

Driver, F. (1999) *Geography Militant: Cultures of Exploration in the Age of Empire*, Blackwell, Oxford.

Duncan, J. S. (1990) *The City as Text: The Politics of Landscape Interpretation in the Kandyan Kingdom*, Cambridge University Press, Cambridge.

Elden, S. (2007) Governmentality, Calculation, Territory. *Environment and Planning D: Society and Space*, 25, 562–580.

Foucault, M. (1965) *Madness and Civilization: A History of Insanity in the Age of Reason*, Vintage, New York.

Foucault, M. (1975) *The Birth of the Clinic: An Archaeology of Medical Perception*, Vintage Books, New York.

Foucault, M. (1979) *Discipline and Punish: The Birth of the Prison*, Vintage Books, New York.

Foucault, M. (1980) Questions on geography, in *Power/Knowledge: Selected Interviews and Other Writings, 1972–1977* (ed. C. Gordon), Pantheon, New York, pp. 63–77.

Gramsci, A. (1971) *Selections from the Prison Notebooks* (eds and trans. Q. Hoare and G. Nowell Smith), Lawrence & Wishart, London.

* Gregory, D. (1994) *Geographical Imaginations*, Blackwell, Oxford.

Gregory, D. (1996) Areal differentiation and post-modern human geography, in *Human Geography: An Essential Anthology* (eds J. A. Agnew, D. Livingstone, and A. Rogers), Blackwell, Oxford, pp. 211–232.

Hacking, I. (1999) *The Social Construction of What?*, Harvard University Press, Cambridge, MA.

Hägerstrand, T. (1967) *Innovation Diffusion as a Spatial Process*, University of Chicago Press, Chicago.

Hägerstrand, T. and Pred, A. R. (1981) *Space and Time in Geography: Essays Dedicated to Torsten Hägerstrand*, CWK Gleerup, Lund.

Hall, S. and Jefferson, T. (1976) *Resistance through Rituals: Youth Subcultures in Post-War Britain*, Hutchinson, London [for] the Centre for Contemporary Cultural Studies, University of Birmingham.

Hannah, M. G. (2000) *Governmentality and the Mastery of Territory in Nineteenth-Century America*, Cambridge University Press, New York.

Hetherington, K. (1997) *The Badlands of Modernity: Heterotopia and Social Ordering*, Routledge, London.

Jackson, P. (1989) *Maps of Meaning*, Unwin Hyman, London.

Lévi-Strauss, C. (1963) *Structural Anthropology*, Basic Books, New York.

Merriman, P. (2012) Human geography without time-space. *Transactions of the Institute of British Geographers*, 37, 13–27.

* Murdoch, J. (2006) *Post-Structuralist Geography: A Guide to Relational Space*, Sage, London.

Parkes, D. and Thrift, N. J. (eds) (1980) *Times, Spaces, and Places: A Chronogeographic Perspective*, John Wiley & Sons, Ltd, Chichester.

* Philo, C. (1992) Foucault's geography. *Environment and Planning D: Society and Space*, 10, 137–161.

Pile, S. and Keith, M. (eds) (1997) *Geographies of Resistance*, Routledge, London.

* Pratt, G. (2008) Reflections of poststructuralism and feminist empirics, theory and practice, in *Feminisms in Geography* (eds P. J. Moss and K. F. Al-Lindi), Rowman & Littlefield, Lanham, MD.

Pred, A. R. (1977) The choreography of existence: comments on Hagerstrand's time-geography and its usefulness. *Economic Geography*, 53, 207–221.

Pred, A. R. (1981) Social reproduction and the time-geography of everyday life. *Geografiska Annaler: Series B, Human Geography*, 63, 5–22.

* Pred, A. R. (1984) Place as historically contingent process: structuration and the time-geography of becoming places. *Annals of the Association of American Geographers*, 74, 279–297.

Rose, G. (1993) *Feminism and Geography: The Limits of Geographical Knowledge*, Polity, Cambridge.

Rose, M. (2002) The seductions of resistance: power, politics, and a performative style of systems. *Environment and Planning D: Society and Space*, 20, 383–400.

Saussure, F. D. and Baskin, W. (1959) *Course in General Linguistics*, Peter Owen, London.

Sayer, A. (2000) *Realism and Social Science*, Sage, Thousand Oaks, CA.

Schwanen, T., Kwan, M. P., and Ren, F. (2008) How fixed is fixed? Gendered rigidity of space-time constraints and geographies of everyday activities. *Geoforum*, 39, 2109–2121.

Sharp, J. P., Routledge, P., Philo, C., and Paddison, R. (eds) (1999) *Entanglements of Power: Geographies of Domination/Resistance*, Routledge, New York.

Soja, E. W. (1989) *Postmodern Geographies: The Reassertion of Space in Critical Social Theory*, Verso, New York.

Soja, E. W. (1999) Thirdspace: expanding the scope of the geographical imagination, in *Human Geography Today* (eds D. Massey, J. Allen, and P. Sarre), Polity, Cambridge, pp. 260–278.

Thrift, N. (1983) On the determination of social action in time and space. *Environment and Planning D: Society and Space*, 1, 23–57.

Toal, G. (1996) *Critical Geopolitics: The Politics of Writing Global Space*, Routledge, London.

Tuan, Y.-F. (1977) *Space and Place: The Perspective of Experience*, University of Minnesota Press, Minneapolis.

Werlen, B. (1993) *Society Action and Space: An Alternative Human Geography*, Routledge, New York.

Williams, R. (1977) *Marxism and Literature*, Oxford University Press, Oxford.

Williams, R. (1980) *Problems in Materialism and Culture: Selected Essays*, New Left Books, London.

Chapter 11

Relational Geographies

One way of thinking about poststructural geographies is as relational geographies. Rather than thinking about the inhabited world as a set of discrete things with their own essences (this place, different from that place), we can think about the world as formed through the ways in which things relate to each other. One popular illustration of this approach is to consider the difference between topography and **topology**. While topography refers to the discrete shape of the land and is often used to denote a discrete place, topology refers to the connectedness of things. A topographic map is useful if you are going hiking. An example of a topological map is the map of the London Underground (or any public transportation map) which only needs to show which points connect to which other points and in which order (see Figure 11.1). Scale and absolute location are no longer important. Relational geographies are, in some sense, topological.

In *Post-Structural Geographies*, Murdoch argues that it is this theme of relationality that lies at the heart of poststructural geographies. It is not spaces or places in and of themselves that are at the heart of such an approach, he argues, but the ways in which they become related. This relational approach to space, he suggests, draws our attention to the ways in which things become related and how this produces **relational spaces**:

> The relational making of space is both a consensual and contested process. 'Consensual' because relations are normally made out of agreements or alignments between two or more entities; 'contested' because the construction of one set of relations may involve both the exclusion of some entities (and their relations) as well as the forcible enrollment of others. In short, relational space is a 'power-filled' space in which some alignments come to dominate, at least for a period of time, while others come to be dominat*ed*. So while multiple sets of relations may well co-exist, there is likely to be some competition between these relations over the composition of particular spaces and places. (Murdoch 2006: 20)

Here, Murdoch outlines a politics of relationality that surrounds the ways in which relations between entities happen or, conversely, fail to happen. For many geographers, space has too long been seen as an inert setting for more active processes to happen. Space, in critical geography, is

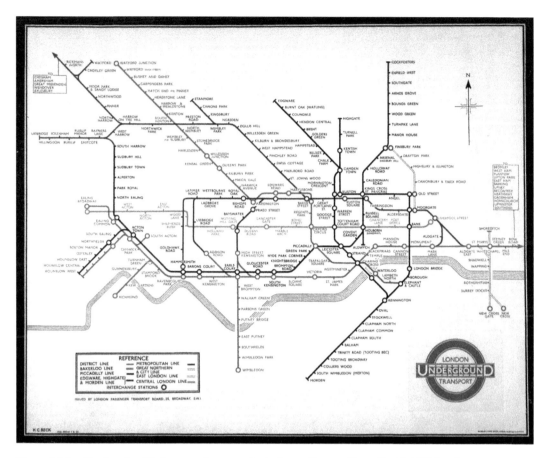

Figure 11.1   The London Underground – a topological map. Source: Harry Beck's 1933 London Underground map © TfL, London Transport Museum Collection.

active, dynamic, and composed of relations. Any sense of spatial solidity or permanence only arises from the coming together of connections and processes. Doreen Massey has argued through many papers and books that a relational conception of space insists that space is a product of interrelations, is a sphere of multiplicity, and is always in process, or *becoming*.

Massey insists that space is made from (and continues to be made, in a constant process of becoming) interrelations that move between and across scales from the local to the global. Space is the arena of multiplicity – it allows things to be different and perceived as such. In this sense, space is constantly active – not dead, fixed, or inert (Massey 2005). Massey's vision of relational space is one of a world where boundedness operates quite weakly in the face of flow and networks and where spatiality is not characterized by a "cartographic" imagination that focuses on ideas such as "territory" and "region" as discrete geographical units. There are many similarities between this view of space as becoming and Soja's Lefebvrian concept of "thirdspace" outlined in Chapter 10. Marcus Doel captures this sense of space in a constant state of becoming in a review of Murdoch's book:

> Since space is *continuously* being made, unmade, and remade by the incessant shuffling of heterogeneous relations, its potential can never be contained and its exuberance can never be quelled. What *becomes* of space always and necessarily eludes the grasp of every will to order. Although space may be stabilized for a time, it cannot be entirely mastered. Relational, ecological, and heterogeneous spaces are essentially

disturbing. Whence the shift from a "topographical" to a "topological" appreciation of space: from an orderly patchwork of surfaces to an unruly skein of relations, from a structuralist to a post-structuralist kind of geography. (Doel 2007: 810)

Doel, Murdoch, Soja, and Massey would, no doubt, have plenty to argue about, but central to their claims are similar interests in the processual and relational nature of space. Each of them resists a sense of space (or place) as finished, rigid, and ordered. Space is *mobilized* in their work.

The relational view of space adds another layer to a complicated array of ways in which geographers and others have thought about space. "Space" as a concept came to the fore in spatial science. Spatial scientists have tended to think of space as "absolute," Cartesian space – an abstract container within which things occur that can be measured and modeled (but see Forer 1978). Other geographers chose to emphasize "relative" space – space that does not exist independently of what is in it but is formed through what happens in it. Space here is the distance between objects rather than some prior already existing thing called space – without the objects there would be no (relative) space (Sack 1980). Relational space collapses the difference between objects and space altogether, arguing instead that "space does not exist in and of itself, over and above material objects and their spatiotemporal relations and extensions" (Jones 2009: 491). Relational space *happens* at the same time as the things that are supposed to occur in space. It is constantly being formed topologically through relations between things. Absolute and **relative space** remain topographical in character. Let us consider two important concepts in the history of geography in relational terms – "place" and "scale."

Relational Conception of Places

We have seen how place (and its sister concept – region) have been thought of both as unique segments of the world (in older forms of regional geography) and as a form of attachment between people and the world (in humanistic geography). Both of these conceptions of place have a sense of places being different from other ones around them and a sense of deep historical rootedness. These are places marked by rootedness and boundedness.

In the 1990s, as poststructuralism began to influence geography, the focus on place began to shift and geographers began to take notions of power, exclusion, and difference seriously. They asked what "other" was being constructed in order to create places where "we" belong. "Home" is a case in point. To humanistic geographers, in particular, home represents an ideal kind of place where we feel most attached and secure. Ideally, it is the most secure and meaningful place in our world (Blunt and Dowling 2006). But home is also a problematic place. Feminist geographers pointed out that home is where the oppression of women is felt most acutely. It is, for instance, where they are most vulnerable to sexual assault. It is also a place where children are often seen as symptom of disorder in a place which is supposed to be ordered – a point made by David Sibley:

> Inside the home and immediate locality, social and spatial order may be obvious and enduring character-istics of the environment. For those who do not fit, either children whose conceptions of space and time are at variance with those of controlling adults or the homeless, nomadic, or black in a homogeneously white, middle class space, such environments may be inherently exclusionary. (Sibley 1995: 99)

Sibley is one of the key geographers who have taken processes of othering and exclusion seriously. Place, in this work, is part of the process by which insiders are differentiated from outsiders – "us" from "them." Place is not simply endowed with meaning by people who are in it in a positive way. Place, here, is constructed through relations to an outside that is always simultaneously part of it.

"Here" is "not there." It was such thinking that informed my doctoral thesis, written in the late 1980s and the early 1990s and later published as *In Place/Out of Place: Geography, Ideology and Transgression* (Cresswell 1996). I was one of a group of geographers reading elements of poststructural theory (mostly Pierre Bourdieu and Michel Foucault) who wanted to reveal how places were constituted as normative places where it is possible for people, things, animals, and activities to be considered "in place" or "out of place" (Sibley 1981; Philo 1987; Valentine 1993; Philo 1995; Valentine 1996; Kitchin 1998). Importantly, all of these accounts depended on the recognition of how the outside of place was every bit as important as the inside – that in some ways the inside included the outside within it.

This was taken further by Doreen Massey in an important set of papers in the early 1990s where she introduced the notion of a "global" or "progressive" sense of place (Massey 1993, 1997). The notions of place in regional and humanistic geography she labeled "reactionary" – based on clear boundaries, deep roots, and singular identities that could be mapped onto them. Indeed, it is possible to see visions of place that are clearly bounded and deeply rooted in all kinds of right-wing, xenophobic propaganda which wants to map singular national and racial identities (British, White) on to clearly bounded bits of the world (Britain). Massey suggests instead that places are relational – that they are constituted as much by their "outside" as by their "inside." In a later essay, she links this relational turn to important work on identities:

> Thus, to take one very established but very different example: 'identity' has over recent decades been subject to dramatic reconceptualization. No longer is identity (on the broader canvas, 'entities') to be theorized as an internally coherent bounded discreteness. Rather it is conceptualized relationally – with implications both internal (in terms of fragmentation, hybridity, decentring) and external (in terms of the extension of connectivity). (Massey 2006: 37)

If identity is no longer discrete and coherent, it is not possible to map "identities" onto places in straightforward ways.

As an illustration of the relationality of place (and identity), she offers a walk down the street she lived near, Kilburn High Road in London. She reflects on the range of identities at play as she wanders down the street encountering a Muslim paper seller, passing a sari shop and an Irish pub with Irish Republican posters. She notes the planes flying out of Heathrow above and the fact that major routes out of London are just around the corner:

> Kilburn is a place for which I have a great affection; I have lived there many years. It certainly has 'a character of its own'. But it is possible to feel all this without subscribing to any of the static and defensive – and in that sense reactionary – notions of 'place' which were referred to above. First, while Kilburn may have a character of its own, it is absolutely not a seamless, coherent identity, a single sense of place which everyone shares. It could hardly be less so. People's routes through the place, their favourite haunts within it, the connections they make (physically, or by phone or post, or in memory and imagination) between here and the rest of the world vary enormously. If it is now recognized that people have multiple identities then the same point can be made in relation to places. Moreover, such multiple identities can either be a source of richness or a source of conflict, or both. (Massey 1993: 65)

This is a lively evocation of a place which is characterized by heterogeneity. This, she tells us, is a place made through its connections with the rest of the world. It is not a place defined by strict boundaries or any easy sense of a singular identity. Its histories are the histories of the connections (to Ireland, the empire, the Commonwealth, the global capitalist economy) that come together in a unique way in Kilburn. To Massey, this provides a model for thinking of place in general. All places are made in horizontal space by their connections, their role in networks that spread across the globe. The inside and outside are no longer easy to identify.

These observations extend even to the material landscape of rocks and mountains. In an evocative essay on the mountain of Skiddaw, in Britain's Lake District, she notes how these rocks were formed thousands of miles away and millions of years ago, south of the equator. One way in which people make claims to the permanence and essential qualities of place is through recourse to nature. This happens in bioregionalism, for instance. By recounting the mobilities of rocks, Massey undermines these arguments, pointing out that even the supposedly natural landscape is made up of trajectories – or relations between inside and outside:

> Maybe instead of, or as well as, the time-embeddedness that enables that relational achievement of the establishment of a (provisional) ground, such histories push a need to rethink our security. Certainly such histories have the potential to be read as removing the absoluteness of such grounding, so that all we are left with is our inter-dependence, a kind of suspended, constantly-being-made interdependence, human and beyond human. (Massey 2006: 43)

Massey's interventions have marked a transformation from thinking of place vertically – as rooted in time immemorial – to thinking of it horizontally – as produced relationally through its connections. This notion of place, rather than being introverted and reactionary, is extrovert and politically hopeful. As she notes above, it forces us to think in terms of interdependence rather than insular identities. It is a relational conception of place.

Another relational way of theorizing place is as an **assemblage** (Anderson and Mcfarlane 2011; Woods *et al.* 2021). The concept of an assemblage comes out of the work of the philosophers Gilles Deleuze and Felix Guattari and was elaborated by DeLanda (2006, 2016). An assemblage, in DeLanda's terms, is a whole that emerges from the interrelations between parts. The individual parts can be replaced but the whole remains. This is an excellent description of a place. Elsewhere, I use the example of my home as place to illustrate the nature of assemblage.

> A familiar example. My home is a place. It is made from red bricks, breeze blocks, terracotta tiles, window panes, copper wires, plastic outlets, wooden floorboards, cotton curtains, a stainless steel hob and oven, mortar and glue, all the things we eat, notes on the fridge. This list could go on for a long time. Together, the full contents of this list make my house. The way that they are assembled make my house different from a supermarket or a football stadium and even, in details at least, different from other houses in the street. It is an assemblage that is always changing too – the food in the fridge is rarely the same as the day before, cracks grow in the plaster, weeds push out from between paving stones in the back garden. I could take out elements of this assemblage. I have, for a while now, considered removing the floorboards and putting in some new ones. The assemblage would remain despite the change in parts. My house is a discrete thing – an assemblage –made by the relation between parts and the kinds of things we do with those parts. The place as assemblage is made by the way the parts are connected. It is more than just a random collection of parts – it is a distinctive assemblage that makes it what it is – my house. All places can be thought of in this way. (Cresswell 2014: 8)

DeLanda refers to two axes that are important to understanding how assemblages come into being – one axis links material dimensions of an assemblage with expressive dimensions – broadly speaking, things and meanings. The other axis links forces that broadly stabilize an assemblage – forces of territorialization – with forces that broadly destabilize an assemblage – forces of deterritorialization. For parts to coalesce into a whole that has some sense of permanence, there needs to be some process of coming together that DeLanda calls territorialization. Territorialization is necessary for a distinct whole (place) to form out its material and expressive parts. As DeLanda puts it: "the concept of territorialization plays a synthetic role, since it is in part through the more or less permanent articulations produced by this process that a whole emerges from its parts and maintains its identity once it has emerged" (DeLanda 2006: 14). Importantly for DeLanda, there are always processes of deterritorialization seeking to pull an assemblage apart.

> One and the same assemblage can have components working to stabilize its identity as well as components forcing it to change or even transforming it into a different assemblage. In fact, one and the same component may participate in both processes by exercising different sets of capacities. (DeLanda 2006: 12)

We can think, for instance, of all the forces of entropy at work in my home that require constant maintenance to ensure that the house remains an effective assemblage. At a larger scale, we can think of socio-economic forces at work that constantly undo places. Forces of gentrification or de-industrialization for instance.

> So, in the first place, processes of territorialization are processes that define or sharpen the spatial boundaries of actual territories. Territorialization, on the other hand, also refers to non-spatial processes which increase the internal homogeneity of a neighbourhood. Any process which either destabilizes spatial boundaries or increases internal heterogeneity is considered deterritorializing. (DeLanda 2006: 13)

Assemblage theory draws our attention to how places are gatherings of both material things and expressive (meaningful, symbolic) elements. In addition, it underlines the degree to which places are in process, becoming and potentially dissolving on a daily basis.

In *Maxwell Street: Writing and Thinking Place* (Cresswell 2019), I used assemblage theory to account for the coming together and falling apart of North America's largest open air market on and around Maxwell Street in Chicago. In addition to DeLanda's material and expressive elements, I also traced how practices – the things we do in combination with the other elements of place – accumulated and dispersed over the twentieth century. Markets are, in many ways, ideal kinds of places to think about assemblage with as they exuberantly bring things (food, clothes, smells, sounds) together alongside stories and symbols and the practices of buying, selling, and performing. They are exaggerated kinds of places. Jon Anderson uses assemblage theory to consider a very different kind of place – the "surfed wave" (Anderson 2012). Drawing on interviews with surfers, he shows how an assemblage of wave, surfer, and surfboard becomes a momentary kind of place that briefly cohere before dispersing.

> As with any assemblage it is how each component part relates, connects, and interacts that forms a particular territorialised or watery place. In one proximate location on the sea a number of different places may occur one after the other: it may first be a place of flat calm then a place of a wave, a place of a surfed barrel, a place of a surfed wipeout, or a place of a ducked-under wave. Each assemblage will produce a different relational agency, risk, and experience before the constituent parts disengage and dismantle. This perspective emphasises the changing nature of place in one location: due to the different combination of components in every instance place is always disassembling and reassembling. (Anderson 2012: 580)

Geographers have used assemblage theory in several ways including using it to rethink the Heideggerian notion of dwelling in the city where dwelling aligns with acts of assembling material in unequal ways. Colin Macfarlane, writing about acts of dwelling and assembling in Sao Paolo and Mumbai argues that "a conception of assemblage can serve to expose which groups and ideologies have the greater capacity to render urbanism in particular ways over others, and therefore offers a ground for thinking how the city might be assembled differently" (Mcfarlane 2011: 667). To Macfarlane, it is important to ask who and what has the capacity to bring assemblages into being as this impacts what kinds of urbanisms are allowed to emerge. Waves and cities are very different kinds of places, but both have proven susceptible to assemblage thinking which helps us to show how places emerge and fade, and how places are related to the worlds around them through the processes of gathering and assembling.

Thinking of the world in terms of places means distinguishing one thing from another. It is way of naming heterogeneity and delineating spatial difference. The world is made up of heterogeneous places. And yet, places are not internally homogeneous. A place is made up of gathered things, meanings and practices that are different from one another. The unique assemblage of differences make the place what it is and differentiates it from other unique assemblage-places. (Cresswell 2019: 175)

The End of Scale?

Relational thinking has also been at the heart of recent thinking on *scale*. Think about the ways in which notions of scale enter into discussions of events. How is something on, say, the global scale interpreted? What about the local scale? Generally, when politicians or journalists report on something being global in scale, they are saying that it is important and likely to have far-reaching consequences. If they say it is local or localized, they are saying it is less important. Here there is a relationship between spatial scope and level of importance in a hierarchy. This is also the way geographers (and others) have most often used scale in explanations (often without much thought). A local thing is particular while a global thing is, well, global! Global forces are often structural forces beyond our control. Politicians and business leaders often talk about globalization as a force that cannot be resisted. It is just something that is happening and we had better get used to it. This way of thinking about scale sees scales as "a nested hierarchy of differentially sized and bounded spaces" (Marston *et al.* 2005: 416):

> ... a 'vertical' differentiation in which social relations are embedded within a hierarchical scaffolding of nested territorial units stretching from the global, the supranational, and the national downwards to the regional, the metropolitan, the urban, the local, and the body. (Brenner 2004: 9)

The global, as a scale, always appears to do more work than, and be more important than, the scales beneath it, despite the fact that anything that happens globally literally has to take place locally. Nothing happens globally without lots of things happening locally. This is analogous to the view of structurationists such as Giddens and Bourdieu that there can be no structure without people doing things – without agency.

In an important and controversial paper, Sallie Marston, John Paul Jones III, and Keith Woodward propose the notion of a human geography without scale. They trace a number of attempts to stop taking scale for granted and to think of scale as a social construct. Political geographer Peter Taylor, for instance, had argued for a three-part definition of a politics of scale that defines the local or urban scale as the scale of experience (our bodies interact with the world in our immediate environment), the national scale as the scale of ideology (it is at the national scale that politics is argued out), and the global scale as the scale of "reality":

> But what exactly do we mean by 'reality' in this context? It is obviously provocative, since all geographers ultimately claim to be studying the 'real world'. In our usage of the term we are not emphasizing our empirical credentials but rather the notion that this is the scale that 'really matters'. (Taylor 1982: 26)

Key to this definition of the global scale as the scale of reality is the idea that the global political economy is where small-scale events can be looked at as a whole. This is more or less the same as Marx's objective economic base – the real conditions – upon which the variety of everyday life is built. It also calls to mind the idea of the global as a space above the local – a point from which it is possible to look down on the world and see it objectively. This God-trick, of course, is exactly the perspective that geographers since the 1970s have been seeking to overturn. It maps on to what Dixon and Jones earlier called the "grid epistemology" – a spatialized way of conceiving of an always impossible objectivity (Dixon and Jones 1998). Think about it for a second. Where do you think "reality" is? In a world political economy or right outside your window? In Taylor's account, the level of the nation-state provides an ideological service that masks the reality of the global economy from the experiential level of the local and urban. For the theorist there is a definite sense that the global scale is simply the biggest and the best site for studying the world objectively. The global scale, it is argued, is the scale at which we can find the best explanations. As Marston *et al.* point out, there is a conflation of size (horizontally) with level (vertically) which provides an ordering frame of ways of knowing the world.

Others have attempted to save something of scale by mixing it with the newer notion of *networks*:

> Whereas the spatiality of a politics of scale is associated with vertical relations among nested territorially defined political entities, by contrast, networks span space rather than covering it, transgressing the boundaries that separate and define these political entities. (Leitner quoted in Marston *et al.* 2005: 417)

Here Helga Leitner is combining a sense of scale as hierarchy with a spatial imaginary of networks that connect points that may exist at any number of scales. It is like imagining a continuous plain crisscrossed by lines of connection. Marston *et al.* reject this too. What they are after is a complete doing away with scale talk. They propose, instead, a **flat ontology**:

> Instead, we propose an alternative that does not rely on any transcendent predetermination – whether the local-to-global continuum in vertical thought or the origin-to-edge imaginary in horizontal thought. In a flat (as opposed to horizontal) ontology, we discard the centering essentialism that infuses not only the up–down vertical imaginary but also the radiating (out from here) spatiality of horizontality. (Marston *et al.* 2005: 422)

One of the benefits of thinking about scale is that most of us have some idea what it means. When we use the words "local" and "global" or "micro" and "macro," we have some commonsense notion of scale which we use to understand them. Getting rid of common sense is a tricky task. So how do Marston and her colleagues go about it? One way forward would be to embrace an ontology and epistemology that focused on movement and flow rather than scale and hierarchy. "According to this approach," they write, "the material world is subsumed under the concepts of movement and mobility, replacing old notions of fixity and categorization with absolute deterritorialization and openness" (Marston *et al.* 2005: 423). It is certainly the case that many writers have been grappling with the issue of mobility both as a fact of the twenty-first-century world and as a theoretical tool for understanding that world (Cresswell 2006a; Sheller and Urry 2006). Marston *et al.*, however, are not impressed:

> We take issue, however, with [this] reductive visualization of the world as simply awash in fluidities, ignoring the large variety of blockages, coagulations and assemblages (everything from material objects to doings and sayings) that congeal in space and social life. It remains difficult to discern what, if anything, takes the place of these negated objects other than the meta-spatial categories that flow thinking was meant to dissolve. Thus the tendency for global, typological categories – here the 'world city' and 'globalization' – to slip in through the back door: concepts placed under erasure that nevertheless found and ground the flows that supposedly make them meaningless. (Marston *et al.* 2005: 423)

Having dispensed with the alternatives to scale, Marston *et al.* go about inventing a spatial ontology that fixes on social sites – "localized and non-localized event-relations productive of event-spaces that avoid the predetermination of hierarchies or boundlessness" (Marston *et al.* 2005: 425).

In some sense, this way of thinking is not unlike the saying that is always popular on long car journeys (when I am driving) – "wherever you are, there you are." It is an approach that insists on being exactly where we are and refusing to immediately jump up or outwards to seek an explanation. The world is made of an infinite series of sites that may or may not be related. These sites are the products of things and events that happen within them and in other places (that are equally just sites). There is no "global" to look to, just sites and heir relations. For Marston *et al.* this entails a particular politics that refuses explanations that locate mysterious forces that are always above and beyond the here and now:

> In this fashion 'the global' and its discursive derivatives can underwrite situations in which victims of outsourcing have no one to blame, a situation possibly worse than blaming oneself. The same macro-mystification is discursively available for managers, who when submitting to interviews about outsourcing, are likewise eager to appropriate 'globalization' in relieving them and their corporation of social responsibility. We do not deny that the contexts for these sorts of corporate decisions are not spatially extensive – indeed, the social sites of boardrooms depend upon a vast distribution of resonating social sites, all 'diversely invested' in practices and orders, employees and ledgers. But the imaginary transposition from boardroom to global corporation obscures those sites of ordering practices, as well as the possibilities for undoing them. (Marston *et al.* 2005: 427).

Marston *et al.*'s paper upset quite a lot of people. Alongside positive responses, a whole series of responses defended scale in one way or another (Collinge 2006; Jonas 2006; Leitner and Miller 2007). The argument against scale, however, was just one part of a larger theoretical field of relational ontologies. On occasion these can be quite contradictory. It is worth comparing, for instance, Marston *et al.*'s insistence (above) on the ability of a flat ontology to allocate blame with the work of political theorist Jane Bennett in her book *Vibrant Matter*. She uses a massive power outage in the North East of the United States in 2003 to consider what she calls "distributed agency":

> … the electrical grid is better understood as a volatile mix of coal, sweat, electromagnetic fields, computer programs, electron streams, profit motives, heat, lifestyles, nuclear fuel, plastic, fantasies of mastery, static, legislation, water, economic theory, wire and wood – to name just some of the actants. (Bennett 2010: 25)

One question that arises, when such a complicated arrangement of things breaks down, is who (or what) is to blame? The answer is very complicated and, like agency, distributed. If the capacity to act (in this case, produce and distribute electricity) is distributed across actors in a network and links between them then so is the blame that follows breakdown:

> The notion of a confederate agency does attenuate the blame game, but it does not thereby abandon the project of identifying (what Arendt called) the sources of harmful effects. To the contrary, such a notion broadens the range of places to look for sources. Look to longterm strings of events: to selfish intentions, to energy policy offering lucrative opportunities for energy trading while generating a tragedy of the commons, and to a psychic resistance to acknowledging a link between American energy use, American imperialism, and anti-Americanism; but look also to a stubborn directionality of a high consumption social infrastructure, to unstable electron flows, to conative wildfires, to exurban housing pressures, and then to the assemblages they form. In each item on the list, humans and their intentions participate, but they are not the sole or always the most profound actant in the assemblage. (Bennett 2010: 37)

Marston *et al.* and Bennett are arguing for very similar ontologies. While Marston *et al.* suggest that a flat ontology will give victims of outsourcing someone to blame (unlike a scalar leap to blaming "globalization"), Bennett argues that a similar ontology will distribute blame to the point where there is no firm basis for saying that corporate greed is any more worthy of blame than some electrons in a wire or someone turning the lights on in Brooklyn. In this sense, and Bennett recognizes this, a distributed sense of agency can be politically immobilizing. It is much easier to have a clear and unequivocal target to blame when things fail.

The notion of a "flat ontology" proved particularly troublesome to political geographers who had been developing critical notions of the territory and region in their work (Jones and MacLeod 2004; Jones 2009). The work of Martin Jones is particularly instructive. He does not completely contradict relational ways of thinking about space and scale but, instead, insists on the continuing salience of territory and region. He sees the relational argument as setting up a straw-person argument of territory and region as static concepts that need to be replaced. In fact, he suggests, political actors continually assume space to be formed territorially and act as though it is. These territories, he argues, are being made and remade all the time through political, economic, and cultural practices. Similarly, networks, the favored spatial form of many relational thinkers, are not always fluid and dynamic but have their own fairly static components or nodes, routes, and moorings. Another question raised by Jones and others relates to the often poorly specified nature of the "relations" that make up space or place. What are these relations? What is it that is being related? How come some things (places, etc.) are rich in relations while others are relationally impoverished? Who gets to make things relate? Is anything nonrelational? Jones suggests that there may even be structural reasons why some things enter into relations and others do not. At the core of Jones's argument is the necessity of taking temporality into account when thinking through the interrelations between relational space and forms of territory and region. He uses the notion of "*phase space*" which he borrows from physics. Phase space describes the way one set of spatial possibilities leads (or does not lead) to a new arrangement of space at some future point. Space, he argues, is "sticky" and things can get stuck or anchored in it. This shapes the unfolding of space and society over time through a familiar interaction of structure (institutions, imperatives, etc.) and agency (individual acts):

> Sociospatial relations … are deeply processual and practical outcomes of strategic initiatives undertaken by a wide range of forces produced neither through structural determinism nor through a spontaneous voluntarism, but through a mutually transformative evolution of *inherited spatial structures and emergent spatial strategies* within an actively differentiated, continually evolving grid of institutions, territories and regulatory activities (…). In short, constructed and always emergent space matters in shaping future trajectories. (Jones 2009: 497–498)

Phase space describes the observation that present spatial arrangements contain within them a multitude of possibilities for futures – different versions of what happens next. It also reintroduces a space of boundedness and territory in a terrain which relational theorists are attempting to dissolve completely:

> … it involves conceding that there *may* be certain circumstances in which, as an object of analysis, practical and bounded spaces that been institutionalized through particular struggles and become identified as discrete territories in the spheres of economics, politics and culture, matter. (Jones 2009: 501)

Nonrepresentational Theory

Perhaps the most recent and contentious version of relational thinking is **nonrepresentational theory** (Thrift 2008; Anderson and Harrison 2010b). Relational geographies in general have sought to move as far away as possible from a view of the world as divided into neatly contained segments formed through deeply structural forces and practices. They are informed by geographical imaginations of connectedness, flow, and networks rather than rigid architectures.

Nonrepresentational theory (NRT) continues in this direction. The term "nonrepresentational theory" is a little bit of a misnomer. It certainly isn't *a* theory. It really describes an attitude to the world or even, as Nigel Thrift has suggested, a *style*. It is also clearly the case that work lumped together under the heading of NRT includes people who disagree with each other about emphasis and substance. Having said that, NRT does name a number of shared approaches to the world. Perhaps most important is a desire to think of the world as *lively* and in a state of *becoming*. Previous approaches, NRT writers suggest, have tended to deaden the world – to make it appear already done. In the sense the world is presented as complete; it has little space for creative moments. It is these creative moments that enliven living in the world and, NRT insists, they are everywhere. One of the things that deadens the world is representation, or rather the emphasis placed on representation by cultural geographers.

NRT is both a product of poststructuralism (Doel 2010) and a critique of some elements of it. It is poststructuralist in its embrace of relationality and its refusal to countenance structural determination. It is critical, however, of some of the key elements of poststructuralism as understood by an earlier generation of poststructuralist geography focused on what we might call the "textiness" of the world. In Chapter 10, we saw, for instance, how Foucault's notion of discourse involves the construction of "truths" discursively. Elsewhere, Derrida has suggested that "there is nothing outside the text" (Derrida 1976: 158), by which he may have meant that there is nothing that is not text – there is no prior thing (often called "reality") that we can rely on to interpret texts that is not already interpreted through text. Context, in other words, is a process of endless (con)textualization. Text always gets in the way of our access to the world. This is a classic poststructuralist statement in its denial of something deeper that we can rely on for explanation. Practitioners of NRT would probably have no quarrel with that. What would upset them, however, is the use of the word "text." They may have similar problems with Foucault's similar construction of "discourse." Both "text" and "discourse" refer to more than our everyday uses of the terms – they do not refer to just words or writing. Nevertheless, the root of both of these terms lies in articulated thought in spoken or written form. They are terms which underline the significance of a particular, and fairly intellectual, way of thinking about the world in terms of words. The idea that the world is discursive or textual has been central to the work of many poststructural geographers (particularly in social and cultural geography) who have insisted that the most important aspects of the world do not have "essences" but are socially constructed. In his paper "Afterwords," Nigel Thrift reflects on the death of his father:

> … my father was a good man who did a lot of good; more than most, I suspect. Almost nothing that he ever did was written down and whereas I once would have seen this as a problem I now think that putting his life in order through text, in order to rescue him from the enormous condescension of posterity, may, in certain senses, be just another form of condescension. I am not sure, in other words, that he needs writing down, or, put in another way, we need a form of writing that can disclose and value his legacy – the somatic currency of body stances he passed on, the small sayings and large generosities, and, in general, his stance to the world – in such a way as to make it less important for him to be written. (Thrift 2000: 213)

Here, and elsewhere, Thrift calls for a nonrepresentational style that looks to other places than the written word (or other forms of representation) for the ongoing making of the world. The world, Thrift insists, is lively and ongoing and (potentially) full of surprise. It is a kind of intellectual and elitist conceit to value things, or think things, through the idea of things like "text" and "discourse":

> In previous books and papers (…) I have pointed to the uses of an alternative 'nonrepresentational' style of work. Note that I use the word 'style' deliberately: this is not a new theoretical edifice that is being constructed, but a means of valuing and working with everyday practical activities as they occur. It follows that this style of work is both anti-cognitivist and, by extension, anti-elitist since it is trying to

counter the still-prevalent tendency to consider life from the point of view of individual agents who generate action by instead weaving a poetic of the common practices and skills which produce people, selves, and worlds. (Thrift 2000: 215–216)

NRT (despite Thrift's claims we are still very much in the world of "theory" here) resists the idea that worlds are made textually and looks instead to moments of creativity and surprise in the way the world is performed. It is not how we think or write that matter, but what we do. This is positioned against (and deriving from) a world in which signs, symbols, and text signify and structure the "world":

> The world and its meanings; this divide is the cost. On one side, over there, the world, the really real, all 'things coarse and subtle', and on the other, in here, the really made-up, the representations and signs which give meaning and value. It's a classic Cartesian divide. Once established there can be no sense of how meanings and values may emerge from practices and events in the world, no sense of the ontogenesis of sense, no sense of how real the really made-up can be. (Anderson and Harrison 2010a: 6)

Signs, texts, and representations are pushed gently aside by NRT and other theoretical constructs come to the fore. One central concept is the **event** – "the fleeting contexts and predicaments which produce potential" (Thrift 2000: 214). Many theories of how the world works pay very little attention to such contexts and predicaments (though humanistic geography gave it a go). Thrift recognizes that the world is not just a series of ongoing surprises and that events have constraints: "Events must take place within networks of power which have been constructed precisely in order to ensure iterability. But what is being claimed is that the event does not end with these bare facts" (Thrift 2000: 217). There is a clear attempt here to escape the clutches of systematic forces which predetermine, mediate, or cause the small moments of life.

The NRT focus on the idea of the *event* attempts to undo a sense of finishedness in the world and refocus our attention on a constant state of *becoming*. This is perhaps most evident, in everyday life, when we are surprised by something that is unpredictable. But the idea of the *event* is at its strongest when it is applied to those very things that *do* appear to be "finished." Often, for instance, the material structure of urban space is presented as a set of structural restraints and determinations. This is how de Certeau presents urban space in his account of the surprising creativity of walking, for instance. In his account, it is the tactical creativity of walking that is full of life while the space of the architecture around the walker is presented simply as an attempt at structuring the walker's actions. But what if we think of the architecture as an event? This is how Jane Jacobs presents the geography of tower blocks (or any architectural space):

> … the materiality of the building is a relational effect, its 'thing-ness' is an achievement of a diverse network of associates and associations. It is what we might think of as a *building event* rather than simply a building. Conceived of in this way, a building is always being 'made' or 'unmade', always doing the work of holding together or pulling apart. (Jacobs 2006: 11)

Perhaps this is best illustrated by the fate of the Pruitt-Igoe estate in St. Louis discussed in Chapter 9. Jacobs focuses on another spectacular failure – the destruction of Ronan Point in London, a 22-story tower-block that collapsed two months after it was "completed" in 1968. Jacobs traces the very big event of the building collapsing as well as a series of small events such as the faulty fitting of a gas cooker as a favor for a friend in Flat 90. The faulty work led to a gas explosion that ruptured a load-bearing wall that finally led to the destruction of the whole building. Jacobs shows how a number of events piled up to produce the big event – the explosion. Integral to the whole process was the weakness in the structure of the building that carried with it the potential for collapse. Ronan Point, a large tower-block, can thus be seen as a layering of events that resulted in its surprising and excessive downfall. To return to de Certeau, it is clear here that the architecture of the city is far from a rigid structure but potentially every bit as surprising as the moves of walkers in the city. What we are talking about is not a rigid stasis set against fluid and cunning flow but rather an elaborate set of variable potentialities happening at variable speeds – a point made by J. D. Dewsbury:

> Take, for example, the building you walk through/within – what is the speed of flux that is keeping it assembled? It seems permanent, less ephemeral than you, but it is ephemeral nonetheless: whilst you are there it is falling down, it is just happening very slowly (hopefully). (Dewsbury 2000: 487)

Thinking in this way also forces us to reconsider what is normally meant by the term "materiality." This term is often used to stand in for a sense of things which are solid. The word "concrete" is often evoked to signify what is meant by "material." Material is then used to contrast things with a "non-material" world that seems more elusive and lacking in substance. Alan Latham and Derek McCormack refute this distinction and instead, in a hypermaterialist way, insist on a pervasive materiality which includes such things as sound and light, and, at the same time, point towards the "excessive potential of the immaterial":

> This is not least because concrete itself, or indeed any other building material, is not 'brute matter'. It is a particular aggregate organization of matter and energy. It is no more (or less) 'real' than apparently 'immaterial' phenomena like emotion, mood and affect, although it has a different duration and threshold of consistency.... Thus, to argue for the importance of materiality is in fact an argument for apprehending different relations and durations of movement, speed and slowness rather than simply a greater consideration of objects. (Latham and McCormack 2004: 705)

To NRT geographers, then, the world is full of sometimes surprising eventfulness. Whatever happens always has the potential to be otherwise and is thus radically underdetermined.

Another key concept in NRT is **affect** (McCormack 2003; Thrift 2004; Anderson 2006). It is not easy to say exactly what the term *affect* refers to. It is often lumped in with more familiar ideas such as emotion and feeling – human capacities that have also been dealt with by humanists, feminists, and geographers engaging with psychoanalysis (Tuan 1974; Davidson *et al.* 2005; Bondi 2008). Ben Anderson uses affect to get away from views of emotions that insist on them being entirely *personal* or, on the other hand, *socially constructed*. Affect, he argues, is a product of relations between things (Anderson 2006). One body (human or otherwise) acts on other bodies. Something passes between them, and this is the moment of affect. In this way "the world is made up of billions of happy or unhappy encounters, encounters which describe a 'mindful connected physicalism' consisting of multitudinous paths which intersect" (Thrift 1999: 302).

Affect differs from emotion in that affect comes in some sense before emotion. Emotions, in this rendering, are ways in which individual bodies makes sense of affect by giving it a meaning that is partly socially and culturally determined. Examples of affect include things like a shiver down the spine that results from a particular encounter – or perhaps a sense of revulsion as it is experienced by a body. These then get translated into an emotion so that they might make sense as fear or disgust. Affect is both relational – as it emerges from encounters – and pre-representational. Derek McCormack illustrates the difference between affect and emotion using the example of dance:

> [T]hink about an occasion when you have entered a dance venue of any kind – it could have been a drum and bass club or an afternoon tea dance. Regardless of what kind of dancing might have been taking place, the affective quality of the space in which bodies move is never only something personal – it is a product of a complex mix between music (although music is not necessary for dance), light, sound, bodies, gesture, and, in some cases, psychoactive substances of various kinds (…). What is clear is

that this affective intensity is felt – you can feel it in your gut (whether you like it or not is a different matter) – and that this felt sense can be modulated by changes in the level of those factors listed above. The extent to which this felt sense is an emotional one depends on the degree to which it can be articulated: so if an interviewer asked you how you felt after your dance, you might articulate this feeling through identifying a specific emotion – 'I feel happy'. And this designation would make sense because we have a collective – if vague – understanding of what it means to feel an emotion such as happiness. (McCormack 2008: 1827–1828)

The event and affect are two key terms at the heart of NRT. These, combined with other ideas, insist on: the fluidity of things – on the practical and processual (posited as opposed to the finished and fixed); on the production of meaning in action (rather than through pre-established systems and structures); on an ontology that is relational (rather than essentialist); on habitual interaction with the world (rather than "consciousness" of it); on the possibilities of things emerging surprisingly (rather than being predetermined); on a wide definition of life as humans/with/plus (rather than strictly humanistic); and on all-inclusive materiality where everything produces the "social" constantly (rather than an already achieved "social" constructing everything else).

So how has NRT been mobilized in the empirical realm? Here I want to focus on just two of a myriad of applications for NRT. The first focuses on a broad realm of creative potentials and the second on the dark side of affective engineering.

In an essay on dance, Thrift argued that there is something about dance that is beyond representation or even meaning. Dance is playful, he argued, and in this playfulness there is something productive (Thrift 1997). Thrift points to what he considers an absence of work on dance in the social sciences and suggests that this is because dance is a form of *play* and that play is exactly that which is not taken seriously owing to its gratuitous nature and its lack of significance. Play does not signify anything, Thrift argues, except itself. The rules of play are internal to the game and not imposed on it. Thrift asked geographers to take the gratuitous as-ifness of dance as play seriously.

In the years following Thrift's essay, dance became the center of a small industry of geographers weighing both the nonrepresentational and representational geographies of dance (Latham 1999; Nash 2000; McCormack 2003; Revill 2004; Cresswell 2006b; Somdahl-Sands 2006). In a review essay, McCormack places this literature in a tradition which includes the phenomenological work of David Seamon and Hägerstrand's time-geography. The politics of movements of dancing bodies, he suggests, are not known or given in advance; there is always something in excess of what can be explained through the contexts of dance and it is in that excess that there is the capacity for surprising and creative moments. McCormack considers the geographies of the tango. Tango, he admits, is heavily implicated in imagined geographies of Argentina and, further, is quite clearly a highly gendered set of practices with men leading women through a set of prescribed moves. And yet, he argues, there is always something else that is present each time the tango is danced that cannot be explained away through recourse to representation:

This is not to say that all ways of dancing are similarly inventive. Some are more heavily policed than others, and this often depends on the cultural context within which dance is enacted. Nor is it to say that inventiveness is the opposite of cultural context; on the contrary, the inventiveness of dancing bodies is facilitated by the generative constraint of the cultural contexts in which they move. The key point is that to dance is not necessarily to unthinkingly reproduce a given cultural identity: it is also a matter of actively reworking, albeit on a microscale, the tangible corporeality of this identity. (McCormack 2008: 1827)

Dance is one of the ways in which space and action are endowed with expressive and creative power in NRT. It is used to point toward the moments of life that are free from structural determination and are filled with joy and the possibility of surprising excess. Alan Latham captures one such moment in striking prose:

> East Berlin, November 1993. A shaft of light cuts through the winter darkness, reaching heavenwards. A film of snow stretches across an open space, gorgeously white and luminous. The corpse of an old Wartburg automobile and other pieces of trash are dusted around. Behind us a ruin hangs gently over the space, the holes in its side melting its solidity.... I am curious, delighted, and a little awed. This is some special place. Softly disorientating, an industrial fairy-scene. Out at the back, over and past a ragged corrugated iron fence, the lights of the nearby hotels and stores fuzz into the night. The snow crunches solidly under my boots as I wander about this lovely place. We dance, my girlfriend and I, upon the crude wooden stage that sits flanked by two weirdly ornamented metal columns (one claims to be a ticket office, although there doesn't appear to be anything for sale), under the shadows of the ruin. Behind the stage is a stripped down, whacked-out van, squatting mouth agape in the snow like an exhausted beast. A smaller creature sits by its side. With wide angular eyes, it looks absently back at me. Everything comes from something else. Everything was something else. A little further away, the skeleton of an old bus, its hinterparts buried deep in the soil, as if under its weight the earth beneath had suddenly turned to quicksand, rears up in a struggle to free itself. We circle around the stricken bus, admire its fossilized grandeur. Then we wander towards the bleary lights of the city in front of us, infinitely touched by the wonder of the space we have just experienced. (Latham 1999: 161)

Latham contrasts this experience of delight and dancing with writing about the city that has described it as increasingly dead space that fails to provide stimulation. Latham, through beautiful evocation of very personal and joyful moments, seeks to explore urban moments where spaces have the capacity to induce everyday forms of surprise and wonder. Such an approach to the city could be accused of a naïve romanticism. It is a view that skirts over the deeply inbuilt rationality of the city and the structural forces that produce exclusion and inequality. It has little to say about the increasing levels of discipline and surveillance that are at play in modern urban space. And yet who is to say that these dark forces are more important or more fundamental than those moments when space comes alive?

In a cogent critique of early forays into NRT, Catherine Nash takes this fixation on dance to task for reinscribing a false dichotomy between thought and represented worlds on the one hand and preconscious, bodily, performed worlds on the other (Nash 2000). Following the work of a string of dance theorists, she suggests that this turn to dance is just the latest attempt to enact a Romantic return to some pre-cognitive state:

> ... desiring to return to an unmediated, authentic relationship to the world, to be like "primitive" others who are unburdened by thought – dancing women, the "exotic people," rural peasants, to reject the modern in favour of the "primitive." The appeal of my dreams of doing other kinds of things is their offer of a kind of transcendent loss of self and a solitary escape from the social in the rhythm of dancing or digging. The asocial implications of this idea of noncognitive embodied practice are somewhat at odds with the deeply social character of coded performances of identity in theories of performativity. But the separation between thought and action is problematic in other ways also. Dance, like music, is a cultural form and practice especially susceptible to essentialist readings of "natural" rhythm and instinctive aptitude often ascribed to racialized people whose "supposed genetic propensity for rhythmic movement rests on an implied division between moving and thinking, mind and body" (Desmond, 1997b: 41). The image of the dancer dancing to a world elsewhere and beyond the reach of words and power does not easily provide a model for effective political strategy nor a useful cultural politics. This is especially the case when the more abstract discussions of dance or performativity lose the sense of the ways in which different material bodies are expected to do gender, class, race or ethnicity differently. Moving "quickly past arms, legs, torso and head on their way to a theoretical agenda that requires something unknowable or unknown as an initial premise," the "body remains mysterious and ephemeral, a convenient receptacle for their new theoretical positions" (Foster, quoted in Wolff 1995: 81). (Nash 2000: 657)

Nash's critique of NRT is not a critique of its interest in dance, but it is a critique of the way in which NRT engages dance. Nash is arguing for continued attention to the importance of representational contexts in which dance happens while simultaneously reminding us that NRT is itself part of a representational history that has often sought to locate an embodied space beyond representation where a certain kind of freedom can be placed. This history, she suggests, is one which is based on a split between the embodied and the representational that cannot be easily sustained. In the end this may just be a matter of emphasis. Few NRT geographers completely reject the influence of representation; they are simply arguing that the emphasis on representation has been pervasive and that it is time to recognize the world beyond (or before) representation. Similarly, there are few geographers who would not admit that there *are* some things that are nonrepresentational. They might, however, be skeptical about their relative importance.

The emphasis on creative and excessive moments (such as those in dance) in NRT may lead us to think of NRT as a rose-tinted kind of style that is always looking for evidence of the joyful and creative. Often, however, NRT has been combined with other theoretical imperatives in sometimes unlikely ways in order to elucidate what we might call "dark affect." Work from within NRT has, for instance, engaged with animal slaughter, the Holocaust, and torture (Anderson 2010; Roe 2010). One particularly fruitful realm has been work on architecture and security. Peter Adey has combined an interest in the nature of *affect* with more Foucauldian concepts such as *discipline* and *biopower* to consider how security is changing in the twenty-first-century airport and beyond (Adey 2008, 2009). Adey accounts for how affect is engineered and prefigured in the architecture of airports as well as in the security procedures that are increasingly deployed in these spaces. He is interested in how seemingly personal realms of affect and emotion are part of the process of controlling and regulating people in airport space, examining how:

> affective expressions of hope, fear, joy, sadness, and many others, as well as the constitutive mundane bodily motions that occupy the airport terminal, may not be as distanced from power and control as we might think. In fact, they are central to their perpetuation as certain triggers – designed-into the terminal space – are intended to excite bodily and emotional dispositions at an unconscious and pre-cognitive register. (Adey 2008: 439)

Adey conducted ethnographic fieldwork at a western international airport, including an extensive set of interviews in order to figure out how airports work. In this work, he combined interests in architecture, mobility, security, and affect. One purpose was to discover how "affectual cues such as texture, feel, lighting, are designed-into spaces to create ethological capacities and potentialities of affectual expression" (Adey 2008: 441). This involved paying attention to the details of airport construction and design:

> At a particular airport, the materials they used for the flooring are intended to give a sense of grounding to people, helping diffuse the anxiety of flight. The architect stated that they wanted: "Good quality finishes, and natural materials […]. The flooring looks nice, it's limestone, but it also has that feeling – solid" (Interview, Airport Architect, 2003). Approximately 10,000 limestone tiles were imported from Italy for this purpose. The floor was highly reflective, bouncing the natural light coming in from the glass facade. It was hoped that the feelings such a floor invoked would stimulate movement forward through the terminal. In fact, the airport believed the flooring would create a "yellow brick road syndrome" that would pull the passengers past the retail environment. (Adey 2008: 446)

But it is not just the architecture of the airport that engineers affect. Adey traces how forms of identification in airports (including biometric procedures such as finger-printing and iris scanning) are based on establishing who you are while new forms of profiling and surveillance focus on what you might become – what you might do. Security forces can now focus on tell-tale give-away signs of anxious bodies through the way that they walk, or through involuntary facial gestures:

> ... behavioural profiling enables a vision of the body-becoming, and, in this context, a body becoming-threatening. They are imminently threatening subjects rather than threatening subjects now, for that is what other forms of security and screening detect – the immediate threat – scrutinised presently and proximately through x-ray screening, metal detectors, and the body pat down. Behavioural monitoring examines a trace of an emotion registered in anticipation of a hostile act or intention to act with hostility. Thus, it acts as a clue to what that body-subject might turn into. (Adey 2009: 286)

Security is moving toward anticipatory actions – trying to predict what will happen next through minute gestural clues that have their origins in the world of affect. This is one of the reasons that Thrift has argued that geographers need to engage with the lively presence of the world – because it is already being engaged with by darker forces seeking to anticipate and then direct our every move (Thrift 2011). And these forces are operating well beyond airport terminals:

> Advances in facial recognition and the detection of honest signals make it increasingly easy to read bodily responses – from the face or more general body movement – in real time. The result is that an old ambition, dating back to the ancient Greek, Physionymas, to be able to read the signs of a person's nature, to read a person's temperament and inclinations from external appearance, is able to begin to be realized en masse as the imperceptible becomes capable of being measured. A new doctrine of bodily signs can come into being which makes its way into official certifications of who we are embedded in software which literally recognizes us. (Thrift 2011: 10)

Thrift describes a world (our world) which is increasingly engineered by an entertainment–military complex which seeks to prefigure our movement and emotions in precise but flexible and creative ways. It is a world in which software and hardware and "wetware" (bodies) are thoroughly integrated. Thrift's suggestion is for us social scientists to be equally experimental in our methodologies, epistemologies, and ways of representing the world.

NRT has been quite controversial because it pokes at the pressure points of a large number of theoretical and political convictions. It has provoked a number of spirited critiques, particularly by feminist geographers but also by others who retain some investment in notions of broadly structural and asymmetrical arrangements of power (Nash 2000; Thein 2005; Tolia-Kelly 2006; Cresswell 2012).

Conclusions

At the heart of this chapter has been the embrace of the idea of "relationality" by a diverse array of broadly poststructuralist geographers. Indeed, to Murdoch, poststructuralist geography *is* relational geography (Murdoch 2006). For something to be relational, it has to be a product of its connections rather than a product of some essential self. Relational thinking is, therefore, anti-essentialist thinking. As connections (relations) change, so those things which are in relation change. We have seen how relational thinking changes the ways we might think about the key geographical concepts of space, place, and scale in this chapter. We have also seen how geographers engaging with (and leading in) NRT have produced a new kind of geography which emphasizes how "events" and "affect" are created relationally both in the production of hopeful moments of creativity and in a dark side of surveillance and affective engineering. These geographers almost certainly disagree on many things. Some here would probably be more comfortable with some level of structural explanation being implicated in accounting for exactly how relations are constituted, for instance. Many others remain unconvinced, insisting on the continued saliency of older notions of more or less fixed and bounded territories. Nevertheless, relational geographies have been busy creating a new topological geography of networks, relations, and flows that has transformed the ways in which many of us think about the production and practice of space and place. This style of thinking is continued in the next chapter which focuses on some attempts to renew links between the interests of human and physical geographers.

References

Adey, P. (2008) Airports, mobility and the calculative architecture of affective control. *Geoforum*, 39, 438–451.

Adey, P. (2009) Facing airport security: affect, biopolitics, and the preemptive securitisation of the mobile body. *Environment and Planning D: Society and Space*, 27, 274–295.

*Anderson, B. (2006) Becoming and being hopeful: towards a theory of affect. *Environment and Planning D: Society and Space*, 24, 733–752.

Anderson, B. (2010) Morale and the affective geographies of the "War on Terror". *Cultural Geographies*, 17, 219–236.

*Anderson, J. (2012) Relational places: the surfed wave as assemblage and convergence. *Environment and Planning D: Society and Space*, 30, 570–587.

*Anderson, B. and Harrison, P. (2010a) The promise of non-representational theories, in *Taking Place: Non-Representational Theories and Geography* (eds B. Anderson and P. Harrison), Ashgate, Farnham, pp. 1–36.

Anderson, B. and Harrison, P. (2010b) *Taking-Place: Non-Representational Theories and Geography*, Ashgate, Farnham.

Anderson, B. and Mcfarlane, C. (2011) Assemblage and geography. *Area*, 43, 124–127.

Bennett, J. (2010) *Vibrant Matter: A Political Ecology of Things*, Duke University Press, Durham, NC.

Blunt, A. and Dowling, R. M. (2006) *Home*, Routledge, London.

Bondi, L. (2008) On the relational dynamics of caring: a psychotherapeutic approach to emotional and power dimensions of women's care work. *Gender, Place and Culture*, 15, 249–265.

Brenner, N. (2004) *New State Spaces: Urban Governance and the Rescaling of Statehood*, Oxford University Press, New York.

Collinge, C. (2006) Flat ontology and the deconstruction of scale: a response to Marston, Jones and Woodward. *Transactions of the Institute of British Geographers*, 31, 244–251.

Cresswell, T. (1996) *In Place/Out of Place: Geography, Ideology and Transgression*, University of Minnesota Press, Minneapolis, MN.

Cresswell, T. (2006a) *On the Move: Mobility in the Modern Western World*, Routledge, New York.

Cresswell, T. (2006b) "You cannot shake that shimmie here": producing mobility on the dance floor. *Cultural Geographies*, 13, 55–77.

Cresswell, T. (2012) Nonrepresentational theory and me: notes of an interested sceptic. *Environment and Planning D: Society and Space*, 30, 96–105.

*Cresswell, T. (2014) Place, in *The SAGE Handbook of Human Geography* (eds R. Lee, N. Castree, R. Kitchin, V. Lawson, A. Paasi, C. Philo, S. Radcliffe, S. Roberts, and C. W. Withers), Sage, London, pp. 7–25.

Cresswell, T. (2019) *Maxwell Street: Writing and Thinking Place*, University of Chicago Press, Chicago, IL.

Davidson, J., Bondi, L., and Smith, M. (2005) *Emotional Geographies*, Ashgate, Aldershot.

Delanda, M. (2006) *A New Philosophy of Society: Assemblage Theory and Social Complexity*, Continuum, London, New York.

Delanda, M. (2016) *Assemblage Theory*, Edinburgh University Press, Edinburgh.

Derrida, J. (1976) *Of Grammatology*, Johns Hopkins University Press, Baltimore, MD.

Dewsbury, J. D. (2000) Performativity and the event: enacting a philosophy of difference. *Environment and Planning D: Society and Space*, 18, 473–496.

Dixon, D. and Jones, J. P. (1998) My dinner with Derrida: or spatial analysis and post-structuralism do lunch. *Environment and Planning A*, 30, 247–260.

Doel, M. (2007) *Post-structuralist geography: a guide to relational space* by Jonathan Murdoch. *Annals for the Association of American Geographers*, 97, 809–810.

Doel, M. (2010) Representation and difference, in *Taking Place: Non-Representational Theories and Geography* (eds B. Anderson and P. Harrison), Ashgate, Farnham, pp. 117–130.

Forer, P. (1978) A place for plastic space? *Progress in Human Geography*, 2, 230–267.

* Jacobs, J. M. (2006) A geography of big things. *Cultural Geographies*, 13, 1–27.

Jonas, A. E. G. (2006) Pro scale: further reflections on the "scale debate" in human geography. *Transactions of the Institute of British Geographers*, 31, 399–406.

* Jones, M. (2009) Phase space: geography, relational thinking, and beyond. *Progress in Human Geography*, 33, 487–506.

Jones, M. and Macleod, G. (2004) Regional spaces, spaces of regionalism: territory, insurgent politics and the English question. *Transactions of the Institute of British Geographers*, 29, 433–452.

Kitchin, R. (1998) "Out of place", "knowing one's place": space, power and the exclusion of disabled people. *Disability and Society*, 13, 343–356.

Latham, A. (1999) Powers of engagement: on being engaged, being indifferent, and urban life. *Area*, 31, 161–168.

Latham, A. and Mccormack, D. P. (2004) Moving cities: rethinking the materialities of urban geographies. *Progress in Human Geography*, 28, 701–724.

Leitner, H. and Miller, B. (2007) Scale and the limitations of ontological debate: a commentary on Marson, Jones and Woodward. *Transactions of the Institute of British Geographers*, 32, 116–125.

* Marston, S., Jones III, J. P., and Woodward, K. (2005) Human geography without scale. *Transactions of the Institute of British Geographers*, 30, 416–432.

* Massey, D. (1993) Power-geometry and progressive sense of place, in *Mapping the Futures: Local Cultures, Global Change* (eds J. Bird *et al.*), Routledge, London, pp. 59–69.

Massey, D. (1997) A global sense of place, in *Reading Human Geography* (eds T. Barnes and D. Gregory), Arnold, London, pp. 315–323.

Massey, D. (2006) Landscape as a provocation – reflections on moving mountains. *Journal of Material Culture*, 11, 33–48.

Massey, D. B. (2005) *For Space*, Sage, London.

* McCormack, D. (2008) Geographies of moving bodies: thinking, dancing, spaces. *Geography Compass*, 2, 1822–1836.

McCormack, D. P. (2003) The event of geographical ethics in spaces of affect. *Transactions of the Institute of British Geographers*, 28, 488–507.

* Mcfarlane, C. (2011) The city as assemblage: dwelling and urban space. *Environment and Planning D-Society & Space*, 29, 649–671.

Murdoch, J. (2006) *Post-Structuralist Geography: A Guide to Relational Space*, Sage, London.

Nash, C. (2000) Performativity in practice: some recent work in cultural geography. *Progress in Human Geography*, 24, 653–664.

Philo, C. (1987) *"The Same and the Other": On Geographies, Madness and Outsiders*. Loughborough University of Technology Department of Geography Occasional Paper, p. 11.

Philo, C. (1995) Animals, geography, and the city: notes on inclusions and exclusions. *Environment and Planning D: Society and Space*, 13, 655–681.

Revill, G. (2004) Performing French folk music: dance, authenticity and nonrepresentational theory. *Cultural Geographies*, 11, 199–209.

Roe, E. (2010) Ethics and the non-human: the mattering of animal sentience, in *Taking Place: NonRepresentational Theories and Geography* (eds B. Anderson and P. Harrison), Ashgate, Farnham, pp. 261–282.

Sack, R. D. (1980) *Conceptions of Space in Social Thought: A Geographic Perspective*, Macmillan, London.

Sheller, M. and Urry, J. (2006) The new mobilities paradigm. *Environment and Planning A*, 38, 207–226.

Sibley, D. (1981) *Outsiders in Urban Societies*, St. Martin's Press, New York.

Sibley, D. (1995) *Geographies of Exclusion: Society and Difference in the West*, Routledge, London.

Somdahl-Sands, K. (2006) Triptych: dancing in thirdspace. *Cultural Geographies*, 13, 610–616.

Taylor, P. J. (1982) A materialist framework for political geography. *Transactions of the Institute of British Geographers*, 7, 15–34.

Thein, D. (2005) After or beyond feeling? A consideration of affect and emotion in geography. *Area*, 37, 450–456.

Thrift, N. (1997) The still point: resistance, expressiveness embodiment and dance, in *Geographies of Resistance* (eds S. Pile and M. Keith), Routledge, London, pp. 124–151.

Thrift, N. (1999) Steps to an ecology of place, in *Human Geography Today* (eds D. B. Massey, J. Allen, and P. Sarre), Polity, Cambridge, pp. 295–322.

*Thrift, N. (2000) Afterwords. *Environment and Planning D: Society and Space*, 18, 213–255.

Thrift, N. (2004) Intensities of feeling: towards a spatial politics of affect. *Geografiska Annaler: Series B, Human Geography*, 86, 57–78.

Thrift, N. J. (2008) *Non-Representational Theory: Space, Politics, Affect*, Routledge, London.

Thrift, N. (2011) Lifeworld Inc – and what to do about it. *Environment and Planning D: Society and Space*, 29, 5–26.

Tolia-Kelly, D. (2006) Affect – an ethnocentric encounter? Exploring the "universalist" imperative or emotional/affectual geographies. *Area*, 38, 213–217.

Tuan, Y.-F. (1974) *Topophilia: A Study of Environmental Perception, Attitudes, and Values*, Prentice Hall, Englewood Cliffs, NJ.

Valentine, G. (1993) (Hetero)sexing space: lesbian perspectives and experiences of everyday spaces. *Environment and Planning D: Society and Space*, 11, 395–413.

Valentine, G. (1996) Children should be seen and not heard: the production and transgression of adult's public space. *Urban Geography*, 17, 205–220.

Wolff, J. (1995) Dance criticism: feminism, theory and choreography, in *Resident alien: feminist cultural criticism* (ed. Wolff, J.), Polity Press, Cambridge, pp. 68–87.

Woods, M., Fois, F., Heley, J., Jones, L., Onyeahialam, A., Saville, S., and Welsh, M. (2021) Assemblage, place and globalisation. *Transactions of the Institute of British Geographers*, 46, 284–298.

Chapter 12

More-than-Human Geographies

In most departments of geography, certainly the ones I am familiar with, human and physical geographers appear to live separate intellectual lives. They meet for coffee, chat about the weather, compare the fortunes of sports teams, and talk about family, but geographical ideas are rarely the subject of conversation. We share buildings and have offices next to each other. We may read the posters on the walls which describe our research projects while idling away some time. We are in departments of geography, in part, because of the history of ideas I have recounted in this book. We share a common lineage of a discipline that emerged as the study of the inhabited earth. Often, however, this seems to be an accident of history that fails to provide any sense of unity to a discipline that ranges from the humanities to the natural sciences. Even the everyday practices of research and publishing are strikingly different. Human geographers wonder at the multitude of authors that appear at the head of most papers in physical geography. Physical geographers wonder at the significance given to writing books by human geographers. Most significant disciplinary journals for geography as a whole are dominated by human geographers who see them as prestigious places to make significant statements. Physical geographers rarely publish in them, preferring their own subdisciplinary or general earth science journals. Major geography conferences are overwhelmingly meeting places for human geographers. The word "theory" means almost completely different things to the two sides of the discipline.

Everything in this book up to the quantitative revolution (Chapters 1–5) is equally relevant to human and physical geography. Early geographers including Greek and Roman geographers, early geographers in the Muslim world, and geographers of the nineteenth century such as Alexander von Humboldt or Peter Kropotkin were able to move between human and physical elements of our world without any need to differentiate one from the other. Kropotkin, for instance, studied glaciers in Finland, reindeer herds in Siberia and came up with the very human idea of mutual aid based on observations he had made of the natural world. A geomorphologist such as William Morris Davis would state that "any statement is of geographical quality if it contains a reasonable relation between some inorganic element of the earth on which we live, acting as a control, and some element of the existence or growth or behavior or distribution of the earth's surface organic inhabitants serving as a response" (Davis 1906: 71). Environmental determinists such as Ellsworth Huntington (a student of

Geographic Thought: A Critical Introduction, Second Edition. Tim Cresswell.
© 2024 John Wiley & Sons Ltd. Published 2024 by John Wiley & Sons Ltd.

Davis) and Ellen Semple made very strong claims about the ways they believed that physical environments determined aspects of human life. Regional geographers of the early twentieth century would provide accounts that started with the bed rock, moved on to soils, described climates and then moved on to human factors of the environments. By the time of the quantitative revolution in geography, arguments were being made for the need to be scientific by both human geographers such as Bunge and Schaefer, and physical geographers such as Strahler.

This was not necessarily the intention. Yi-Fu Tuan, for instance, in a spirited attempt at defining geography as the study of "earth as the home of people" (the old idea of the global *ecumene*), was insistent that physical geography was very much part of this humanistic account:

> No human group can survive unless it makes sense of its environs – its Airs, Waters, and Places, as Hippocrates put it. Physical geographers have built on this basic curiosity and need. They have tried to understand the earth as the physical or natural entity on which humans live. Physical geography can be a pure physical science that scarcely mentions people and their works (…). Yet, for all its rightful claim to being a physical science, it remains a science tied, at a fundamental level, to the human scale (…). Thus, whereas physical geographers may study the earth as a whole or its parts, the parts they study rarely, if ever, reach the microscale of, say, the molecular structure of minerals or the turbulences of air over leaves of different shape. Physical geographers examine the surface and the upper crust of the earth but almost never its core, which is too remote from ordinary human interest. They study land-forms, climates, and biological organisms of the last two million years, but rarely those of more distant geological times. They restrict themselves to the Quaternary period, with special emphasis on the Holocene (…), because it is then that the earth has become the home of the human species. (Tuan 1991: 99–100)

At the heart of Tuan's humanistic account of the core of geography is this interest in the ways in which humans and the natural world interact – how the earth is made into a home. The role of "nature" in the constitution of place is also at the center of Robert Sack's theorizations of place. He has long argued that the concept at the heart of geography – place – is too often reduced to a by-product of some field or other of thought from outside of geography. He writes of the concept of place as the coming together of three realms. These are the realms of meaning, the social, and nature. Place, he argues, is constantly weaving these three things together in particular ways. Implicit in his argument is the observation that contemporary theory, particularly radical social theory, has tended to privilege the social realm over the other two:

> Indeed, privileging the social in modern geography, and especially in the reductionist sense that "everything is socially constructed," does as much disservice to geographical analysis as a whole as has privileging the natural in the days of environmental determinism, or concentrating only on the mental or intellectual in some areas of humanistic geography. While one or other may be more important for a particular situation at a particular time, none is determinate of the geographical. (Sack 1997: 2)

His argument is that a properly geographical approach to place insists on the interweaving of all three of these realms so that place reemerges as a synthesis of natural, social, and cultural worlds. It is geography's history, as a discipline that always brings these things together, that should allow us to start with place and theorize outward from there.

Marxist geographers start from a very similar place to Tuan. People are confronted with first nature (the wild) and have to transform it through labor in order to live. This is the starting point for the development of productive forces and relations of production. The Brazilian geographer, Milton Santos, had a similar starting point for his conception of geographic space.

> Geographic space is nature modified by man through labor. The idea of a natural nature, in which man either does not exist or is not central, cedes its place to the idea of a permanent construction of artificial, social nature, synonymous with human space. (Santos 2021: 88)

In each of these cases, the definition of geographical enquiry is based on a dualism of "nature" on the one hand, and "people" on the other. Geography's subject matter emerges when one (humans) transforms the other (nature).

Within different subdisciplines of human geography there have been energetic attempts to link human and physical (natural) worlds. A significant body of work now exists on the role of nature in the city, for instance. Often this is placed within either broad cultural geography or political economy frameworks. One source of inspiration is environmental history. William Cronon (both an environmental historian and a geographer) considered the role of nature in the production of the city of Chicago in his landmark book, *Nature's Metropolis*, which traces a number of bits of nature (timber, corn, etc.) as they are transformed into urban commodities while at the same time transforming the emerging metropolis of Chicago (Cronon 1991). A number of geographers emerging from the Marxist political economy tradition have focused on the politics of water in urban environments (Swyngedouw 1999; Kaika 2005; Loftus and Lumsden 2008). Matthew Gandy has done for New York what Cronon did for Chicago by writing an account of the role of nature in the city. Rather than following bits of nature into the city, Gandy shows how a metropolitan nature is produced both through its spaces (parks, water systems, etc.) and through its relations to the world outside which are connected through infrastructures that connect the city to the raw materials that make it. All of this is framed by both the circulation of capital and the development of ideas (ideologies even) about nature and its role in urban life (Gandy 2002).

For the most part, though, human geographers, wary of being labeled as environmental determinists, have subordinated the natural world to the social (human) one. Nature is said by most to be "socially constructed" (Smith 1991; Castree 2005). One of the most striking ways in which recent relational approaches to geography have contributed to longstanding theoretical debates in the discipline is in their approach to the familiar dualism of nature and culture. With the turn to relational ways of thinking, human geographers began to return to the physical and natural world in a way that attempted to get beyond the nature/culture dualism. Simultaneously they attempted to bypass the cul-de-sac of thinking of nature as either "real" or "constructed." In this chapter (and the next), I consider some of the attempts that have been made to reengage the realms of nature and culture across the discipline of geography. Before that, however, it is helpful to account for the general philosophical approach of **post-humanism** that these more-than-human geographies form part of. As we saw in Chapter 4, post-humanism has arisen as a critique of the centering of humans in humanism. While humanism posited humans as the center of experienced worlds, and underlined the importance of seemingly human characteristics of reason and imagination, post-humanists, since around 2000, have revealed how agency is formed through relations between humans and other species and things where reason and imagination are only two among many attributes that lead to different outcomes in an interconnected world. Post-humanism thus includes nonhumans as legitimate subjects of social enquiry with the capacity to (in relation with others) enact agency in the world (Braun 2004; Castree and Nash 2006). Post-humanism is not just about leveling the playing field of agency between human and nonhuman subjects but also about recognizing how the label of humanism has masked differences ascribed to those humans labeled either closer to nature or less-than-human – particularly women and Black and Brown people. As Castree and Nash put it:

> There is a considerable body of work within feminist and other critical traditions that addressed the ways the category of humanity as separate from and in control over nature has historically also involved hierarchical models of difference amongst humans (…). Humanism's history is a history of race and sex being used to define some humans as more human through their distance from nature than others…. (Castree and Nash 2006: 501)

Post-humanism, then, asks very old questions but in new ways. Post-humanists, like humanists, want to know what makes a human a human. But also, as Bruce Braun has put it:

> How are boundaries between humans and animals, bodies and machines drawn? How are human bodies produced and transformed? What does it mean to admit non-humans into our understanding of humanity and society? And what are the consequences of doing so for ethics and politics? (Braun 2004: 269)

Braun considers the links between the internal differentiation of humans into categories of the more or less human, but also the links between humans and the animals that surround us suggesting that rather than thinking of the classic singular humanist subject, we think of our bodies as permeable and connected.

> Here in the United States, evening news broadcasts mess up the category further: 'barely human' others (Iraqis, Rwandans, Muslims), and 'almost human' companions (monkeys, dogs and cats), are discussed alongside accounts of 'inter-species' exchange (bird flu, SARS) in which the boundaries of the human are suddenly porous and mobile. Conversely, the ascendance of rights discourse has produced a demand to 'fix' the body, to give the human determinate form and content, confronting the continuous differentiation of bodies with an earnest desire for reassembly, consistency, and universality. (Braun 2004: 269)

Post-humanism recognizes that that some humans have historically been considered more fully human than others, and that humans always exist in relation to other animals and things. It also acknowledges that the human body itself is internally differentiated. We know, for instance, that other organisms within our bodies outnumber the cells of the human bodies by 10 to 1. Over 10,000 microbial species occupy our body as an ecosystem and we would not be able to live without them. In addition to that, our body is laced with chemicals including plutonium from nuclear testing and micro plastics that have been found in the placentas of mothers with unborn babies. Increasingly our bodies include technologies such as metal plates, artificial lenses, pacemakers, and the like. Our bodies too are relational entities that combine what might conventionally be thought of as human and nonhuman components. Geographers have mobilized the ideas of the post-human to produce a "more-than-human" geography "…displacing the hubris of humanism so as to admit others into the calculus of the world" (Braun 2004: 273). In the remainder of this chapter, we will explore three forms of these "more-than-human" geographies: animal geographies, actor-network theory, and hybrid geographies.

Animal Geographies

Biogeography is a subfield of geography usually recognized as a part of physical geography. Its focus is on living, nonhuman, things that inhabit the biosphere and it represents the intersection of geography and biology as disciplines. It is one place where the human and physical realms are likely to be theorized together. In the 1990s, however, a slightly different field of "animal geography" emerged which brought together a range of geographers from across the cultural geography/biogeography divide.

Animals are perhaps the closest part of the nonhuman to humans. We recognize them as sentient beings that share the earth with us, we endow them with human-like characteristics in movies and cartoons from *Donald Duck* to *Madagascar*. We live with them as pets and visit them in zoos. At the same time, we experiment on them (because they are *like* us but *not* us), farm them, and kill and eat them on an industrial scale. They are one of the first bits of "nature" to be included in a wider understanding of the social, as it is not too much of a stretch to think of animal *subjectivities*. Braun draws on the insights of philosopher Jacques Derrida to consider the arbitrary nature of the division between humans and animals (Braun 2004).

> It is less a matter of asking whether one has the right to refuse the animal such and such a power – …than of asking whether what calls itself human has the right to rigorously attribute to man, which means therefore to attribute to himself, what he refuses the animal, and whether he can ever possess the pure, rigorous, indivisible concept, as such, of that attribution. (Derrida 2003: 138 quoted in Braun 2004)

Derrida questions using the attribute of human intellect as a worthy dividing line between humans and animals by suggesting both the possibility that nonhuman animals have their own forms of intelligence and whether we should even give intelligence the kind of priority we have given it when forming our hierarchies.

Insights such as these have led geographers to explore the interface of human geography and the world of animals in order to break down and problematize the culture/nature dualism. These geographers were inspired by philosophical questions such as those posed by Derrida as well as work in poststructuralism, cultural geography, and environmental ethics and within the political movements calling for the humane treatment of animals. Some work had previously been done on this. Most notable perhaps was Tuan's book *Dominance and Affection: The Making of Pets*, which explored how humans exhibit both negative practices of domination and positive practices of affection in their attitude to both animals (pets) and nature in general (trimmed hedges in formal gardens, for instance) (Tuan 1984).

Cultural geographers explored the human/animal divide through the changing ways in which the *idea* of the animal has changed over time and the way in which it has been used to designate both part of us (our animal natures) and human "others." This geography focused on representations of animals and animality and their relations to our identity as humans (Anderson 1995, 1997). While early work tended to focus on human representations of the animal, later work turned to the agency of animals themselves and the ways in which animals impacted on human lives in a variety of settings. While early work looked to animals as human-like in their ability to exhibit subjectivity, the later work decided that subjectivity and intentionality were not actually the most important things in acknowledging agency (Wilbert 1999). Several edited books have appeared charting this new engagement between the human and the animal (Wolch and Emel 1998; Philo and Wilbert 2000; Gillespie and Collard 2015).

You might think that the likeliest place to look for work on the human/animal interface would be the countryside, with its farms, ranches, and nature reserves. A lot of the most intriguing work, however, has focused on the city as a space shared by animals and humans. Kay Anderson, for instance, looked to the zoo in Adelaide, Australia, as a site where humans inscribed a range of beliefs on its animal inhabitants in order to, in part, construct a national Australian identity as a colonial space where Indigenous people were excluded or oppressed (Anderson 1995). Another approach to animals is that taken by Chris Philo in his examination of markets and slaughterhouses in nineteenth-century London and Chicago. He considers the possibility of animals as a kind of social group which occasionally transgresses the borders of acceptable behavior in the city:

> It seems to me that … many animals (domesticated and wild) are on occasion transgressive of the socio-spatial order which is created and policed around them by human beings, becoming "matter out of place" in the process; and it is in this respect that animals often squeeze out of the places – or out of the roles that they are supposed to play in certain places – which human beings envisage for them. (Philo 1995: 656)

Philo's paper focuses on the existence of livestock in urban areas until comparatively recently, noting the various acts that were passed in Britain in the nineteenth century to remove livestock markets, milking facilities, and slaughterhouses from public view. Philo suggests that this was part and parcel of wider processes going on in the nineteenth-century British city including, especially, the "long-term splitting apart of the urban and the rural as distinctive entities conceptually associated with particular human activities and attributes (the industrial and civilized city, the agricultural and barbarian countryside)":

> My argument is simply that animals as a social group have become inextricably bound up in these stories, much as have certain outsider human groups, and as a result animals have become envisioned in particular ways with particular practical consequences: one of which is that some animals (cats and dogs) have been turned into pets valued as an element of the urban world whereas other animals (cows, sheep, and pigs) have become matter that should be expelled to the rural world. (Philo 1995: 666)

Philo supports his argument with reference to a multitude of historical documents recounting the various reasons why contemporary observers in cities such as London and Chicago objected to livestock in the city, despite the fact that they had been part of the urban life for as long as there had been cities. One set of commentators objected on health and safety grounds, arguing that large numbers of cattle moving along an urban road could not mix safely with humans. Another set of objections were more cultural, focusing on the perceived immorality of animal behavior in full view of women and children. The sites of animal processing in the city, particularly meat markets, were associated with immorality and degeneracy in the human population around them. Philo reports one observer of London's Smithfield Market, a Mr. J. T. Norris, as writing:

> Can you speak as to the morality of the neighbourhood of Smithfield, whether it is as pure as the rest of the city of London? – I am afraid it is not; the language I hear used as I pass, and the frequent visits of a good many concerned in the business of the market to the gin shops, the fighting and the disturbances that occur thereabout, make me say that it exhibits the lowest state of morals to be found anywhere in the metropolis, excepting perhaps St. Giles. (Quoted in Philo 1995: 669)

Philo's point in recounting these processes is to open up the possibility of a new animal geography based on the re-inclusion of animals in the theory construction of human geographers.

The process of exclusion Philo describes is never complete. The city is a space that humans always share with animals, ranging from the chickens people keep in their back yards to pets and the wild animals that find urban and peri-urban spaces rich in sustenance discarded by humans. Examples of the latter include foxes, coyotes, deer, and cougars (Gullo *et al.* 1998). Far from being a totally human space – the most social of social spaces – the city turns out to be a space in which nonhuman animals are constantly circulating and inhabiting:

> … cities are replete with animate, sentient beings with legs, wings, antennae and tails – namely animals. Yet, despite the fact that explaining relations between nature and human society is ostensibly a primary goal for geographical research, animals rarely figure in urban geographical studies. We see them as parts of urban ecosystems, raw materials powering the growth of industrial cities, or symbols of urban popular culture. Mirroring larger trends in human geography, particularly over the last 25 years, urban geography has largely ignored animals as a topic for serious scholarly attention. (Wolch 2002: 722)

One animal that makes its home in and around the homes of humans in Australia is the possum (see Figure 12.1). Emma Power, in an ethnographic study of Australian homeowners, has shown how the possum ruptures the border between the domestic space of the home and the wild outside in a way that destabilizes the home and provokes a level of anxiety in homeowners. The possums inhabit the cavities and ducts that exist in the borders of the home (inside the walls, for instance). Many of the participants found their presence (mostly experienced as sound and smell) unsettling – a

Figure 12.1 Common Bushtail Possum, Tasmania, Australia. Source: Photo by J. J. Harrison, CC-BY-SA-2.5. http://en.wikipedia.org/wiki/File:Trichosurus_vulpecula_1.jpg (accessed May 29, 2012).

presence that detracted from their sense of home. On the other hand, many recognized that the possums were inhabitants of the land before they were, and in some sense "belonged." Paradoxically, many thought that possums did not belong in an urban environment:

> Participants' encounters with possums were framed by discussions about whether possums belonged, or did not belong in the urban and urban-bushland environments that participants lived in. Participants' beliefs about whether possums belonged in these environments shaped their own sense of hominess in these spaces and in the house-as-home. Although urban environments have traditionally been framed as human-only spaces, participants expressed a more complex conception of possum-belonging that drew on narratives of nativeness, invasion, colonialism and contemporary urban development. (Power 2009: 38)

The contradictory sense of possums both belonging and not belonging in urban space was also manifested in the participants sometimes claiming that the presence of possums in and around their homes led to a feeling of co-habitation and hominess. The presence of possums also increased their sense of a shared human–nonhuman space:

> Possums, perceived as wild, native animals, extended the home into nature and brought nature into the home to make the house more homey. These experiences conflict with views of home as a human-space and illustrate some ways that wildness and hominess can come together within the house-as-home. They extend discussion of designed, integrated living spaces that blur inside and outside, because here blurring was a product of uninvited ruptures in home. It went beyond human design to instead reflect the particular impact of possum-agency on home and homemaking. (Power 2009: 45)

Power's work shows how these animals are part of urban life and that they have "possumagency."

This reflects a wider concern with animal agency in the city explored by Jennifer Wolch. Wolch asks us to take animals seriously as moral and political agents in city life. She asks what might happen if animal species were allowed to be agents in the production of urban space through regulation. What kind of cities would we build if we took animals seriously?

> If animals are granted subjectivity, agency and maybe even culture, how do we determine their survival opportunities in the city? To what urban arenas should our moral compass direct us – homes, businesses, streets, parks and open spaces, restaurants and supermarkets? The implications of training our moral gaze on such urban places are enormous. For example, federal law mandates water-quality standards for all waterways, but these standards are designed for humans; what is tolerable for humans is not necessarily tolerable for, say, frogs. Does this mean that the US EPA needs amphibian water-quality standards for urban watersheds? (Wolch 2002: 733)

Wolch's answer is "yes." And if this is the case then there is clearly a need for increased research on and theorization of the interaction of animal life and urban regulation. Wolch asks us to recognize the fact that cities are spaces which humans share with animals and we should therefore factor this into political decision making and urban design so that the metropolis might be thought of as a zoöpolis (Wolch 1996). This concern with cities as "more-than-human" spaces is focused on the agency of animal life and points toward the existence of animals as sentient beings as an important reason for us to take them seriously.

Another example of work on human–animal interactions in the city is Krithika Srinivasan's work on street dogs in Chennai in India (Srinivasan 2019). Dogs that are not pets but, instead, free living, are a common part of street life in the Indian city. Srinivasan uses these street dogs to think through some of the ethical complications of more-than-human geographies. The street dogs, she argues, are neither the pure wild nor the pure domestic but live instead as part of spontaneous unintentional landscapes – spaces made by humans but inhabited by nonhuman animals in unintended ways. The dogs eat waste from humans and use human objects, such as cars, for shelter. In this sense, the dogs are like rats, or weeds in the pavement, or kestrels hovering beside the motorway waiting for roadkill. Not wild and yet not fully urban. Srinivasan charts how street dogs live alongside and among humans. The dogs are allocated a certain kind of local citizenship, known as denizenship, where they are seen, in legal terms, as having rights to inhabit the city streets without being killed thanks to ABC (animal birth control) laws. These rights are, of course, different from the rights of human citizens. Due to fear of rabies, for instance, the dogs are immunized. Similarly, they are prevented from reproducing in uncontrollable numbers through sterilization. Nevertheless, their right to exist was confirmed in law demonstrating a kind of expanded cosmopolitanism that is open to kinds of nonhuman difference. While the dogs were sometimes annoying, with occasional incidents of biting for instance, they were also often semi adopted and fed by the residents of streets. The dogs are part of a zoöpolis with residents, even when they do not particularly like the dogs, acknowledging that the urban space is as much the dogs' space as it is human space.

Srinivasan argues that the existence of these street dogs and their uneasy alliance with humans challenges aspects of the nature/society binary. Modern states, she argues, are based on the policing of unruly nature, creating purified human spaces. Liminal animals like rats and Chennai's street dogs challenge this idea of modernity. We saw earlier how Philo traced the gradual exclusion of slaughterhouses from the city as part of the work of modernity that clearly delineates human space and the space of nature (Philo 1995). Srinivasan's example challenges this. Her example of canine cosmopolitanism balances coexistence and conflict with the demands for a safe and sanitized urbanity.

Part of Srinivasan's purpose in thinking with the street dogs of Chennai is to ask us to consider the ethics of dealing with the nonhuman animal. While humans can grant denizenship to dogs, the dogs do not have the capacity to do the reverse. While dogs may occasionally attack humans, humans can do so much worse to the dogs.

> Street dogs, on the other hand, continue to be subject to social violence that usually goes inadequately challenged. What's more, the law guaranteeing their secure residence continues to be contested in Indian courts. In some ways, street dogs are seen to be too social to be granted the ethical status and protections that other nonhuman organisms that are viewed as nature have, but not social enough to have the ethical status attributed to humans. (Srinivasan 2019: 387)

Here, Srinivasan puts her finger on the in-between nature of the street dogs. They are social but not social enough to get all of the rights that come with being a human in India. They are natural, but not natural enough to be protected or revered in the way, say, a rare tiger might be protected and revered. This raises questions about the ethics of dealing with the nonhuman that goes beyond notions of purity and impurity.

> By contrast, those nonhumans that are seen as not being "natural" enough – hybrids of socio-nature – have ethical statuses inferior to not only the purely social (the human) but also the purely natural (wildlife). These include organisms like free-living dogs, pigeons, rats, cockroaches, and weeds that create niches for themselves in the midst of human life but are then persecuted, controlled, and eradicated because they are undesirable or risky to people and/or valued wildlife. They also include organisms that are conceptually "tainted" by the human, as exemplified by invasive alien species that are exterminated because they were introduced by humans in particular regions at some point in history. In other words, material or conceptual intertwining with the social, the human, results in the "denaturalisation" of some nonhuman and concomitant ethical disprivileging. All these are just some instances of how ethical dualisms operate. As such, it is vital that geographical scholarship on nature and the more-than-human go beyond the label of non-dualism to explicitly engage with the prevalence of ethical dualisms in society-nature interactions. (Srinivasan 2019: 387)

Srinivasan acknowledges that more-than-human approaches to animals have often succeeded in breaking down various kinds of dualistic thinking based on the nature/society divide. Her key point, though, is that we have not done this in the realm of ethics and that deeply entrenched beliefs persist in the way humans view animals in ethical or nonethical ways. The binary of nature/society and the accompanying binary of animal/human still guides ethical human–animal relations. Our understanding of "alien invasive species," for instance, is, as Srinivasan points out, tainted by their association with us humans. Our ethical relations with animals are still guided by a dualistic foundation.

Arguments for the nonhuman, however, go much further than human–animal relations in the wider project of actor-network theory, which denies that sentience and subjectivity are any kind of special qualities when it comes to considering the agency of the nonhuman world.

Actor-Network Theory

Since the 1980s, geographers have been struggling, in different ways, with the problem of dualistic thinking. Three of these dualisms (themselves central to and outcomes of forms of structuralist thinking) are structure/agency, culture/nature, and realism/social constructionism. We have seen how the issue of structure and agency has been at the heart of, first, humanistic and radical responses to spatial science and, second, the various attempts of structuration theory to map a third way through individual action and structural imperative. The dualism of culture and nature has run right through all of the theoretical approaches in this book. The distinction between realism and social constructionism is another key set of debates that often distinguishes the natural and physical sciences (where the independent existence of an external world is taken for granted) from the social sciences (where the dominant view is that much that passes as "reality" is produced by social forces). This has been mapped on to physical and human geography more or less wholesale. Physical geographers, by and large, have remained wedded to the idea that a real, solid, tangible world exists out there and that their science explains it with ever-increasing accuracy (Inkpen 2005). Human geographers, for most of the past 30 years, have been equally happy accounting for the inventive capacities of humans to socially produce everything from oceans to nature itself (Smith 1991; Steinberg 2001). It is not that surprising that conversation has been difficult. Recently, however, there has been a return to philosophical realism within human geography. While the older forms of realism were framed by positivism and were characteristically essentialist, new forms of realism advertise themselves as anti-essentialist and "speculative." Perhaps the most significant of these theoretical turns has been the emergence of **actor-network theory**.

Actor-network theory (ANT) emerged in geography in the late 1990s (Bingham 1996; Murdoch 1997a, b; Hinchliffe 2002, 2010; Greenhough 2006). It has made tentative inroads into physical geography (Allen and Lukinbeal 2010). It arose from the work of the French thinker Bruno Latour and his work on the history of science, and particularly the site of the laboratory (Latour 1993, 2005). The most important insight of ANT for our purposes here is that the agency of humans in the production of the world is matched by and enabled by the agency of the nonhuman world. This is the world of animals to be sure, but it is also the world of rocks and trees. And it is not just "nature" that has nonhuman agency – so does the world of manufactured objects such as books, test-tubes, and sparkplugs. Latour is concerned to show how the space of the laboratory (for instance) forms part of a network of people, ideas, and things that brings things (such as antibiotics or public transport systems) into being. Rather than suggesting that individual agents (humans or animals) have the ability to accomplish things in and of themselves or, conversely, that some overarching structure compels certain outcomes, Latour asserts that agency is the achievement of things (including people) in networks. Agency is the product of this connectedness rather than either atomistic humans or systematic structures.

Just as Bourdieu stated his intention to escape the poles of subjectivism (phenomenology and ethnomethodology) and objectivism (structuralism and Marxism), so Latour charts a course between the local and global (or perhaps micro and macro):

> Let us admit that the ethnomethodologists are right, that there exist only local interactions, producing social order on the spot. And let us admit that mainstream sociologists are right, that actions at a distance may be transported to bear on local interactions. How can these positions be reconciled? An action in the distant past, in a faraway place, by actors now absent, can still be present, on condition that it be shifted, translated, delegated or displaced to other types of actants, those I have been calling nonhumans. (Latour 1994: 50)

In other words, local actions, including those of individual humans, always contain something of the far away and long ago within them. Other local actions are connected to the one in question and these have influence. Structural explanations, on the other hand, frequently forget to say *how* a macro-scale phenomenon (the economy, society, etc.) can reach into and influence a small, local action:

> Everything in the definition of macro social order is due to the enrolment of nonhumans … even the simple effect of duration, of long-lasting social force, cannot be obtained without the durability of nonhumans to which human local interactions have been shifted. (Latour 1994: 51).

In this argument about duration we might consider, for instance, how buildings, books, regulations, codes of practice, or whole landscapes are needed to make sure that individual or collective human intentions are made to endure. For Latour, society is not a thing that exists in advance of the world and continuously produces it. Rather society (or the social) is what is always being made in the relations between people and things.

Latour's project in the development of ANT is to navigate the dualisms of micro/macro, structure/agency, and subject/object by asking us to focus on the way in which things (including us humans) *connect* in order to produce forms of agency. In this sense, objects (sparkplugs, test-tubes, guidebooks, microbes – anything) are not seen as "products" of humans collectively acting as the "social" but as active agents in the ongoing constitution of the world. Agency itself is not something that humans come pre-endowed with but is an outcome of the establishment of networks of connections between heterogeneous objects. Latour insists that objects are full participants in the production of both "agency" and the "social":

> At first, bringing objects back into the normal course of action should appear innocuous enough. After all, there is hardly any doubt that kettles 'boil' water, knifes [sic] 'cut' meat, baskets 'hold' provisions, hammers 'hit' nails on the head, rails 'keep' kids from falling, locks 'close' rooms against unintended visitors, soap 'takes' the dirt away, schedules 'list' class sessions, prize tags 'help' people calculating, and so on. Are these verbs not designating actions? How could the introduction of these humble, mundane, and ubiquitous activities bring any news to any social scientist? (Latour 2005: 70–71)

Here, Latour is making his central claim that objects have agency in such a way that it seems uncontroversial. Within a social science framework, however, these are controversial claims. To say that a hammer, for instance, hammers in nails is to make a mistake, as it is people that hammer in nails, using hammers. From a critical theory perspective, to endow things with agency is to indulge in **fetishism** and **reification**. A fetish, in anthropological terms, is a belief that something has powers that it does not. An example might be a voodoo doll or a rainstick. This term is applied in social theory terms to mistaken attributions of agency. In geography the most well-known accusations of fetishization are made against those who fetishize space – who believe that space (in and of itself) has agency. It is not space that has agency, the critical theorist argues, but people, arranged into social groups, who use space to do things. It is society that has agency. In all of Latour's examples above, it is possible to argue that the things in the list are incapable of doing anything without the inputs of humans and their intentionality. Latour wants to take our attention away from human intentionality and ask instead: "Does it make a difference in the course of some other agent's action or not?" (Latour 2005: 71). In other words, if an object was removed from a network, would it make a difference? Latour's answer is clear:

> The rather common sense answer should be a resounding 'yes'. If you can, with a straight face, maintain that hitting a nail with and without a hammer, boiling water with and without a kettle, fetching provisions with and without a basket … running a company with or without bookkeeping, are exactly the same activities, that the introduction of these mundane implements changes 'nothing important' to the realization of tasks, then you are ready to transmigrate to the Far Land of the Social and disappear from this lowly one. For all the other members of society, it does make a difference under trials and so these implements, according to our definition, are actors, or more precisely, *participants* in the course of action waiting to be given figuration. (Latour 2005: 71)

Latour insists on a range of forms of action between intentional human action and complete irrelevance and argues that it is not possible to draw a line dividing the important from the inconsequential. One possible consequence of ANT for geographers is to enable us to think of our objects (humans and the physical world) on a level playing field:

> Thus Latour believes a focus on the exchange of properties between the social and the material problematizes prevailing ideas of 'social' and 'society' as these terms only conjure up one side of the traditional dualism – the world of human subjects and social interaction – and leave out the picture objects and materials. He, therefore, calls for the use of new concepts – such as 'actor-network' – which permit the *symmetrical* analysis of subjects and objects and actors and things. (Murdoch 1997b: 329)

Jonathan Murdoch has argued that ANT is inherently geographical as the dependence on nonhuman resources as part of actor-networks means that both space and time are seamlessly entangled into the ongoing analysis of social life. At the same time, ANT challenges geography-as-usual as it forbids us from thinking of spatial scales in a hierarchy of tiers where some tier is more important and "determines" what goes on at lower levels (usually bigger scales determining smaller scales):

> … it is only the mobilization of nonhumans across space and time that distinguishes the local from the global, the micro from the macro. Thus, theory puts space 'in its place': there can be no purely spatial processes, for the use of nonhuman resources to facilitate action-at-a-distance not only binds space into social, natural and technical processes but is also a means of ensuring these actions become historical, that is, of rendering them permanent and stable. Time and space become seamlessly entwined. (Murdoch 1997b: 329)

In ANT, heterogenous entities are continually being brought into association with each other through networks that are always spatial. As networks are created, discrete spaces (or places) are drawn together. Most of ANT focuses on science, and the kinds of spaces we encounter tend to be laboratories (or centers of calculation) or the "field" (Greenhough 2006). Actor-network theorists, in geography and elsewhere, have accounted for how scientific knowledge and power emerge from within networks and are not something imposed from some bigger and more important "elsewhere" (often called the "social"). The constituent parts of the network include people (scientists, for instance), but also all the objects that have to be enrolled into the network to make things happen. Agency is thus distributed and attributed to things as much as it is to humans. Possibilities for action arise from the connections between humans and nonhumans.

Here, ANT is firmly "realist" in its ontology. Unlike much of poststructuralism and all of social constructionism, actor-network theorists insist on the reality of the world beyond our construction of it. Things (microbes, animals, rocks, etc.) cannot be reduced to the creations of humans – they are more than social constructions. At the same time, however, things only become active when they are enrolled into networks – so ANT is constructivist. In this theorization of the world, the "social" cannot be an explanatory factor. It cannot be assumed. Rather it is always being made through connections:

> … when social scientists add the adjective 'social' to some phenomenon, they designate a stabilized state of affairs, a bundle of ties that, later, may be mobilized to account for some other phenomenon…. Problems arise … when 'social' begins to mean a type of material, as if the adjective was roughly comparable to other terms like 'wooden', 'steely', 'biological', 'economical', 'mental', 'organizational', or 'linguistic'. At that point, the meaning of the word breaks down since it now designates two entirely different things: first, a movement during a process of assembling; and second, a specific type of ingredient that is supposed to differ from other materials. (Latour 2005: 1)

ANT insists that everything takes place at ground level. There is no need to jump scales in order to provide an explanation. You simply need to follow the network as far as it will go and account for it. In this sense ANT is quite empiricist in its formulation. Latour makes this clear though asking whether a railroad is global or local:

> … it is local at all points, since you always find sleepers and railroad workers, and you have stations and automatic ticket machines scattered along the way, Yet it is global, since it takes you from Madrid to Berlin or from Brest to Vladivostock. However, it is not universal enough to be able to take you just anywhere. It is impossible to reach the little Aubergnat village of Malpy by train, or the little Staffordshire village of Market Drayton. There are continuous paths that lead from the local to the global, from the circumstantial to the universal, from the contingent to the necessary, only as long as the branch lines are paid for. (Latour quoted in Murdoch 2006: 71)

To understand and account for a rail network (and by analogy, any network), it is not necessary to leave the immediate and local in order to look for explanation somewhere else (in abstract structures such as the "economy" or the "social"). Instead, Latour urges, we should stay in the local but realize that the local only makes sense in wider networks of other "locals":

> So, it is perfectly true to say that any given interaction seems to *overflow* with elements which are already in the situation coming from some other *time*, some other place, and generated by some other agency. This powerful intuition is as old as the social sciences, ... Action is always dislocated, articulated, delegated, translated. Thus, if any observer is faithful to the direction suggested by this overflow, she will be led *away* from any given interaction to some *other places, other times* and *other agencies* that appear to have molded them into shape. It is as if a strong wind forbade anyone to stick to the local site and blew bystanders away; as if a strong current was always forcing us to abandon the local scene. (Latour 2005: 166)

Clearly ANT, while realist, has little room for top-down views from nowhere. It is committed to always being located in the particular but recognizes the importance of all the connections to other particular sites that form a network.

ANT is not without its critics. Indeed, there are plenty of them. Perhaps the most telling point of critique surrounds the key point of **agency**. ANT creates a level playing field for agency – making no distinction between forms of agency based on intentionality and subjectivity (human and possibly animal agency) and forms of agency that are based on brute materiality. This raises questions of whether there is a need to differentiate agencies from one another. One of Latour's key questions for a network is whether it would make any difference if something was removed from a network. If the answer is yes, he suggests, then the thing has agency. Most people would, in some limited sense, agree with this. But there are other questions to ask. One might be "what elements in a network can be replaced and what elements have the power to do the replacing?" Consider the hammer and nail example. In our house it is frequently the case that a hammer cannot be located. Bricks, ends of screwdrivers, mallets, and shoes have often been used to do some hammering. These are then enrolled into a network to produce a particular kind of agency. In each of these cases the one element it would be impossible to replace is not the hammer but the human. Another way of putting this is to ask what elements in a network have the power to replace other elements. This question clearly points to human (or possibly animal) agency. The point of these questions is to insist on varieties of power and agency in networks – a point that Latour and ANT practitioners would be quick to refute.

Hybrid Geographies

Relational geographers of one kind or another start from the intuition that dualistic thinking results in us dividing the world up and thus missing out on connections between things. The world, they tell us, is actually not divided up in this way, rather it is seamless and connected in multiple ways. In addition to missing out on the ways in which things on either side of a dualism are connected, we also miss out on the multitude of differences that exist within that which is supposed to be the "same." In terms of the culture/nature divide, for instance, there are just as many differences within the elements of the world labeled "nature" as there are between "nature" and "culture." What, for instance, do an elephant and a waterfall have in common? For this reason, it is argued, it is better for geographers (and others) to think relationally and to use another set of terms – hybrid, mixed up, and networked. Heavily informed by ANT, some have begun to write and think in a way that navigates a different path from either strict physical science or social constructionism – a more-than-social, less-than-natural geography:

> Relational thinkers argue that phenomena do not have properties in themselves but only by virtue of their relationships with other phenomena. These relations are thus *internal* not external, because the notion of external relations suggests that phenomena are constituted *prior* to the relationships into which they enter. (Castree 2005: 224)

Most of geography has started from an assumption that something called society or something called nature (or environment) causes other things to happen. Relational, more-than-human geographies are attempting to produce a new version of how the world is connected to a new sense of ethics we might call an *ethic of connection*. By focusing on connections, rather than blocks of stuff labeled "society" or "nature," we will be able to see beyond enduring dualisms.

It might help to consider a well-known bit of "nature" to think about the ways in which these perspectives change the way we approach the world. Consider an American National Park – Yellowstone or Yosemite. One approach to these amazing landscapes is that of a realist. The realist would argue that the park is a wonderful and unique piece of nature that needs to be protected from society. They may describe it in biological or ecological terms as an ecosystem, bioregion, or catchment area. They might point to the ecological diversity of the region or, conversely, its unique natural attributes. Along comes the social constructionist. "Hold on a minute," she says, "this park only exists because of historical conceptions of wilderness which exclude human (indigenous) influence and because it is economically worthless to the forces of capitalism. It is a human image of nature and social through and through." The relational geographer, on the other hand, sees the park as a "network" made up of a multitude of links between heterogeneous parts. All these parts, he insists, are interrelated "actants" – trees, waterfalls, laws, rangers, tourists, artists, guidebooks, etc. The existence and capabilities of each depend on connections to all the others. There is no need to appeal to larger things such as "society" or "nature" – each actant is the cause and consequence of each other. "Things" have effects just as people do and they frequently escape the intentions of the "social."

Arguments such as these have most frequently been made with reference to things produced in laboratories. One prominent example is Oncomouse (Castree 2005). Oncomouse is a genetically modified mouse used in cancer experiments. Is Oncomouse a part of nature? It is certainly a mouse with fur and a tail and DNA and all the things that go to making up animals and mice in particular. But it is a mouse that has been deliberately engineered so maybe it is a material construction of scientists in just the same way as a test tube or an MRI scanner is. Relational thinking, Castree tells us, would lead us to think of Oncomouse as a product of an array of human and nonhuman actants including "sophisticated laboratory equipment, scientific papers containing information on how to modify mice genetically, and electronic flows of money to fund Oncomouse's ongoing production" (Castree 2005: 231). Following ANT we can see how Ocnomouse is a part of a network in which each part's role is determined by a combination of its own properties and its position in a network relative to other "actants" in that network. Oncomouse, like Yosemite National Park, is thoroughly relational.

Perhaps the most influential statement on this relational view on nature/culture is Sarah Whatmore's book, *Hybrid Geographies* (Whatmore 2002). In this book, Whatmore develops a notion of culture-natures in quite complicated ways through a series of examples ranging from elephants to soybeans. In her account of soybeans, for instance, Whatmore places the long history of soybeans in the context of recent food scares in the United Kingdom. These food scares include outbreaks of E. Coli, listeria, and salmonella as well as the more spectacular panics around mad-cow disease and foot and mouth disease. These scares, Whatmore suggests, present us with questions about culture/nature:

What begins as a catalogue of errors by accident – 'rogue' bacteria and proteins whose presence signals a failure in the clinical production and distribution of milk and meat, chickens and eggs, becomes a catalogue of errors by design – the traces of scientific and economic rationalizations of plant and animal bodies as crops and livestock that, in their multifarious incarnations as human foods become incorporated into our own. (Whatmore 2002: 121)

As with Oncomouse and Yosemite, it is hard to separate the social and the natural in these events, or to say that one is the cause of the other. It was in the context of a series of food scares that genetically modified (GM) soybeans began to arrive in the United Kingdom. While many "experts" insisted on the safety of these and other GM foods, people no longer trusted them.

Whatmore describes how soybeans have been part of human culture for over 3,000 years, valued as nutritious sources of protein that also help to fix nitrogen in the soil. Before the advent of engineering at the micro scale of genetics, farmers and others had been breeding soybeans for different habitats throughout that period (see Figure 12.2):

The dense fabric of socio-material relations between plants and farmers, soils and bacteria accumulated in Asia has engendered thousands of genetically … variable land races of soybean with different environmental and disease tolerances … associated with their particular ecological nexus. (Whatmore 2002: 126)

This history of breeding and the enlisting of soybeans into the social networks and practical knowledge of Chinese peasant farmers resulted in the sub-species, Glycine Max, which provided the basis for the modern cultivated soybean.

In addition to the process of breeding over centuries, the soybean was also part of a network of trade, science, and culinary culture. In 1880, soybeans were proven to be nutritionally valuable in a laboratory in France. By 1930, Chinese cooking techniques were being exported from China and universities in the United States were conducting experiments to identify the most likely varieties to succeed in the United States. In the mid twentieth century, US agribusiness was busy producing F1

Figure 12.2　Soybeans. Source: Photo by Annie Mole, CC-BY-2.0. http://commons.wikimedia.org/wiki/File:Edamame_Annie_Mole.jpg (accessed May 30, 2012).

hybrids of corn for optimum production levels in particular environments. F1 hybrids are the off-spring of two genetically distinct varieties that have been cross-bred. F1 hybrids have the advantage, for agriculture, of being homogenous and predictable – making them ideal for mechanized forms of harvesting, for instance. On the other hand, they do not produce useful seeds for replanting and they are relatively expensive to buy. They are good for agribusiness because farmers have to buy the seeds every year. Soybeans proved to be more difficult to hybridize than corn. They were recalcitrant objects.

In the 1970s, there was a shift from sexual intervention (controlled breeding) to cellular modification and what would become known as genetic modification. The US 1970 Plant Varieties Protection Act made it possible, for the first time, to patent a plant variety. We tend to think of patents as claims on the ownership of an invention (a silicon chip, a combine harvester, etc.) and not a living thing. This Act scrambled culture and nature. It was one of the actants in this increasingly complex network. It helped lead to more development of soybeans as technology advanced and, eventually, to a massive increase in the soybean crop and use of soybeans in a multitude of processed food products.

As the soybean used in farming became progressively more uniform and genetically impoverished, it became more susceptible to disease. Weeds also became more of an issue as a threat to soybean monocultures. This made the idea of a soybean that was genetically modified to be resistant to herbicide attractive. Farmers could spray crops with a herbicide and kill the weeds but not the crop. The Monsanto Corporation produced the soybean "Roundup Ready" which could be sprayed with the herbicide "Roundup." This variety of soybean, which had been produced with the help of tobacco DNA, was not produced in order to increase yield but to make the use of "Roundup" easy and unavoidable, thus increasing Monsanto's profits and making farmers dependent on Monsanto. "In their Roundup Ready incarnation soybeans are hybrid agents of corporate science in which the entanglement of technical and business practices is incorporated in the seed" (Whatmore 2002: 132).

When GM crops, including soybeans, were being introduced to the United Kingdom, they were met with stubborn resistance from consumers who no longer believed assurances by experts of food safety. The litany of recent food scares had removed trust. Soybeans and other GM foods were labeled "Frankenstein food" by activists – producing what Whatmore calls "an anti-disciplinary configuration of the soybean" (Whatmore 2002: 141) in which the soybean was enrolled into a different kind of network.

Whatmore's story of the soybean is an illustration of the way in which particular culture-natures are produced through diverse acts, things, and relations. The soybeans hold within them farmers, activists, legal frameworks, Monsanto, scientists, consumers, and others, and:

> … emerges as a lively presence that agglomerates very diverse acts and complicates the distribution of powers and knowledges in the precarious business of growing and eating.… Journeying in multifarious guises, the soybean fleshes out the interval between these distant but simultaneous moments, tracing rents and folds, currents and frictions in the topological performance of producing and consuming global and local, 'humans' and 'things'. (Whatmore 2002: 142)

Whatmore's conception of **hybrid geographies** asks us to think again about culture-nature. While early geographers looked at the role of culture in relation to nature with one always determining the other, later radical geographers insisted on the social production of "second natures." Now, Whatmore and others are asking where we might draw the line between culture and nature. Where does nature end and culture (or the social) start for a soybean? This complicated story suggests that we need new ways of thinking beyond two discrete sets of stuff labeled nature and culture. A hybrid, relational approach asks us to trace the relations between diverse objects in the world that act to assemble wholes that are irreducible to their component pieces and cannot simply be labeled culture or nature.

Whatmore's work has been extended, along with colleagues', into the complicated, "more-than-human" world of flood risk management. In this work, Whatmore enrolls both specialist broadly scientific expertise as well as the local expertise of residents in flood-affected areas to open up new way of thinking about flood risk and management that bring together a wide array of human and nonhuman actants (Lane *et al.* 2011).

Conclusions

In this chapter, we have explored a number of ways in which contemporary geographers are seeking to bring together human and nonhuman aspects of the world in a geography that crosses the divide between what we know as human and physical geography. These attempts to collapse the nature/culture dualism and replace it with a flat ontology of connections and networks has not been without its critics. One line of critique, for instance, is that by distributing agency beyond the human world we also distribute blame. When asking questions about climate change, or species extinction, or plastic pollution, for instance, how does it help to point out how humans and things are connected and interrelated? How does this help us get out of the trouble we have caused? (Büscher 2022). Nevertheless, many human geographers and an increasing number of physical geographers are working to close the divide that opened up as human geography embraced an array of theoretical approaches that were critical of the positivism inherent in much of spatial science – an approach that treated human and physical processes as more or less analogous. We have seen how human geographers and physical geographers have sought to engage each other and how animal geography has sought to include the world of animals (part of the domain of biogeography) in its orbit. In the latter half of the chapter, we explored how ANT and hybrid geographies provide theoretical approaches to worlds in which divides between human and nonhuman, or sentient and nonsentient, cease to make sense. As with spatial science, human and nonhuman are seen as broadly equivalent and the human characteristics of reason, imagination, and subjectivity are removed from the pedestal constructed by humanists and others. In many ways it is no longer easy to say whether a practitioner of ANT or hybrid geography is "human" or "physical." Such a divide does not make much sense in such a world. Perhaps we are returning to a world that would be recognized by Humboldt or Ritter in which nature and culture are part of the same intellectual arena. It remains the case, however, that the vast majority of practitioners of ANT are geographers formally known as human geographers. As we will see in the next chapter, though, physical geographers are increasingly joining them. Theoretical engagements such as the ones discussed in this chapter do open up the possibility of a "more-than-human" and "more-than-physical" geography in which we may have more than sports, the weather, and family to talk about in the corridor. These are encouraging signs.

References

Allen, C. D. and Lukinbeal, C. (2010) Practicing physical geography: an actor-network view of physical geography exemplified by the Rock Art Stability Index. *Progress in Physical Geography*, 35, 227–248

Anderson, K. (1995) Culture and nature at the Adelaide Zoo: at the frontiers of "human" geography. *Transactions of the Institute of British Geographers*, 20, 275–294.

Anderson, K. (1997) A walk on the wild side: a critical geography of domestication. *Progress in Human Geography*, 21, 463–485.

Bingham, N. (1996) Object-ions: from technological determinism towards geographies of relations. *Environment and Planning D: Society and Space*, 14, 635–657.

Braun, B. (2004) Querying posthumanisms. *Geoforum*, 35, 269–273.

Büscher, B. (2022) The nonhuman turn: critical reflections on alienations, entanglement and nature under capitalism. *Dialogues in Human Geography*, 12, 54–73.

Castree, N. (2005) *Nature*, Routledge, London.

Castree, N. and Nash, C. (2006) Posthuman geographies. *Social and Cultural Geography*, 7, 501–504.

Cronon, W. (1991) *Nature's Metropolis: Chicago and the Great West*, Norton, New York.

Davis, W. M. (1906) An inductive study of the content of geography. *Bulletin of the American Geographical Society*, 38(1), 67–84.

Derrida, J. (2003) And say the animal responded, in *Zoontologies: The Question of the Animal* (ed. C. Wolfe), University of Minnesota Press, Minneapolis, MN.

Gandy, M. (2002) *Concrete and Clay: Reworking Nature in New York City*, MIT Press, Cambridge, MA.

Gillespie, K. and Collard, R.-C. (eds) (2015) *Critical Animal Geographies: Politics, Intersections, and Hierarchies in a Multispecies World*, Routledge, London.

Greenhough, B. (2006) Tales of an island-laboratory: defining the field in geography and science studies. *Transactions of the Institute of British Geographers*, 31, 224–237.

Gullo, A., Lassiter, U., and Wolch, J. R. (1998) The cougar's tale, in *Animal Geographies: Place, Politics, and Identity in the Nature–Culture Borderlands* (eds J. R. Wolch and J. Emel), Verso, London.

*Hinchliffe, S. (2002) Inhabiting—landscapes and natures, in *The Handbook of Cultural Geography* (eds K. Anderson *et al.*), Sage, London, pp. 207–226.

Hinchliffe, S. (2010) Working with multiples: a non-representational approach to environmental issues, in *Taking Place: Non-Representational Theories and Geography* (eds B. Anderson and P. Harrison), Ashgate, Farnham, pp. 303–320.

Inkpen, R. (2005) *Science, Philosophy and Physical Geography*, Routledge, London.

Kaika, M. (2005) *City of Flows: Modernity, Nature, and the City*, Routledge, New York.

Lane, S. N., Odoni, N., Landstrom, C., Whatmore, S. J., *et al.* (2011) Doing flood risk science differently: an experiment in radical scientific method. *Transactions of the Institute of British Geographers*, 36, 15–36.

Latour, B. (1993) *We Have Never Been Modern*, Harvester Wheatsheaf, New York.

Latour, B. (1994) On technical mediation—philosophy, sociology, genealogy. *Common Knowledge*, 4, 29–64.

Latour, B. (2005) *Reassembling the Social: An Introduction to Actor-Network-Theory*, Oxford University Press, New York.

Loftus, A. and Lumsden, F. (2008) Reworking hegemony in the urban waterscape. *Transactions of the Institute of British Geographers*, 33, 109–126.

Murdoch, J. (1997a) Inhuman/nonhuman/human: actor-network theory and the prospects for a nondualistic and symmetrical perspective on nature and society. *Environment and Planning D: Society and Space*, 15, 731–756.

*Murdoch, J. (1997b) Towards a geography of heterogeneous associations. *Progress in Human Geography*, 21, 321–337.

*Murdoch, J. (2006) *Post-Structuralist Geography: A Guide to Relational Space*, Sage, London.

*Philo, C. (1995) Animals, geography, and the city: notes on inclusions and exclusions. *Environment and Planning D: Society and Space*, 13, 655–681.

Philo, C. and Wilbert, C. (eds) (2000) *Animal Spaces, Beastly Places: New Geographies of Human–Animal Relations*, Routledge, New York.

Power, E. R. (2009) Border-processes and homemaking: encounters with possums in suburban Australian homes. *Cultural Geographies*, 16, 29–54.

Sack, R. D. (1997) *Homo Geographicus*, Johns Hopkins University Press, Baltimore, MD.

Santos, M. L. (2021) *For a New Geography*, University of Minnesota Press, Minneapolis, MN.

Smith, N. (1991) *Uneven Development: Nature, Capital, and the Production of Space*, Blackwell, Oxford.

*Srinivasan, K. (2019) Remaking more-than-human society: thought experiments on street dogs as "nature". *Transactions of the Institute of British Geographers*, 44, 376–391.

Steinberg, P. E. (2001) *The Social Construction of the Ocean*, Cambridge University Press, Cambridge.

Swyngedouw, E. (1999) Modernity and hybridity: nature, regeneracionismo, and the production of the Spanish waterscape, 1890–1930. *Annals for the Association of American Geographers*, 89, 443–465.

Tuan, Y.-F. (1984) *Dominance and Affection: The Making of Pets*, Yale University Press, London.

Tuan, Y.-F. (1991) A view of geography. *Geographical Review*, 81, 99–107.

* Whatmore, S. (2002) *Hybrid Geographies: Natures, Cultures, Spaces*, Sage, Thousand Oaks, CA.

Wilbert, C. (1999) Anti-this-against-that: resistances along a human non-human axis, in *Entanglements of Power: Geographies of Domination/Resistance* (eds J. P. Sharp, P. Routledge, C. Philo, and R. Paddison), Routledge, London, pp. 238–255.

Wolch, J. R. (1996) Zoöpolis. *Capitalism Nature Socialism*, 7, 21–48.

* Wolch, J. R. (2002) Anima urbis. *Progress in Human Geography*, 26, 721–742.

Wolch, J. R. and Emel, J. (1998) *Animal Geographies: Place, Politics, and Identity in the Nature–Culture Borderlands*, Verso, New York.

Chapter 13

More-than-Physical Geographies

We saw in Chapter 12 how more-than-human geographies emerged from (mainly) human geographers seeking to break down the binary of nature and society as well as the divisions between human and physical geography. In this chapter, we will explore the knotty problem of theory and physical geography and how physical and human geographers have worked together to question these long-established divisions from the perspective of physical geography and aligned disciplines such as geology. The starting point is physical geography's long-standing antipathy toward theory and philosophy. We saw in Chapter 12 how as we moved beyond the 1950s, geographers became more specialized and human and physical geography began to take very different paths. Human geographers, as we have seen throughout this book, became fixated on a range of theoretical and philosophical debates while physical geographers resolutely did not. In 1994, Rhoads and Thorn wrote that

> Geomorphologists generally receive little, if any, formal training in philosophical topics at the undergraduate or graduate level. Moreover, in many cases what little training they do receive often comes from human geographers, who simply inform the physical geographers that they are empiricists and then proceed to survey the myriad philosophical perspectives in human geography. (Rhoads and Thorn 1994: 91)

If we generalize geomorphologists to physical geographers, then this is the problem I am trying to avoid. It is my job to convince readers who identify as physical geographers that theory and philosophy are an important part of your toolkit just as they are for your human geography peers and colleagues.

Reengaging Human and Physical Geography

One concerted attempt to reengage human and physical geography has been made by Doreen Massey. She saw a common backdrop of new ways of thinking about space and time that might transcend the major split in disciplinary thinking. To Massey, geographical imaginations of **space–time**

---

provide a common thread for all parts of the discipline. The particular version of space–time she had in mind was a deeply relational one:

> In brief, a number of human geographers are now trying to rethink space as integrally space-time and to conceptualize space-time as relative (defined in terms of the entities 'within' it), relational (as constituted through the operation of social relations, through which the 'entities' are also constituted) and integral to the constitution of the entities themselves (the entities are local time-spaces). (Massey 1999: 262)

While Massey was not the first geographer to point to a unification of space and time as a basis for the reunification of geography (see Unwin 1992), her argument is framed with reference to relational conceptions of space and recent developments in physics. Massey is on tricky ground here. Spatial science, as we have seen, sought to gain some of the cachet of hard science by dealing with imported epistemologies. It was widely critiqued for this misapplication of natural science techniques to humans. It was not only spatial scientists who looked to physics for legitimation, though: "Cultural geographers may cite chaos theory, urban theorists turn to formulations from quantum mechanics, anyone arguing about the nature of knowledge might draw on the thinking of Heisenberg" (Massey 1999: 263). Massey makes it clear that she does not suffer from such "science envy":

> In human geography and related disciplines, for instance, what precisely is the status of appeals to quantum mechanics or chaos theory? What, *really*, are the grounds for evocations of fractal space? As provocations to the imagination they may be wonderfully stimulating; as implicit assertions of a single ontology they need justifying; as invocations of a higher, truer science they may be deeply suspect. (Massey 1999: 264)

One of the issues Massey raises is that geography (whether human or physical) is dealing with extremely complicated systems while classical physics, in its rush to find universal statements, deals with simple systems. This may seem like news to many. Part of the capital that physics accrues as the pinnacle of science comes from a sense that it is difficult. Things like gravity and quantum mechanics are presented to us through dazzling equations and with the help of unimaginably large and expensive pieces of equipment. In fact, though, the kinds of things geographers deal with in particular places – from an eroding desert landscape to a global city – are massively complex systems with no clear edges or logic. They are beyond computation. Scientifically speaking, they are "open systems" subject to sudden and inexplicable transformations due to either factors imported from elsewhere or through human reflection and action. This was the mistake of spatial science – to think of the geographical world as simple. Massey suggests that subjects such as human/physical geography that deal with open and complex systems could now lead the way. Both physical and human geography, Massey argues, are "sciences of the complex and the historical, which are badly served by looking to (…) physics as a model" (Massey 1999: 266):

> … for there to be multiple trajectories – for there to be coexisting differences – there must be space, and for there to be space there must be multiple trajectories. Thus, I want to argue, a more adequate understanding of spatiality for our times would entail the recognition that there is more than one story going on in the world and that these stories have, at least, a relative autonomy. (Massey 1999: 272)

Massey, as with other prominent geographers, had spent a few decades arguing for the importance of space in social thought. Since the nineteenth century, she argued, time had taken precedence over space in social thought and philosophy. While time appeared to be lively and full of possibility, space was taken to be a frozen slice across time where everything is held in situ. Part of Massey's program was to enliven space. By thinking mainly in temporal terms, Massey argued, theorists have thought about "difference" in a particular way. Difference has been seen as points along a temporal continuum. Perhaps the clearest example of this is thinking of the world as divided into developed and developing worlds. In this sense all countries are basically the same, just at different points along a (temporal) path. A similar logic applies to the Marxist theory of history that sees an inevitable progression from feudalism to capitalism to socialism, etc. What this fails to take into account is the possibility of spatial difference, where existing real differences between places leads to different temporal possibilities. Countries may develop differently. This is termed "path dependence":

> … for time genuinely to be held open, space could be imagined as the sphere of the existence of multiplicity, of the possibility of the existence of difference. Such a space is the sphere in which distinct stories coexist, meet up, affect each other, come into conflict or cooperate. This space is not static, not a cross-section through time; it is disrupted, active and generative. It is not a closed system; it is constantly, as space-time, being made. (Massey 1999: 274)

It is in this kind of theorization of space–time that Massey saw a possibility for the emergence of a renewed dialogue between human and physical geographers who can unite in their interest in the relationships between the historical and geographical in particular places that are different from one another. They are united also by their engagement with open and complex systems that cannot easily be bounded and closed without doing violence to actuality. Sharp readers will notice, yet again, a return to the central geographical interest in the theoretical import of singularity in life on earth.

For the most part it is human geographers who have attempted to bridge the widening divide between human and physical geography (Demeritt 1996; Massey 1999, 2006). But physical geographers, traditionally disinclined to engage with what human geographers refer to as theory, have begun to tackle this problem from their side (Sugden 1996; Clifford 2001; Phillips 2004; Inkpen 2005). The physical geographer Jonathan Phillips, for instance, argues that many physical geographers have spent a great deal of time and effort on what are essentially problems of measurement. These problems, he suggests, have become less urgent as new techniques of measurement have been developed:

> With new dating techniques, for instance, we are now able to test some hypotheses regarding long-term landscape evolution, such as whether landforms are (or can be) in steady-state equilibrium, and mitochondrial DNA analysis allows biogeographers to examine evolutionary biogeography in new ways. In another example, new ice-core data, enabled by improved ice-drilling techniques, allows the evaluation of hypotheses regarding climate changes. (Phillips 2004: 38)

Once the new methods have been developed, the main fetter on the development of physical geography, Phillips argues, is the identification of new problems. "The effort to advance theoretical physical geography in the near future," he writes, "will be less about algorithmic or laboratory skill and more about generating ideas about soils, ecosystems, climates and so on" (Phillips 2004: 38). One place he looks for the generation of these new ideas is the boundary between human and physical geography. Sophisticated theoretical work on this boundary is comparatively rare, however. It is widely recognized at a commonsense level that human and physical processes are related, but this recognition is missing in theory development. Phillips makes a strong case for the interrelatedness of the traditional domains of human and physical geography, noting that "even if one is only interested in hillslopes, ecosystems, or evapotranspiration, with no professional concern with human activities or behavior, human agency must still be engaged, because there is often no way to understand hillslopes, ecosystems, or evapotranspiration – for example – without it" (Phillips 2004: 39). Phillips points toward work in geoarchaeology and cultural ecology as bodies of knowledge where the interaction is most developed but notes that we are still "a long way from – for example – a fluvial project in which stream power and political power are both brought to bear on studies of channel change" (Phillips 2004: 39). Here we have a physical geographer urging an understanding of stream flow that includes within it a consideration of politics – the realm of the human geographer. This works both ways. Consider the words of the human geographer Bruce Braun. In a review of various writings by human geographers on the political economy of water, he writes:

> A great deal is written about water, but nary a word is said about the *properties* of water, and how these might influence the sociospatial development of cities. Water flows. It reacts with certain chemicals and dissolves others. Often these dissolved chemicals are invisible, and diffuse rapidly and uncontrollably. Water evaporates when warmed, condenses when cooled, and, as any homeowner in Minneapolis knows, expands when it freezes. It obstructs movement and enables movement. It serves as a pathway for viruses and bacteria, but is also used to cleanse. It seeps into porous materials, but flows across those that are nonporous. Do these properties matter to the material form of the technological networks and bureaucracies that control its movement, or to the narratives, hopes and fears that circulate around it? Can water be 'mobilized' in just any old way? Must the real actors in urban political ecology be always already social? (Braun 2005: 645–646)

Here a human geographer is asking why it is not the case that other human geographers writing about water do not consider its physical properties. If you consider Phillips's point and Braun's together, it is possible to see the potential of work that bridges the two sides of the discipline.

Returning to Phillips, he goes on to ask his physical geography colleagues to engage with the old thorny issue of particularity. As we have seen, Science (with a big S) has traditionally been associated with the universal and the **nomothetic**. It was the perceived backwardness of **idiographic** regional geography that spurred on the advocates of the quantitative revolution. We have seen how an investment in the importance of the particular has been a constant in various versions of human geography. Physical geography, for the most part, has maintained its commitment to the generation of law-like certainty. Phillips's embrace of particularity, and even the very human idea of "place," is therefore striking (see also Allen and Lukinbeal 2010). Phillips argues that historical and geographical contingencies are beginning to be recognized as important across the natural sciences – perhaps even just as important as universal laws. This particularity cannot be magicked away by the refinement of models. Rather contingency and particularity have to become part of the theoretical explanation of physical/human processes:

> The operations and manifestations of earth systems are often characterized by contingent factors that may have significant – even predominant – influences on system states and outcomes. That is, contingencies are not necessarily noise superimposed on patterns determined by general laws, or complications that can eventually be described by general laws. They may be factors which are irreducibly place- and/or time-dependant as or even more important than general laws in determining how the world works. (Phillips 2004: 40)

At the core of Phillips's arguments is the belief that physical geography (and, by extension, geography in general) is uniquely placed to meet the challenge of integrating nomothetic and idiographic approaches to the world. Indeed, part of the reason that physical geography might be able to tackle these issues is its long-time association with human geography and the old debates about the value of regional geography (which was always partly physical). Part of this embrace of the particular is also an embrace of the entanglement of physical and human worlds:

> We will proceed with understanding of and engagement with realities such as, for example, the fact that ecosystem restoration goals have as much or more to do with cultural values as with environmental factors, or that water-resource economics is likely to influence the flow of some streams more than precipitation. And we will find new ways to integrate nomothetic science with historical, interpretive research in ways that preserve the information and insight of both approaches. (Phillips 2004: 42)

Here we have a physical geographer grappling with the two most consistently important questions in the history of geographic theory – the relationships between the natural and human worlds and between the general and the particular. The arguments for a greater theoretical conversation between human and physical geographers have grown in recent years. It is possible to see a productive relationship between aspects of human and physical geography in the work of political ecologists (Blaikie and Brookfield 1987; Peet *et al.* 2010; Robbins 2012).

Why Theory and Philosophy (Should) Matter to Physical Geographers

The physical geographer, Richard Chorley, famously said, "Whenever anyone mentions theory to a geomorphologist, he instinctively reached for his soil auger" (Chorley 1978: 1). Chorley was not thinking of himself here – his work clearly did engage with theory. He was thinking of his colleagues who thought of their work as primarily practical rather than conceptual. To Chorley's geomorphologist, the world was out there waiting to be recorded – evidence was something in the world in slopes, and profiles, ice flows, and riverbeds. Research was something that happened in the field and not sitting behind a desk. Physical geographers have classically seen themselves as practitioners of empirical forms of enquiry, and as suspicious of any non-empirical investigation. This reluctance to engage with the world of theory is reflected in books about geographical theory and philosophy that are overwhelmingly written by human geographers. Whatever Chorley's colleagues think, however, the practice of physical geography is just as theory-laden as human geography. It is not possible to pick your soil auger, of any other instrument and simply go into the world and start measuring stuff at random.

Philosophy and theory are necessary to any science because they remind scientists that they are part of the social world, not neutral observers gifted with the mythical view-from-nowhere. Philosophy and theory provide a form of inquiry into the strengths and weaknesses of science as a practice that seeks valid and certain knowledge of the world. Theory can provide the foundation for claims to knowledge while guarding against the illusions of certainty and truth. Philosophy and theory are important bedrocks for any science – including physical geography.

There are a number of philosophical approaches to science that help us to understand the practice of physical geography. **Logical Positivism** dominated science in early twentieth century – it was based on the elimination of the metaphysical and the promotion of knowledge based on direct, unmediated experience – that which can be observed. It insisted that empirical knowledge can be theory neutral – perhaps the philosophical underpinning of the geomorphologist's desire to reach for his or her soil augur. Logical positivism became so prevalent that it became known as the "received view." Despite this success, it was generally discredited by the 1960s as it was challenged by a series of other philosophies of science.

At the other end of the philosophical spectrum from logical positivism is **constructionism** or **constructivism**. Constructionism insists that observations are always theory dependent and often linked to both personal beliefs and opinions as well as wider societal structures that make certain kinds of enquiry legitimate and others illegitimate. Constructivism acknowledges the value judgments of scientists and uses less formalized and mathematical versions of "theory." Constructionism insists that the context of discovery and context of justification are related. In other words, scientific justification is not pure and removed from values but thoroughly wrapped up in them. While logical positivism states that scientific views are acceptable when they are in some sense objective, constructionism states that theories are acceptable not because of objectivity but because of agreement between scientists. Science does not move toward *truth* but toward *agreement*.

A third approach to the philosophy of science is **scientific realism.** Scientific realism, like positivism, insists on the reality of a world beyond our individual or social conceptions of it. Unlike positivism, however, it recognizes the imperfect fit between our theories of the world and the truth of the world out there. Scientific realists argue that theories move toward truth and that when scientific theories are approximately true then they tend to be empirically successful.

Let's return to Richard Chorley's geomorphologist and his soil auger. Someone mentions philosophy in his study, he reaches for his soil auger, throws on his outdoor physical geographer gear, and heads for the field. What is he going to do with his auger? I want to focus here on two things that physical geographers generally do, with or without soil augers – they classify and they measure. Let's start with classification. Geomorphology is the study of landforms. This is based on acts of classification. But what are landforms? How do we know when one landform ends and the other begins? Do they just exist in nature (natural kinds) waiting to be discovered, or are they artifacts made up by scientists? Are they natural or nominal? The search for **natural kinds** is the attempt to discover the true nature of the world and is epistemologically privileged.

Why is a banana like a computer? It isn't! There is little we can say that bananas share with computers. There are, however, possible lists of things which include both bananas and computers – such as "solid things," "or things that can be found in my house." But are these groupings based on things that matter? Or are they shallow and superficial resemblances? The group of things that can be found in my house is certainly a group of things – but the connections seem arbitrary and trivial. But if we say all these things are fruit then we seem to be saying something deeper and more important – that they contain the seeds of the plant. We can divide up reality any way we want – but some divisions seem to make more sense – seem natural. Chemical elements are good example of natural kinds. They have a logic that divides them – the number of electrons and atomic weight – these allow us to guess the properties of all members of a group. If I discover something about a single hydrogen atom – it will be true of all of them. The kind, hydrogen atom, is seen to be part of reality and how the universe behaves. Groups like Italian cheeses, and culinary vegetables make sense – but are clearly groups that reflect human interests. They are kinds that make sense – but not natural kinds.

One question for geomorphology (and physical geography more generally) is "are its kinds natural?". One argument is that things like cliffs, rivers, glaciers, and volcanoes do not have essences defined by either their morphology, or shape, or processes at work – but are instead the result of more essential kinds at a more micro level – the kinds in physics and chemistry. So, is physical geography just applied physics and chemistry? Does that matter? Can explanations at the scale of a landscape be reduced to explanations at the level of atoms – or does the accumulation, or assemblage, of a landscape lead to the need for other kinds of explanation?

Physical geographers characteristically engage in processes of classification – dividing the world up into things that are different from each other. Here's an example given by Blue and Brierley (2016). Fluvial geomorphologists distinguish braided and anastomosing rivers. Braided rivers form a network of many branches surrounding small, temporary islands or sand bars. Anastomosing rivers also include multiple channels but these are semi-permanent and surround parts of a flood plain. These are not separate things like hydrogen and helium. There is a continuum of instances between them – and the line that separates them is in some sense arbitrary. They are clearly different things but there is no clear division. In fact, fluvial scientists invented another type to sit between them – a so-called transitional state. This is true for instance of the Upper Yellow River at Dari in Western China, which has a transitional form between a braided and anastomosing river. These two forms have different processes associated with them. Calling a third state transitional recognizes that not all rivers fit normal classification boundaries. These are not natural kinds. Some kinds are not "natural" but loose convenient groupings. This might be true of pretty much all "kinds" in physical geography. These kinds are, at least in part, products of human interests and perceptions rather than the underlying structure of the universe. This is important as the idea of natural kinds undergirds the notion of scientific realism – the idea that successful theories represent theory-independent phenomena – the idea that scientific laws and principles describe the universe as it actually is.

If we cannot be certain about the kinds of physical geography, then we might turn instead to another approach – such as constructivism or conventionalism. This, you will recall, treats kinds as human inventions based on whatever purpose we bring to the natural world. Cooks have kinds and so do geomorphologists – and their kinds are equally real (or unreal). In its extreme form, constructivists argue that all kinds are the product of human interest and discourse. Not "real" and independent of us but just useful for our purposes. Constructivists don't care that a kind is not real – just that it works and is in some sense useful.

A third approach to the problem of classification and kinds is known as **promiscuous realism** (Dupre 1993). Promiscuous realism believes in the reality of the world but insists that our classifications are countless and depend on our interests. A common-sense definition of fruit is different from our biological classification of fruit. Bananas as cultivated may have seeds but they are sterile and very small. Does this mean they are not a fruit anymore? Tomatoes are fruit biologically speaking but we do not conventionally think of them that way. Knowledge is knowing a tomato is a fruit. Wisdom is knowing not to put it into a fruit salad. Each way of thinking is equally legitimate.

This philosophical musing about kinds is important for physical geography. In conversations between physical geographers and human geographers, physical geographers often play the role of scientists against the soft logics of social and cultural enquiry. The status as science gives it status – the very status that human geographers looked for from the 1950s onwards in the quantitative revolution. Compared to physicists and chemists however, physical geography looks like a soft science that focusses on superficial kinds that some insist can only really be investigated by reducing things to the hard science of natural kinds in physics and chemistry. It comes down to this – if the kinds of physical geography are not natural kinds, then they are in some sense human constructs that are context dependent. This undermines the privileged position of physical geography as a science of seemingly closed systems.

If we return to Chorley's geomorphologist and his soil auger, one thing he will have to decide is what kind of soil he is examining when he gets his soil sample. Soil scientists have very elaborate taxonomies based on the division of clay soils, sandy soils, and silt soils. Because these are not natural kinds like hydrogen and helium, but messy mixtures, there are intermediate types like silt loams that mix clay, sand, and silt. And then hundreds of other subdivisions. Taxonomy is difficult and somewhat dependent on the interests of the scientist. The Russians were famously good at classifying soils and many of names for soils are actually Russian – like chernozem and podzol. The Russian geographer Vasily V. Dukuchev is widely credited as being the father of soil science. They were so good at naming soils that the Americans decided they wanted their own names and the US Department of Agriculture came up with a completely different set of soil types based on different properties of soil. They did not want their soils with Russian names. In the American system, a chernozem is approximately equivalent to a mollisol. This suggests that the philosophy of constructivism makes some sense here. Soil type naming is in some sense geopolitical. It also suggests that the idea of promiscuous realism might be useful. There are an infinite number of soils and any way that we choose to divide them into types has some basis in reality but is also in some sense arbitrary. We could, and do, classify them differently depending on our interests and how useful those classifications are.

Once we have dealt with the issue of classification and kinds, the second question we have to face is measurement. Let's presume that Chorley's physical geographer is likely to be measuring things when they go to the field to run away from philosophy. They might weigh some samples, they might measure the depth of a soil horizon or the size of individual grains of soil. On the topic of measurement, Blue and Brierley (2016) start from the observation that science has an elevated position in society that is partly based on measurement and quantification which are practices equated with objectivity, fairness, and intellectual debate. Quantification gives a sense of accuracy, precision, credibility, and replicability. It makes something known. Blue and Brierley reveal how seemingly innocent quantification can obscure the contextual nature of knowledge production. They discuss river channels and use the example of the Yellow River in China. We know that river channels are shaped by the transport of sediment by water. Perfect knowledge, they argue, would produce mechanistic understandings of river channel formation right down to the level of particles. We could use this knowledge to manage the flow of rivers through areas inhabited by humans. But we do not have perfect knowledge and have to make models using statistical methods. However hard they try, scientists of things that flow, like water, cannot account for turbulence – an open process subject to chance and contingency. As rivers exist in the world and not in a lab, we have to make decisions about where to draw a line around our investigation and this line belies a messy world. It is another act of classification that we have to do before we take measurements. What we decide to measure reflects judgments we make about what is important – judgments made within social and political contexts. These decisions can become fossilized – accepted knowledge and then part of management schemes for river systems. River management generally aims for improved ecological conditions which, it is generally agreed, means a diversity of physical habitats. But how do you measure diversity? What we choose to measure leads to future management practices. The choice of what to measure is thus both scientific and political. Decisions of what to measure reflect knowledges, experiences, and the mindsets of researchers as well as the social structures that surround them. All of this is influenced by the theory we employ, language we use, and the models and tools (including soil augers) available to us. Measurement, Blue and Brierley insist, involves the co-production of social and physical worlds, and as acts of measurement instill and embed the values of those who make those choices, they become part of the landscape.

> Identifying and naming particular morphologies sets normative expectations as to how the river 'should' be, and how it should or should not change over time. If the river subsequently begins to adjust, whatever the cause, perceptions of environmental 'degradation' may seem to justify particular interventions in the river and/or surrounding land use, removing features (…) or even people (…) that 'do not belong'. Rather than protecting geodiversity such interventions might inadvertently, and ironically, act to restrict it to particular sets of named or easily measured variants. (Blue and Brierley 2016: 193)

Normative expectations about what rivers should be like might lead to certain kinds of intervention that remove features or people in the name of protecting easily measurable variants of diversity – and thus both our classifications and our measurement become part of the landscape.

Chorley's physical geographer with their soil auger is thus not really escaping theory and philosophy. As soon as they start doing anything with the auger, they are involved in philosophical and theoretical puzzles. How they classify soil, and what they measure, are all based on theoretical decisions and connect back to the social world.

Approaching Geomorphology

In this section, we move away from the philosophical speculation on the scientific nature of physical geography and, instead, focus on the history of ideas in the discipline in order to show how different approaches to the same elements of the physical world have shifted. The focus will be on geomorphology. We will start by considering the two different approaches to geomorphology proposed by William Morris Davis and Grove Karl Gilbert.

William Morris Davis (1850–1934), a President of the Association of American Geographers was the leading figure in geomorphology in early twentieth century. We were introduced to Davis and his ideas as examples of the influence of Darwin in Chapter 3. At the center of his approach was his theory of the cycle of erosion. Davis's approach to geomorphology was largely historical, focused on the evolution of landscapes. His focus was on the orderly evolution of landscapes from one form to another as they approached old age. This approach was based on Davis' conviction that geomorphology as a science should be based on observation, description, and generalization. Davis was a geographer at Harvard and supervised many students who went on to be influential in the discipline, ensuring that his ideas continued to hold sway long after his death in 1934.

Davisian geomorphology was challenged by the work of Grove Karl Gilbert, who lived between 1843 and 1918 and worked at the same time as Davis. Gilbert, however, did not hold an official position in a University and did not supervise students. His ideas remained submerged until late in the twentieth century. Rather than thinking of landscapes in an historical evolutionary way, Gilbert approached landscapes from what he called a process perspective – using the scientific method of hypothesis testing in order to identify what processes were at work behind the formation of observed landscapes. To understand landscapes around rivers, therefore, he sought to understand the mechanics of water. The difference between Davis's approach to fluvial landforms and Gilbert's approach can be illustrated by two quotations.

> The forces by which structures and attitudes have been determined do not come within the scope of geographical inquiry, but the structures acquired by the action of the forces serve as the essential basis for the genetic classification of geographical forms. (Davis 1899: 482)

> The first result of the wearing of the walls of a stream's channel is the formation of a flood-plain. As an effect of momentum the current is always swiftest along the outside of a curve of the channel, and it is there that the wearing is performed; while at the inner side of the curve the current is so slow that part of the load is deposited. In this way the width of the channel remains the same while its position is shifted … (Gilbert and Dutton 1877: 126)

In the first quote, Davis specifically states that thinking about the forces that lead to structures is not geographical – that geography should focus on the morphology of the structures themselves – which can be classified as geographical forms. Gilbert, writing in 1877, made a very different claim focusing on the very things – the "forces" – that Davis stated were outside the remit of geographical enquiry. While Davis argued for a focus on forms, Gilbert argued for a more microscopic focus on processes.

The geographer Dorothy Sack has argued that Davis's view held sway for so long because: (i) it was informed by a life-cycle analogy from biology and from Darwin in particular – it linked to "big theory" and to all the other fields that were importing Darwin at that time, (ii) that it was not dependent on specialist scientific knowledge – it was understandable to many people at the time and, (iii) Davis taught at Harvard and influenced many students who followed his methods while Gilbert did not. Science is social! (Sack 1992).

Gilbert's process approach became influential when it too could be linked to "big theory" – the tide of positivist scientific geography that swept geography in the 1950s. Gilbert had influenced Strahler – who insisted in the 1950s that geomorphology should combine empirical numerical data subjected to statistical techniques with rational mathematical modeling – and that over time the two would converge. Gilbert's approach thus only became popular in the 1950s, long after his death. If we return to our discussion of what constitutes a natural kind, we can see how Strahler, following Gilbert, wanted to get close to the level of physics, chemistry, and mathematics to produce a properly scientific geography – one that turned away from historical evolutionary theory of forms that could not be clearly defined as natural kinds – as scientific objects. This view would come to haunt physical geography much later, in its explorations of the East Antarctic ice-sheet.

In Massey's attempt to promote a dialogue between human and physical geography (discussed above), she draws on the work of geomorphologist David Sugden. He provides an account of conflicting ideas about the long-term history of the East Antarctica ice-sheet. One group of scientists has suggested that the ice-sheet has been subject to dramatic changes over the past 14 million years, while another argues that it has been largely stable. This is important because the Antarctic ice-sheet is a major component of the earth's climate system and major changes to the size of the ice-sheets are logically related to changes in sea level, temperatures, and ocean currents, among other things. At the heart of Sugden's paper is the observation that the two theories (dynamic and stable) are the result of different ways of looking at the evidence. He argues for the importance of geomorphology in making sense of long-term history in landforms such as the ice-sheet. Unusually for physical geography, he draws on philosophies of science to make his case. Geomorphology, he suggests, in an attempt to be more like physics, has tended to focus on short-term processes rather than long-term landscape evolution:

> Viewed in this light and driven by the aspiration to be scientific, it is perhaps understandable that geo-morphology has stressed reductionism, short-term process studies and experimentation as the optimum route to knowledge. In contrast, long-term studies seem untestable and dominated by speculation. Seen as unscientific, landscape evolution has remained in a *cul de sac* and been easy to ignore. (Sugden 1996: 451–452)

Sugden's suggestion is that just as it is inappropriate for human geography to ape physics, so it is inappropriate for physical geography. If geomorphology is practiced in this way, he argues, it will just seem like a lesser and imperfect science. Geomorphology, and physical geography in general, he argues, are better thought of as an *interpretive* science:

> Viewed in this light, the study of landscape evolution is seen as one important scientific plank in geomor-phology and one which is necessarily strong on interpretation at the expense of experiment. Such a view of geomorphology as an interpretive science makes it complementary to physics rather than necessarily subservient. (Sugden 1996: 452)

Here, Sugden is rehearsing the arguments of Davis and Gilbert. In many ways, the dynamists were following the process model of Gilbert – looking at the micro scale of diatoms and pollen grains as evidence. The stabilists on the other hand were more Davisian – looking at a wider scale of a land-scape with a history. Sugden makes a powerful argument for a **hermeneutic** geomorphology with a strong focus on the interpretive and the historical – things that are often missing from experimental and lab-based sciences (and missing from the perspective of the dynamists). The take home message here is that facts do not speak for themselves – that there is always a layer of interpretation. In the case of the East Antarctic ice-sheet as elsewhere, different theories relate to different sets of facts and what we make of them. Physical geography is not really based on certainty and falsification. It is not that kind of science. There is a constant circling between evidence and theory, which Davis was certainly aware of.

> It is evident that a scheme of geographical classification that is founded on structure, process, and time, must be deductive in a high degree. This is intentionally and avowedly the case in the present instance. As a consequence, the scheme gains a very "theoretical" flavour that is not relished by some geographers, whose work implies that geography, unlike all other sciences, should be developed by the use of only certain ones of the mental faculties, chiefly observation, description, and generalization. But nothing seems to me clearer than that geography has already suffered too long from the disuse of imagination, invention, deduction and the various other mental faculties that contribute towards the attainment of a well-tested explanation. It is like walking on one foot, or looking with one eye, to exclude from geography the "theoretical" half of the brain-power, which other sciences call upon as well as the "practical" half. Indeed, it is only as a result of misunderstanding that an antipathy is implied between theory and practice, for in geography, as in all sound scientific work, the two advance most amiably and effectively together.

Surely the fullest development of geography will not be reached until all the mental faculties that are in any way pertinent to its cultivation are well trained and exercised in geographical investigation. (Davis 1899: 483–484)

The other main take-home from the debates around the East Antarctic ice-sheet is that physical geography, like human geography, is an interpretive exercise concerned with specific causal circumstances in a particular spatial context – not the search for law-like generalizations.

Critical Physical Geography and Versions of the Anthropocene

The division between human and physical geography reflects a deep-seated binary in Western culture between nature and society. There are several areas where geographers have sought to get beyond this way of thinking. One of these is **Critical Physical Geography (CPG).**

> We term this integrative intellectual practice *critical physical geography (CPG)*. Its central precept is that we cannot rely on explanations grounded in physical or critical human geography alone because socio-biophysical landscapes are as much the product of unequal power relations, histories of colonialism, and racial and gender disparities as they are of hydrology, ecology, and climate change. CPG is thus based in the careful integrative work necessary to render this co-production legible. (Lave *et al.* 2014: 3)

We explored the established approach of "more-than-human" geography in Chapter 12. In a sense, CPG is a similar enterprise – just starting from the other end of a human–physical continuum. We might also call it "more-than-physical" geography. The biggest prompt for CPG is probably the announcement of a new geological age – the **Anthropocene**. The Anthropocene is the name given to the current age by geologists and others who believe that mankind is now the driving force in geological change. While they argue over the exact point at which the Anthropocene began, ranging from around 8,000 years ago when humans took up farming and sedentary life to the colonization of the Americas around 1600, to the advent of the Industrial Revolution starting around 1750, to the first atomic tests in the 1940s. In any case, it is clear that some future geologist or geomorphologist will find evidence of the human influence on the earth's surface in any future record – whether it be increased levels of carbon, or radioactive isotopes from atomic explosions. The general label of the Anthropocene is accompanied by the current era of post 1850 global heating caused by the activities of humans as well as what has been termed the sixth mass extinction caused not by natural forces such as meteor strikes, but by the activities of humans. Currently, rates of species extinction are around 1,000–10,000 times higher than would occur without humans. There are also arguments about whether the label "Anthopocene" is appropriate. The label seeks to point out the irresponsible ways that humans have interacted with the earth – becoming a geological force in our own right. One problem with this is that by naming an age of humans we (humans) are making the same mistake of hubris that led us to this position in the first place. By suggesting that humans are the driving force of geological change we are, perhaps, overestimating our importance. Afterall, the earth as such will not care if and when we are no longer able to live on it. In addition, many critical scholars point out that it is misleading to blame all humans equally for our current multi-layered environmental crisis which is overwhelmingly caused by the affluent inhabitants of the Global North and the populations of colonial powers. Other suggestions include, for instance, the **Capitalocene**, a term that suggests that it is the system of capitalism that is to blame rather than human life in general, and the **Plantationocene** that has been coined to indicate the central role of plantation agriculture, racism, and chattel slavery (Haraway 2016; Davis *et al.* 2019).

One important critical account of the idea of the Anthropocene is geographer Kathryn Yusoff's book *A Billion Black Anthropocenes or None* (Yusoff 2018). She makes the case that the idea of the

Anthropocene is based on the idea of a universal liberal subject who is both the cause of the current condition and the one who will suffer from its impacts. This humanist subject is, of course, a myth, as real personhood is fractured in multiple ways, and particularly in terms of race. This appeal to a universal humanity ignores the actual histories of both the processes that led to the label of Anthropocene and the histories of the discipline and practice of geology.

> If the Anthropocene proclaims a sudden concern with the exposures of environmental harm to white liberal communities, it does so in the wake of histories in which these harms have been knowingly exported to black and brown communities under the rubric of civilization, progress, modernization, and capitalism. The Anthropocene might seem to offer a dystopic future that laments the end of the world, but imperialism and ongoing (settler) colonialisms have been ending worlds for as long as they have been in existence. The Anthropocene as a politicly infused geology and scientific/popular discourse is just now noticing the extinction it has chosen to continually overlook in the making of its modernity and freedom. (Yusoff 2018: xiii)

Here, Yusoff refers to the ongoing nature of world ending. There is a certain kind of global, scientific, world ending that has been labeled the Anthropocene yet, for many Black and Brown people the world has been continually ending, over and over, since the advent of colonialism and the invention of the idea of race. Yusoff shows how the seemingly remote hard science of geology has been historically entangled with, and complicit in, the histories of extractive colonialism and racism – where both geological material such as gold and coal, and the bodies of Black people, could be exchanged as equivalents. The seeming universality as a science, erases these complicated complicities.

> As the Anthropocene proclaims the language of species life – *anthropos* –through a universalist geological commons, it neatly erases histories of racism that were incubated through the regulatory structure of geological relations. The racial categorization of Blackness shares it natality with mining the New World, as does the material impetus for colonialism in the first instance. (Yusoff 2018: 2)

Yusoff traces the way "White Geology" shared its birth with the invention of Blackness and yet pretended, as science, to be a neutral humanist enterprise. The Atlantic slave trade, for instance, was born in 1441 when the first slaves were sold in Lisbon in order to replace Indigenous people in the mines of Brazil. Yusoff reflects on the idea of the **Golden Spike**, a name given to a boundary in time between one geologic era and another thus denoting a new phase in planetary history. As we have seen, there are a number of potential "Golden Spikes" being argued over as the origin point of the Anthropocene. These include 1610 as the beginning of what is euphemistically called the "collision" of the New and Old worlds where flora and fauna were first "exchanged." Yusoff points out that these words, "collision" and "exchange" erase the actual history of colonial subjugation they refer to. Meanwhile 1800 (the origin point of the Industrial Revolution in Britain) replaces the agency of the White explorer Columbus with the actions of White British industrialists. Once again, Yusoff shows how this move conceals the necessary "pre-capitalist" labor of enslaved Black people on plantations that made capitalism possible – including the compensation paid to slave owners (not the enslaved themselves) following the abolition of slavery in the United Kingdom. These reparations became an important part of the capital that financed the new fossil fuel based industries. The final potential Golden Spike was the great acceleration of the 1950s that had its origin point in the first nuclear test in Almogordo in 1945 – leaving traces of plutonium in soil and rock strata the world over. The nuclear age rested on the erasure of Indigenous people both in the southwest of the United States (the site of both testing grounds and Uranium mining), and in Pacific islands such as Bikini Atol where Indigenous Pacific Islanders were removed, dispersed, and then irradiated for years.

Whichever point the geologists decide upon as the pertinent Golden Spike, it will be haunted by its erasure of Black and Indigenous people. This recent pursuit of geology, Yusoff argues, is a continuation of a history of geology's insistence on its own scientific neutrality while being central to the processes of extraction that are at the core of colonialism. It pretended to be asocial, neutral, and rooted in the nature of rocks, all the while performing the role of a technique of resource extraction.

> Geologic classification enabled the transformation of territory into a readable map of resources and organized the apprehension of extraction and the designation of extractable territories. Geology was the science of material dispossession but also a social technology of naturalization. The motivation of colonialism was as an extraction project. The consequence of this formation of inhuman materialism was the organizing of racializing logics that maps onto and locks into the formation of extractable territories and subjects. (Yusoff 2018: 83)

Central to Yusoff's arguments are that it is necessary to ask questions of the humanist, liberal, universal "we" that is blamed for the Anthropocene and who are expected to experience the negative consequences of massive environmental catastrophe.

> If the imagination of planetary peril coerces an ideal of "we," it only does so when the entrappings of late liberalism become threatened. This "we" negates all responsibility for how the wealth of that geology was built off the subtending strata of indigenous genocide and erasure, slavery and carceral labor, and evades what that accumulation of wealth still makes possible in the present – lest "we" forget that the economies of geology still largely regulate geopolitics and modes of naturalizing, formalizing, and operationalizing dispossession and ongoing settler colonialism. (Yusoff 2018: 106)

Geology, like any science, is not innocent and neutral, but complicit in a fractured past that lives on in the present and in attempts to name and date the new geologic era.

However it is defined or configured, the identification of the Anthropocene highlights the entanglements of human and physical worlds that so often form opposite poles of a binary way of thinking. The Anthropocene (and Capitalocene, Plantationocene, etc.) name an era in which the objects of physical geography (rocks, rivers, ice, the weather, etc.) are thoroughly entangled with the objects of human geography (political, economic, social, and cultural systems and spaces). CPG is an approach to physical geography that directly responds to the Anthropocene by recognizing that science needs to be understood as knowledge which is produced in space and time – in historical and geographical context (Lave *et al.* 2014; Biermann *et al.* 2018). It is not the disembodied knowledge produced from nowhere that science has often pretended to be.

We have already seen CPG in action when we considered the philosophical implications of classification and measurement above. In an introduction of CPG, Rebecca Lave and colleagues return to problems of classification noting that existing definition of biomes such as savannas and rainforests are far from innocent descriptions of portions of the earth's surface. As with the case of braided and anastomosing rivers, they become political issues when they are used as categories for management practices. As Lave *et al.* put it, as soon as policy mobilizes these classifications, they become part of the landscape and the boundaries between biomes move from socially constructed lines to lines which play a role in constituting society. Some human activities are allowed, encouraged, or forbidden in a savanna which are not allowed, encouraged, or forbidden in a rainforest. "Because governance and conservation frameworks are organized around these shifting distinctions" Lave *et al.* write, "new points of collaboration are necessary to reassess a wide variety of boundaries, their (mis)use in the policy realm, and their consequences for social justice and ecological health" (Lave *et al.* 2014: 5). In addition to the biomes example, Lave and colleagues consider the impacts of climate change on glacial-fed rivers in the Peruvian Andes, noting how while it was important to do the science in order to show that they are in fact drying up (conventional physical geography), it is also important to recognize "who manages that water, how stakeholders' objectives and power vary, and how hydrological research to date has benefitted hydroelectric companies more than peasants" and that this "is also a crucial step toward producing more accurate, practical, and relevant knowledge" (Lave *et al.* 2014: 6).

CPG attempts to get beyond the human/physical divide within the discipline by paying attention to how and why environmental questions are asked and, importantly, to who gets to ask these questions. Traditional science, based on the ideas of objectivity and the "view from nowhere," generally avoids questions about the positionality of the scientists. For the most part, these scientists have traditionally been White men based in the Global North where there is the kind of funding necessary for sophisticated labs, equipment, and technicians. CPG recognizes that these facts matter when producing environmental knowledge. One way of addressing the politics of knowledge in physical geography has been labeled **ethnogeomorphology** by Deirdre Wilcock, Gary Brierley, and Richard Howitt (Wilcock *et al.* 2013). In their paper, they consider how moving beyond the human/physical divide allows them to consider physical questions of forms, processes, and evolutionary change in the physical landscape alongside human questions of home and meaning. One way of doing this is through an exploration of the Indigenous approaches to landscape which insist on a human–physical totality. One of the cases they explore is the relationship between the Stuart Lake Dakelh (Carrier) holders of traditional keyohs (landholdings) of interior British Columbia and the limestone karst landscape around them. In traditional geomorphological terms, the landscape consists of glaciated, volcanic, and karst landscapes marked by an intricate set of drainage channels connecting surface lakes through underground channels. What geomorphology would not usually consider is the way that the landscape is embedded in stories and practices among the First Nation people who live and walk along trails which run parallel or perpendicular to the underground waterways. Wilcock and colleagues show how walking the landscape entangles multiple space–times as it connects the First Nation walkers to the paths of ancestors. The following interview segment illustrates these connections between people and land, past and present.

> INTERVIEWER: So the land defines you both in a time … with the ancestors, as well as the way in which the land is situated – so where the lakes and the trails are connected?
> JIM: Yeah, it's part of it – it's not just – you can't just look at it from one perspective. It's like, everything's linked together. It's all connected. *Every way.*
> INTERVIEWER: Time, space – everything?
> JIM: Every way. Yeah. Like, Larry or Kenny, and Victor [other keyoh holders], it's everything, you know, it's just like he's lost, eh. Like they cut one more block [of forest], and he's *more lost* – every time. Goes spiritually and physically, not just – you know, it's not just physically. It's not just 'cause it's gone, there's no more trail, there's no more reference points. *Spiritually too.* (Wilcock *et al.* 2013: 588)

For the First Nations inhabitants of the land, the separation of the natural from the social that is at the core of the split between physical and human geography is unimaginable. For Wilcock and colleagues it is instructive to encourage other forms for knowledge into the field to geomorphology in order to recognize how biophysical and cultural entities are mutually constituted as well as to explore the processes by which power hierarchies between forms of knowledge are formed and exclusions enacted.

CPG scholars Mark Carey, M. Jackson, Allessandro Antonello, and Jaclyn Rushing have recently argued for a feminist glaciology (Carey *et al.* 2016). Such a call immediately seems strange as we have been made to believe that earth objects such as ice are outside of the remit of ideologies. Ice, we might think, is just ice. Carey *et al.* carefully show how ice carries with it contested sets of meanings and narratives not least because ice and glaciers provide evidence for global heating. Carey *et al.* ask four questions about glaciology that were rarely asked in traditional glaciological research:

1. How does the gender identity of knowledge producers influence glaciology?
2. How is glaciology as a science gendered?
3. How have power, domination, and colonialism shaped glaciology?
4. How might glaciologists embrace alternative representations and forms of knowledge about glaciers – such as folk glaciologies?

The knowledge producers of glaciology have most often been men. Glaciology is associated with rugged fieldwork in remote locations that are hard to get to. The practice of studying glaciers is wrapped up in ideas of masculine heroic adventurous pursuits in extreme climates and often up mountains. Carey *et al.* argue that there is a need to diversify who is producing knowledge about glaciers in order to include knowledges that emerge from different places. One reason this has proved difficult for women is that sites of research such as Antarctica have been unsafe for gendered reasons as well as climatic ones. A recent news story in Australia highlighted how women working in the Australian Antarctica Division's research stations were consistently subjected to sexual harassment, displays of pornographic images, inappropriate drinking culture, inadequate facilities for menstruation, and homophobic culture.[1] Such a damning report indicates how it is not just the question of who is doing ice science that matters. Even when women are involved in glaciology, they have to conform to masculine expectations combining the traditional "view from nowhere" of normal science with the respect that comes from physical feats of daring, and demeaning views of women.

Carey *et al.*'s third question concerns systems of domination. Glaciology, like most sciences, is linked to and often funded by resource extractive industries and the military. Glaciology in the United States saw a massive growth in funding in the 1950s when the military were alarmed by a possible attack via Greenland. In both Arctic and Antarctic regions, glaciology was enfolded into the larger geopolitical play of power during the Cold War. Glaciology is part of the same process as the "White Geology" that Yusoff wrote of. Carey *et al.* argue that physical science itself is a master narrative heavily concentrated in the Global North where there are the funds and resources to conduct technology-dependent science. The fieldwork of glaciology, especially when it takes place outside of the Global North, looks suspiciously like the colonial exploits of earlier times. It is for this reason that Carey *et al.* suggest that glaciologists might pay heed to what they call "folk glaciologies," such as the multi-sensory knowledge of Indigenous people who live around glaciers. This is an argument for the kind of ethnogeomorphology discussed above (Wilcock *et al.* 2013).

> The goal is neither to force glaciologists to believe that glaciers listen nor to make indigenous peoples put their full faith in scientists' mathematical equations and computer-generated models (devoid of meaning, spirituality, and reciprocal human-nature relationships). Rather, the goal is to understand that environmental knowledge is always based in systems of power discrepancies and unequal social relations, and overcoming these disparities requires accepting that multiple knowledges exist and are valid within their own contexts. (Carey *et al.* 2016: 782)

This is Carey *et al.*'s fourth point – the importance of alternative representations of ice and glaciers that are derived from the local knowledge of people who live with ice rather than just study it.

---

[1] Josh Butler, Measures to Stamp out Sexual Harassment on Australia's Antarctic Stations After Damning Report, *The Guardian* 30 September 2022 https://www.theguardian.com/world/2022/sep/30/measures-to-stamp-out-sexual-harassment-on-australias-antarctic-stations-after-damning-report (accessed 18/012023).

Storytelling, narrative, and the visual arts, they insist, are valid ways of representing knowledge that can help us to more fully understand ice as more-than-physical. They give the example of the Scottish artists, Katie Paterson, who in her 2007 work, *Langjökull, Snæfellsökull, Solheimajökull,* recorded the sounds of glaciers the artwork was named after and then transferred them to LPs made of ice from the glaciers. These LPs were played simultaneously on record players while the ice melted and the resulting sounds melted away with them over a ten minute period.

> Climatic data from ice core records are often imported into climate models, while rates of glacier retreat chronicling meters melted per year are usually taken directly at face value, with policy implications. Both the ice cores and ice loss measurements feed homogenizing global narratives of glaciers with somewhat restricted views of the cryosphere, lacking emotional and sensory interactions with the ice that occurs in Paterson's artworks. Paterson and other artists thus intervene in such 'truths' by presenting purposefully imprecise social and scientific methodologies and works. (Carey *et al.* 2016: 784–785)

The work of Paterson, and others, decenters the knowledge of largely western, largely White, and largely masculine glaciology. Alternative representations, Carey *et al.* tell us:

> … reveal entirely different approaches, interactions, relationships, perceptions, values, emotions, knowledges, and ways of knowing and interacting with dynamic environments. They decenter the natural sciences, disrupt masculinity, deconstruct embedded power structures, depart from homogenous and masculinist narratives about glaciers, and empower and incorporate different ways of seeing, interacting, and representing glaciers – all key goals of feminist glaciology. (Carey *et al.* 2016: 786)

Conclusions

In this chapter, we have addressed the sometimes fraught relationship between physical geography and theory noting how impossible it is for Chorley's noted geomorphologist to escape philosophy armed only with a soil auger. Physical geographers typically classify and measure, and these cannot be done in a way that does not include some philosophical presuppositions. Neither can they be done in ways that do not recognize the unavoidable links between our knowledge of the physical world and the social, cultural, and political worlds that geomorphologists, like the rest of us, are part of. We saw how theoretical presuppositions ran through debates around the nature of geomorphology between acolytes of Davis and Gilbert and how the relative success of these two approaches was tied to the social circumstances under which knowledge was produced, and how these approaches continued to influence the way contemporary scientists have approached the East Antarctica ice-sheet. We examined, with Kathryn Yusoff, how the geological label of the Anthropocene is a far from innocent label which falsely universalizes both the causes of environmental change and its actual and potential victims. Once again, physical geography and geology are intertwined with social and political hierarchies of power in ways that cannot be disentangled. It is this recognition that led to the advent of CPG. Questioning the established ways of doing science can be a dangerous exercise. The combination of ideas of objectivity alongside the deeply engrained belief in western cultures that nature is separate from society are strongly held norms. They appear to many, particularly to scientists, as common-sense. Questioning common-sense always comes with dangers. Carey *et al.*'s paper on feminist glaciology was subjected to massive levels of ridicule by public figures and in the largely conservative media. It seems they touched a nerve by suggesting that ice might not be just ice and that science might benefit from contact with the humanities and social sciences. Some scientists called the paper "gobbledygook" and, according to an article in the Canadian *National Post,* Dallas Weaver, a researcher of biofilters in aquariums and hatcheries (clearly not a glaciologist!) said that "the social science and humanities have gone functionally insane."[2] It is highly unusual for papers in geography journals to receive this kind of attention. As I was writing this (January 2023), the paper had been viewed and downloaded over 16,000 times. It became part of the culture wars. Part of that is due to the label "feminist" which still causes extreme reactions in some commentators. The other part, though, is surely the collapse of the nature/society divide that the title suggests – and the related idea that science, and the things scientists study, are inescapably political and ideological. The kind of critique suggested by this example of more-than-physical geography, as well as other work referenced in this chapter, seems transgressive and heretical because it asks the reader to question what scholars and others in the West have considered to be common-sense. The fury of the critique is a direct response to the power of the authors' arguments.

References

Allen, C. D. and Lukinbeal, C. (2010) Practicing physical geography: an actor-network view of physical geography exemplified by the rock art stability index. *Progress in Physical Geography*, 35, 227–248.

Biermann, C., Lane, S. N., and Lave, R. (eds) (2018) *The Palgrave Handbook of Critical Physical Geography*, Springer International Publishing: Imprint: Palgrave Macmillan, Cham.

Blaikie, P. M. and Brookfield, H. (1987) *Land Degradation and Society*, Methuen, London.

*Blue, B. and Brierley, G. J. (2016) 'But what do you measure?' Prospects for a constructive critical physical geography. *Area*, 48, 190–197.

Braun, B. (2005) Environmental issues: writing a more-than-human urban geography. *Progress in Human Geography*, 29, 635–650.

*Carey, M., Jackson, M., Antonello, A., and Rushing, J. (2016) Glaciers, gender, and science: a feminist glaciology framework for global environmental change research. *Progress in Human Geography*, 40, 770–793.

Chorley, R. J. (1978) Bases for theory in geomorphology, in *Geomorphology: Present Problems and Future Prospects* (eds C. Embleton, D. Brunsden, and D. K. C. Jones), Oxford University Press, Oxford, pp. 1–13.

Clifford, N. J. (2001) Physical geography – the naughty world revisited. *Transactions of the Institute of British Geographers*, 26, 387–389.

Davis, W. M. (1899) The geographical cycle. *The Geographical Journal*, 14, 481–504.

Davis, J., Moulton, A. A., Van Sant, L., and Williams, B. (2019) Anthropocene, capitalocene, … plantationocene?: a manifesto for ecological justice in an age of global crises. *Geography Compass*, 13, e12438.

Demeritt, D. (1996) Social theory and the reconstruction of science and geography. *Transactions of the Institute of British Geographers*, 21, 484–503.

Dupré, J. (1993) *The Disorder of Things: Metaphysical Foundations of the Disunity of Science*, Harvard University Press, Cambridge, MA.

Gilbert, G. K. and Dutton, C. E. (1877) *Report on the Geology of the Henry Mountains*, Govt. Print. Off., Washington.

Haraway, D. J. (2016) *Staying With the Trouble: Making Kin in the Chthulucene*, Duke University Press, Durham.

*Inkpen, R. (2005) *Science, Philosophy and Physical Geography*, Routledge, London.

*Lave, R., Wilson, M. W., Barron, E. S., Biermann, C., Carey, M. A., Duvall, C. S., Johnson, L., Lane, K. M., Mcclintock, N., Munroe, D., Pain, R., Proctor, J., Rhoads, B. L., Robertson, M. M., Rossi, J., Sayre, N. F., Simon, G., Tadaki, M., and Van Dyke, C. (2014) Intervention: critical physical geography. *Canadian Geographies/Les géographies canadiennes*, 58, 1–10.

*Massey, D. (1999) Space-time, science and the relationship between physical geography and human geography. *Transactions of the Institute of British Geographers*, 24, 261–279.

Massey, D. (2006) Landscape as a provocation – reflections on moving mountains. *Journal of Material Culture*, 11, 33–48.

---

[2] See https://nationalpost.com/news/world/heres-why-an-article-about-feminist-glaciology-is-still-the-top-read-paper-in-a-major-geography-journal (accessed 19 January 2023).

Peet, R., Robbins, P., and Watts, M. (2010) *Global Political Ecology*, Routledge, Milton Park, Abingdon, Oxon, England; New York.

* Phillips, J. (2004) Laws, contingencies, irreversible divergence and physical geography. *The Professional Geographer*, 56, 37–43.

Rhoads, B. L. and Thorn, C. E. (1994) Contemporary philosophical perspectives on physical geography with emphasis on geomorphology. *Geographical Review*, 84, 90–101.

Robbins, P. (2012) *Political Ecology: A Critical Introduction*, John Wiley and Sons, Chichester, West Sussex; Malden, MA.

Sack, D. (1992) New wine in old bottles: the historiography of a paradigm change. *Geomorphology*, 5, 251–263.

* Sugden, D. E. (1996) The East Antarctica ice sheet: unstable ice or unstable ideas? *Transactions of the Institute of British Geographers*, 21, 443–454.

Unwin, P. T. H. (1992) *The Place of Geography*, Longman Scientific & Technical, Harlow.

Wilcock, D., Brierley, G., and Howitt, R. (2013) Ethnogeomorphology. *Progress in Physical Geography: Earth and Environment*, 37, 573–600.

* Yusoff, K. (2018) *A Billion Black Anthropocenes or none*, University of Minnesota Press, Minneapolis, MN.

Chapter 14

Postcolonial, Decolonial, and Anticolonial Geographies

It should be clear from earlier chapters of this book that the discipline of geography is tightly connected to the history of largely Western colonialism and imperialism. Even in its radical moments there are still traces of this history in the way geography is taught and learned in the West. The relationship between the kinds of knowledge that pass for geographical theory, and the kinds of knowledge that do not, can be thought of as a process of exclusion that happens in a systematic way undergirded by sets of power relations that are centuries old. There is a direct relationship between the knowledge that is included in a survey of geographic thought such as this one and the kinds of geographical theory that are ignored. Western theory has been complicit in erasing other kinds of knowledge as part of a colonial project. There is, in other words, a whole world of geographical theory beyond the theory that we teach as the discipline's canon.

This chapter focuses on a range of theoretical approaches – postcolonial, decolonial, and anticolonial – that seek to challenge this process of exclusion from within the framework of academic disciplines such as geography. It relates a tense set of positions that attempt to navigate the position of being within a discipline that is clearly colonial in its origins while, at the same time, critiquing it.

One starting point for this discussion is to recognize that despite the violence of colonialism, both material and epistemological, other ways of knowing, other geographies, continue to exist. This point is made forcefully by Indigenous scholar, Vanessa Watts (Anishnaabe and Haudenosaunee), who asks us to take the place-thought of Indigenous knowledge literally (Watts 2013). Watts describes how Western scholars have appropriated Indigenous knowledge systems as stories or myths in ways that extract them from the place-based Land systems that they are part of. She begins by recounting how Sky-Woman fell through the skies on to the back of a turtle.

> According to Haudenosaunee, Sky Woman fell from a hole in the sky. John Mohawk (2005) writes of her journey towards the waters below. On her descent, Sky Woman fell through the clouds and air towards water below. During her descent, birds could see this falling creature and saw she could not fly. They came to her and helped to lower her slowly to waters beneath her. The birds told Turtle that she must

Geographic Thought: A Critical Introduction, Second Edition. Tim Cresswell.
© 2024 John Wiley & Sons Ltd. Published 2024 by John Wiley & Sons Ltd.

need a place to land, as she possessed no water legs. Turtle rose up, breaking through the surface so that Sky Woman could land on Turtle's back. Once landed, Sky Woman and Turtle began to form the earth, the land becoming an extension of their bodies. (Watts 2013: 21)

There are many ways to respond to this account. One is to dismiss it as simply untrue – to assert that science has more convincing accounts of how the earth was formed. Another is to see this as a story that tells us about a belief system that we are either part of (if we are Indigenous) or not (if we are not). As a story it has value and might provide important insights into how we might think of ourselves (nonindigenous scholars) in relation to "nature." Finally, we might simply believe it. In her paper, Watts insists on the literal truth of this and other stories and argues for the importance of thinking of them as true as a practice that resists colonialism. "Before continuing," she writes, "I would like to emphasize that these two events took place. They were not imagined or fantasized. This is not lore, myth or legend. These histories are not longer versions of "and the moral of the story is …." This is what happened" (Watts 2013: 22). Watts compares what she calls Indigenous "place-thought" with Euro-Western thought that divides ontology and epistemology. "Place-Thought," she writes, "is the non-distinctive space where place and thought were never separated because they never could or can be separated. Place-Thought is based upon the premise that land is alive and thinking and that humans and nonhumans derive agency through the extensions of these Thoughts" (Watts 2013: 21). Watts traces how various theories within the Euro-Western tradition have extended agency to nonhuman actors – including in Bruno Latour's Actor-Network Theory – but only by redefining agency so that it does not include knowing will. Agency, she argues, simply becomes the recognition that things (she uses the example of soil) have effects either by themselves or in relation with other things. This is very different from the Place-Thought of Indigenous knowledge systems that Watts takes literally.

Coloniality can be seen, according to Watts, in how a "foreign epistemological-ontological-divide" describes Place-Thought as stories or myths and, in so doing, does violence to Indigenous ways of knowing and being. Taking Sky-Woman literally means refusing the division between knowing and reality, between ontology and epistemology and insisting on the indivisibility of theory and praxis.

> When Sky Woman falls from the sky and lies on the back of a turtle, she is not only able to create land but becomes territory itself. Therefore, Place-Thought is an extension of her circumstance, desire, and communication with the water and animals – her agency. Through this communication she is able to become the basis by which all future societies will be built upon – Land. (Watts 2013: 23)

As Sky-Woman becomes Land, agency is distributed among all things living and non-living: "… non-human beings choose how they reside, interact and develop relationships with other non-humans. So, all elements of nature possess agency, and this agency is not limited to innate action or causal relationships" (Watts 2013: 23). While this agency shares something with accounts of distributed agency in Actor-Network Theory, it goes well beyond that by keeping a view of agency that a Euro-Western worldview normally reserves for humans. Euro-Western belief systems, Watts argues, have corrupted Indigenous forms of knowledge but not defeated, erased or eradicated them. Indigenous knowledge (like Indigenous people) persists despite 500 years of colonial violence.

> When an Indigenous cosmology is translated through a Euro-Western process, it necessitates a distinction between place and thought. The result of this distinction is a colonized interpretation of both place and thought, where land is simply dirt and thought is only possessed by humans. If we operationalize this distinction, we as Indigenous peoples risk standing in disbelief of ourselves. Even amongst ourselves it can be easy to forget that our ability to speak to the land is not just an echo of a mythic tale or part of a moral code, but a reality. (Watts 2013: 32)

Indigenous people believe the world to be largely a living thing filled with significance and willful agency. In much of the Western world "we" only consider sacred space as a distinctive space divided from the everyday or profane space. This is how the sociologist Emile Durkheim accounted for the notion of the sacred – as based on division and boundedness (Durkheim 2008). Thus we put aside spaces such as churches or temples as "sacred." In Indigenous worldviews, no such clear distinction exists and the landscape is full of sacred significance.

In the Australian context colonial views of territory, property and mapping, which allowed land to be owned and traded, exists in sharp distinction with Aboriginal way of being in Country. Just as Land (with a capital L) denotes the importance of being in the world in some North American Indigenous traditions, so the word Country (with a capital C) denotes a living, breathing character in Aboriginal geographies. Australian anthropologist, Deborah Bird Rose, put it this way:

> In Aboriginal English, the word 'Country' is both a common noun and a proper noun. People talk about Country in the same way that they would talk about a person: they speak to Country, sing to Country, visit Country, worry about Country, grieve for Country and long for Country. People say that Country knows, hears, smells, takes notice, takes care, and feels sorry or happy. Country is a living entity with a yesterday, a today and tomorrow, with consciousness, action, and a will toward life. Because of this richness of meaning, Country is home and peace: nourishment for body, mind and spirit; and heart's ease. (Rose 1996: 7)

This relationship to Country is very different from colonial stories about land. Aboriginal scholar Jill Milroy and nonindigenous scholar Grant Revell frame the clash between colonial and Aboriginal worldviews as one of the different "stories."

> The colonizer, when gazing on the "new" land, sees the place and the story he wants to make, the colonial fiction he will create: a story about property and value, not land and spirit, about the "nation" to come, not the "Country" that is. The colonial story is usually man made, though women will be complicit, and then active, in the story. The "new" story will be told in a foreign language. (Milroy and Revell 2013: 3)

While the colonial story is one of new land, the Aboriginal stories are ones that tie Country to the travels of ancestors. The Dreamtime refers to the time when the ancestors brought the world into being through the songlines – paths and stories that tied the people into a network of land, sea, and sky. In Aboriginal culture, the stories need to be kept alive not through a process of archiving but though a more living sense of practice – by walking the songlines. Milroy and Revell give an example from the Goolarabooloo people.

> The Song Cycle is an oral heritage map. Its songs contain codes of behaviour fundamental to sustaining the balance and well-being of the land and its people and are still sung today.... A Song Cycle has a birth-place and an end place, it has physical (landmass) length and width but it is not just a track. It is made up with "sites" – places, grounds, increase sites, ceremonial grounds, seasonal food places, vegetation for ceremonial usage, trees, shrubs, plants, ochre, land and water within the Song Cycle land, providing all that is needed to sustain life both for humans and animals. (http://www.goolarabooloo.org.au/song_cycle.html accessed January 29, 2023)

It is not just the case that colonists and Aboriginal people had different stories, but that the colonists actively attempted to erase the Aboriginal stories as part of the process of colonization. Milroy and Revell recount the history of colonization in the area around present-day Perth – traditional Noongar land. A grid was laid over the land. The river the Noongar people called the Derbarl Yerri-Yerrigan was the track left by the Waugal (Rainbow) serpent – a spirit home where people would go to visit their ancestors. It was renamed the Swan River.

> For the next one hundred years, explorers and surveyors were the vanguard of colonization, mapping out the new territory suitable for land-hungry settlers, pastoralists, miners, and speculators. The key was to find water supplies, and the mapping of the Canning Stock Route …. As the explorers and colonists moved outward, the English language moved with them, and as Country was "discovered," it was "named" accordingly. The renaming of Country disguised its nature, obscured its meaning, and stole its voice. New maps showed nothing of the meaning of Country or the ancient stories embedded in it. Aboriginal peoples and Country were enclosed within the fictional boundaries and borders of Western cartography: a fictional place with lots of names copied from other places and people, neither imaginative nor original and mostly in English. (Milroy and Revell 2013: 4)

The consequences of the attempted erasure of Indigenous stories by colonial ones are serious. Milroy and Revell tell how the stories are linked to mental and physical health and self-esteem. How the absence of stories – not "knowing one's Country" is experienced as grief and loss.

These different worldviews become clear when the geographical imaginations of Aboriginal people come into contact with settler visions of what parts of the landscape merit preservation and count as heritage (Gelder and Jacobs 1998). In a postcolonial nation, Aboriginal claims to the sacred frequently come into contact with the "modern" state, producing a complicated entanglement of different and opposed geographies. In their book *Uncanny Australia*, Gelder and Jacobs (1998) explore this entanglement in a range of arenas, including the role of museums in collecting and portraying Indigenous heritage and Indigenous opposition to mining. In all of the cases, the authors complicate a binary view of indigenous "traditional" knowledge and settler "modern" knowledge as Indigenous people are not stuck in some distant past but are, instead, part of a changing and often hybrid present.

For the most part, the geographical theory of Aboriginal people in Australia has not made its way into the kinds of geographical theory that form the tradition of geography as it is taught in the West. It does appear occasionally. It is mobilized, for instance, in Sarah Whatmore's *Hybrid Geographies* in order to suggest a certain mutability to practices of Western colonial territoriality in Australia (Whatmore 2002). Outside of geography, Aboriginal conceptions of space have been used by Stephen Muecke to ask why "indigenous philosophy" has not been accorded the status of philosophical knowledge but, instead, consigned to "history" and anthropological "culture" (Muecke 2004). He links the Songlines to Deleuze and Guatarri's vitalist philosophy of nomads and rhizomes. While the presence of Aboriginal/Indigenous thinking is welcome in both cases, it remains the case that their geographical thought is valued because it resembles or supports some acknowledged or emerging body of Western theory – Actor-Network Theory or vitalist philosophy. It is not the same as taking Sky-Woman literally. It is not Indigenous Place-Thought. Holding onto Indigenous forms of knowledge and practice is not easy for Indigenous people who are also academic geographers. Indigenous (Kwakwaka'wakw) geographer, Sarah Hunt, reflected on the challenging nature of being both Indigenous and a geographer, inhabiting spaces that make those joint identities difficult. She compares her first experience of being at an academic conference in Seattle with her childhood experience of attending her first potlatch – a traditional ceremony of the Pacific Northwest coastal Indigenous peoples (Hunt 2014). Hunt describes how academic geography represents what she calls "epistemic violence" against Indigenous forms of knowledge – how Indigenous people in North American have been constantly subjected to the logics of grids, reservations, and property, that are not Indigenous logics but the products of colonial geographic imaginations. She struggled with ways to ask questions within the context of an academic conference while doing justice to her own ways of knowing. And those ways of knowing are also fluid and relational – not a romanticized image of past Indigeneity: "The future of Indigenous rights and political struggles" she writes, "depend on the ability of Indigenous knowledge to retain its active, mobile, relational nature rather than the fixity it is given in colonial law, stuck at the point of contact with colonizers" (Hunt 2014: 30). Her memory of her first potlatch involved a different kind of learning. She recalls joining in the dance, thrilled and smiling, and being told to go try again without the smile.

> There was a productive confusion in this way of learning, one which would not have been possible had I been told in a linear way how to dance at a potlatch. No guidebook or PowerPoint, no essay or instructional video could have given me this type of knowledge. Even though I have since read many books and articles about the potlatch, none of them have captured what I know the potlatch to be. The ontological differences are difficult to explain yet that is where their power lies – in the spaces between intellectual and lived expressions of Indigeneity. I would propose that these gaps in regimes of knowledge provide sites where ontological shifts are possible. So how do we better expose and explore these gaps? (Hunt 2014: 30)

Hunt struggles with how to live between these two worlds and these two ontologies and epistemologies – how to be an Indigenous geographer given the histories that the discipline of geography, and academia in general, comes freighted with.

> The situatedness and place-specific nature of Indigenous knowledge calls for the validation of new kinds of theorizing and new epistemologies that can account for situated, relational Indigenous knowledge and yet remain engaged with broader theoretical debates within geography. There is a danger in ghettoizing Indigenous geographic knowledge as 'other' or a curiosity, rather than engaging this knowledge in broader efforts to actively decolonize geography, navigating among differing power relations at the scales of both the individual academic and the broader discipline. (Hunt 2014: 31)

There are many forms of geographical thought out there in the world that remain largely ignored by the tradition I have been schooled in. Another example of an excluded alternative geography is the Chinese art/science of *Feng Shui*. Feng Shui became popular in the West as a set of rules for interior design during the 1990s. There are whole magazines devoted to its use when you are arranging furniture in your Manhattan apartment. To many Chinese people it represents a profound set of beliefs about the natural world and the place of humans in it. Gleaming modern skyscrapers in Hong Kong or Shanghai are planned around the geographical ideas of Feng Shui.

In a rare example of engagement with these ideas, Chuen-Yan David Lai recounts attempts to locate an old Chinese cemetery in Victoria, British Columbia. Some of the Chinese residents claimed that there has never been one and yet there were late-nineteenth-century references to the existence of such a cemetery. Lai proposed that the best way of discovering its location was to use Feng Shui as that was the form of knowledge that Chinese residents would have used to find an appropriate place the bury the dead:

> *Feng Shui* is a pseudophysical science of climatology and geomorphology originating in fear of strong desert winds, devastating typhoons, thunderstorms, floods, droughts, and other forces of Nature which the ancient Chinese could not explain or control. It may also have been derived from their admiration of the magnificent work of Nature which, through its agents of erosion, wind and water, has carved out lofty mountains and hills which form the greater proportion of the land of China. (Lai 1974: 507)

Traditional Chinese beliefs dictate that it is important to bury ancestors in just the right spot, in harmony with nature, so that their souls could rest and their descendants be prosperous. Feng Shui is based on the idea of the natural world breathing:

> The third parameter of the model is related to the concept of *Feng Shui*, wherein Nature is regarded as a living organism that breathes unceasingly. When it moves its breath produces the *Yang* or male energy, and when it rests its breath produces the *Yin* or female energy. Nature's breath is a twofold element consisting of *Yang* and *Yin* energies which interact continuously and produce all forms of existence on earth.

> Mountain chains are indications of these two life-giving energies. The *Yang* energy is expressed as a lofty mountain range, symbolically called the "Azure Dragon," and the *Yin* energy as a lower ridge called the "White Tiger." The most auspicious model of *Feng Shui* topography is a secluded spot where these two energies converge, interact vigorously, and are kept together in abundance and in harmony by surrounding mountains and streams. (Lai 1974: 508)

Once such a spot is found, there is further work to be done. The body of the dead should ideally face south on a south-facing slope – a direction associated with life and wellbeing. The site should not be on the top of a hill or on a low plain. All of these features of the world that we might account for in geomorphological terms have particular meanings that make the siting of a grave a complicated business with an array of geographical implications.

It is not just "traditional" forms of knowledge that have been excluded in the construction of a narrative of geographical theory:

> Every human group, every local community and every nation has its proper history of the evolution of geographical knowledge and, in the context of intellectual history, its own history of indigenous geographical thought. (Takeuchi 1980: 246)

As Keiichi Takeuchi reminds us, most nations have universities and schools that teach some form of geography in an academic sense and yet our accounts of geography focus on the Anglo-American and, to a lesser extent, European versions of the discipline. We (Western geographers) would be hard pressed to say much about the rich tradition of Japanese historical and regional geography that preceded Japan's imperialist phase (Takeuchi 1980).

One example of a neglected geographic tradition is the Indian one. The geographer Rana Singh has spent much of his career insisting on the importance of traditional Hindu ways of imagining the earth combined with what he calls the rigors of Western thinking:

> What is now needed is to enrich geographic research in India through a skillful blending, in appropriate cultural contexts, of the rigor of the western scholarly paradigms and greater application of culturally-rooted and relevant concepts to [sic] which the East and the West both can easily co-share, co-operate and lead to unified march in making geography as a way of understanding, awakening and cohesiveness. (Singh 2009: 11)

Singh insists that the ways in which humanity and nature are imagined in Hindu thought can lead us to live more eco-ethical lives. Such modes of thinking, he argues, have much to teach us in the West who, generally, have not done such a great job of looking after the planet. Singh is well aware of the ways in which the kinds of thought he is advocating have been all but absent from Western accounts of geographical thought. He summarizes, for instance, the contents of five key collections in the development of the field of cultural geography (which you might think would be keen to explore different cultures of geographical thought) and notes that:

> In none of these readers and anthologies is India's rich cultural heritage given even marginal space. In fact, the western hegemony (especially Anglo-American and British) used India as a resource for their own theoretical test, but never given honour to consider it as part of the curriculum. (Singh 2009: 163–164)

Postcolonialism and Geography

One body of theory that helps to explain the process of exclusion and erasure at work in accounts of geographical theory is **postcolonialism**. Broadly speaking, postcolonialism is a body of knowledge that seeks to critique colonialism and its lasting impacts. It looks to account for the impact of colonialism both in the past and in the present, in former colonies and in the metropolitan spaces at the heart of empires. In this sense, London is every bit as postcolonial as Delhi. Postcolonialism as a body of thought has been heavily influenced both by Marxism, with its critique of imperialism, and by poststructuralism, with its insistence on the deconstruction of difference. Many of the key figures in the development of postcolonialism, unlike any other body of theory considered in this book thus far, are notable for being scholars of color. These include the Palestinian cultural theorist Edward Said, Indian literary scholar Homi Bhabha, and Indian theorist Gayatri Chakravorty Spivak (Said 1979; Bhabha 1994; Spivak 1999). The descriptors "Indian" or "Palestinian" used here are themselves problematic as they point toward an established and settled sense of origins which postcolonial theorists would be the first to undermine, preferring the notion of hybridity (associated with the work of Bhabha in particular). Hybridity is an idea which points to the importance of mixing (ethnically and racially) in order to undermine essentialist notions of identity. Nevertheless, the fact that people of color are central to this body of work is significant, as it indicates that voices which are often excluded from debates about theory are being heard.

Postcolonialism refers to two, sometimes contradictory, ideas. As with postmodernism, there is a first sense which is temporal. Postcolonialism was first used in the relatively simple sense of *after* colonialism. In this sense it refers to an understanding of the world after places such as India or Singapore have ceased to be colonies governed by metropolitan centers such as London. Thus, contemporary India, Singapore, and London are postcolonial. Such a definition rests on ideas of imperialism and colonialism. Imperialism is the domination of a distant territory by a metropolitan center. Colonialism more specifically refers to the settlement of those distant territories by the metropolitan center. Colonialism follows from imperialism, which both bind a subordinated people and land into an unequal relationship with a dominant power.

There are a number of problems with this first definition of postcolonialism as *after* colonialism. The first is that it suggests that colonialism is over in its strictest sense when, in fact, many would argue that there are ongoing colonialisms all over the world. Plaid Cymru, the nationalist party of Wales, for instance, might argue that Wales is still being colonized by England. The inhabitants of Puerto Rico might well believe that they are a colony of the United States. Even in places that are no longer formally colonized, there are new forms of international relations that maintain relations of domination and subordination. Persistent inequalities might be referred to as neocolonial and can be traced through the activities of large corporations as much as through the actions of nation-states. Derek Gregory has made a persuasive case for a "colonial present" in Iraq, Afghanistan, and Palestine (Gregory 2004). More conceptually, the continual referrals to colonial and postcolonial worlds make the fact of colonization appear to be the central defining moment in any place's history, thus placing western expansion at the heart of the definition process. This makes the West appear to be the "actor" or "agent" of world history while everywhere else merely has things done to them. This has the ironic effect of intellectually colonizing the world. A further problem lies in the sheer complexity of colonialism on the ground. Not all colonies are remote peripheries to some distant center. There are, for instance, varieties of "internal colonialism" such as the case of Wales in the United Kingdom or Tibet in China. Some places can be thought of as both colonies and colonizers. The most dramatic of these is the United States which started its life as "the colonies" and then became the world power that practices colonialism and neocolonialism. Consider the case of Canada described by Alison Blunt and Cheryl McEwan, drawing on the work of Moore-Gilbert. In Canada, they argue, there are at least five different colonial/postcolonial contexts to account for:

These contexts are: (1) the legacy of the dependent relationship with Britain; (2) the relative (cultural, strategic, and economic) US domination of North America; (3) the issue of Québec; (4) the relationship of the indigenous inhabitants to the various white (Québecois and Anglo) settler colonialisms; and (5) the arrival, role and status of migrants from Asia. (Blunt and Mcewan 2002: 21)

In such a situation, any straightforward diagnosis of who is the colonizer and who is the colonized, what is colonial and what is postcolonial, is impossible.

The second use of the term postcolonialism, while referring to the actual process of colonialism, has a much more theoretical basis. Postcolonialism in this sense refers to a general critique of colonialism in all its forms as well as an attempt to go *beyond* imperialism and colonialism. "In this case, the 'post' in 'postcolonialism' refers more to a *critical* than to a *temporal* aftermath as postcolonial perspectives explore and resist colonial and neo-colonial power and knowledge" (Blunt and Wills 2000: 170). A postcolonialist geography, then, has a number of aims which include:

> ... the unveiling of geographical complicity in colonial dominion over space; the character of geographical representation in colonial discourse; the de-linking of local geographical enterprise from metropolitan theory and its totalizing systems of representation; and the recovery of those hidden spaces occupied, and invested with their own meaning, by the colonial underclass. (Jonathan Crush quoted in Blunt and McEwan 2002: 2)

Postcolonialism in geography clearly rests on the recognition of geography's complicity in the erasure of other forms of knowledge. The ambitious list of aims presented by Crush seeks to provide some kind of corrective to this complicity by both revealing the previously unheard voices of the colonized and demonstrating the ways in which the "center" has been involved in this silencing. A good deal of work has focused on the involvement of geography and geographic thought in the process of colonialism. There is some irony, then, in the idea of postcolonial geographies, a point made in a key paper by James Sidaway:

> The prospect of 'exploring' postcolonial geographies ... is intentionally contradictory and ironic. Therefore, like calls for a postcolonial history (...), any postcolonial geography 'must realise within itself its own impossibility', given that geography is inescapably marked (both philosophically and institutionally) by its location and development as a western-colonial science. It may be the case that western geography bears the traces of other knowledges (...) but the convoluted course of geography, its norms, definitions and closure (inclusions and exclusions) and structure cannot be disassociated from certain European philosophical concepts of presence, order and intelligibility. Feminist and poststructuralist critiques may have sometimes undermined these from within, but they could never credibly claim to be *straightforwardly* outside or beyond those institutions and assumptions that have rooted geography amongst the advance-guard of a wider 'western' epistemology, deeply implicated in colonial imperial power. (Sidaway 2000: 592–593)

Limitations such as these can seem disabling. We could become paralyzed in self-critique (though, it has to be said, this is not a widespread problem!). Just as it may seem too soon to talk of a postcolonial world (because of ongoing colonialism and neocolonialism), it may also be too soon to write of postcolonial theory when that body of theory may be seen as just another Western hegemonic gesture. Soon "we" (Western scholars) will be teaching "them" (scholars and others from outside the West) about postcolonialism and thus a form of colonialism continues as the West retains its hegemony. Despite such conundrums, Sidaway insists on the radical potential of postcolonial theory which may help to reveal and overturn geography's exclusions:

> Yet, at their best and most radical, postcolonial geographies will not only be alert to the continued fact of imperialism, but also thoroughly uncontainable in terms of disturbing and disrupting established assumptions, frames and methods. Between the encouragement to rethink, rework and recontextualize (or, as some might prefer, to 'deconstruct') 'our' geographies and the recognition of the impossibility of such reworked geographies entirely or simply escaping their ('western') genealogies and delivering us to some postcolonial promised land, are the spaces for forms and directions that will at the very least *relocate* (and perhaps sometimes radically dislocate) familiar and often taken-for-granted geographical narratives. (Sidaway 2000: 606–607)

Defining Decoloniality

One place where the limits of postcolonialism have been revealed is in the work of scholars working on **decoloniality** and decolonial geographies. Decolonial scholarship is, in general, more politically assertive than postcolonial scholarship – drawing particularly on Indigenous geographies and Black geographies (see Chapter 15) and posited against White Supremacy and all forms of colonialism. Decolonial scholarship in geography and beyond is both a critique of colonial spatialities and imaginations and an affirmation of different spatialities that cannot be reduced to, or explained by, colonialism. Recognizing that postcolonialism, despite its critical intent, continues to center both colonialism and colonialist powers in its interpretations and expectations, decolonial critique – also called decoloniality – pushes to interrogate how knowledge-making practices contribute materially and discursively to marginalize people, places and thinking, and thereby reproduce the norms and privileges of Western, "universal" knowledges and institutions.

Decolonial theory has its origins in the thought of those who are colonized – particularly in Latin America and particularly among Indigenous people. It has a specific purchase in situations where colonialism never really went away and so it is very difficult to talk about postcolonialism. This is clearest in places marked by settler-colonialism – places no longer ruled by an imperial center but dominated by people (settlers) not indigenous to the area. Examples include the USA, Canada, Australia, and New Zealand. A key term for understanding decoloniality is **coloniality**. *Coloniality* is shorthand for *coloniality of power* and originates in the work of Quijano (2000). It has been taken up and developed by a broad network of scholars most often connected to or from Latin America and Indigenous communities (Schiwy and Ennis 2002; Mignolo and Walsh 2018). It is important that the idea of coloniality arose outside of the Euro-Western world. In this sense it is different from ideas like capitalism, modernism, or post-modernism – all ideas which limit or marginalize the importance of the history of colonialism. Coloniality (and thus, decoloniality) arose out of the lived experiences of people in the colonized worlds of the Americas and thus addresses their concerns. Coloniality refers to the ongoing effects, the afterlives, of colonialism expressed through ongoing modes of thought and action based on social, cultural, and racial hierarchies that valued some people and things while devaluing others. While coloniality is linked to capitalism and modernism, it is not reducible to them. Walter Mignolo links modernity with coloniality. While modernity describes Europe's self-image of progress towards a future horizon, coloniality describes its dark underside – the "Third World" where inhabitants felt the negative consequences of the First World drive for progress. Modernity, Mignolo argues, cannot be understood without coloniality. Most of the content of this book is an example of coloniality as it accounts for a discipline that arose out of Europe with roots in the classical Greek and Roman past.

> The vocabulary of any of the existing disciplines, words that denote the field of investigation or are concepts you use to approach the field, have two semantic dimensions. You will find, first, they have been derived from Greek and Latin. You will find, secondly, that most, if not all, of the words/concepts you use in your discipline and even in everyday conversation were translated and redefined around the sixteenth and seventeenth centuries in Europe. (Mignolo and Walsh 2018: 113)

Decolonial scholars insist that the process of decolonization – the process of becoming independent from colonial power – left the coloniality (of knowledge) in place. They use the term **colonial matrix of power** to describe this process.

> The colonial matrix of power (the CMP) is a complex structure of management and control composed of domains, levels, and flows. Like the *unconscious* in Sigmund Freud or *surplus value* in Karl Marx, the CMP is a theoretical concept that helps to make visible what is invisible to the naked (or rather the

nontheoretical) eye. Unlike Freud's unconscious or Marx's surplus value, though, the CMP is a concept created in the Third World – in the South American Andes, specifically. It is not a concept created in Europe or in the U.S. academy. (Mignolo and Walsh 2018: 142)

Decolonial theory undoes the CMP in part by insisting on the locatedness of theory production – insisting on the links of thought and Land epitomized by the Place-Thought that Watts wrote about above. Decolonial theorists use the term **vincularidad,** derived from the work of Indigenous Andean thinkers, that names the linking of living things to the land in place – another term for Place-Thought. By acknowledging vinicularidad, Mignolo and Walsh argue, we enter a **pluriverse** of theory where theoretical practice is specific to place and resists the impulse to universalize ideas and actions. Thinking of theory in this way resists the universal impulses and abstractions of Euro-Western theory. Importantly, it is not only decolonial theory that is place specific – Euro-Western theory, while often pretending to be universal is also place specific – arising out of places that were at the heart of imperial and colonial systems. Mignolo and Walsh make this case with reference to the philosopher of being, time, and place, Martin Heidegger.

> No-one would claim that Martin Heidegger's writings were German studies. He was German, and what he thought had a lot to do with his personal history and language. But he thought what he deemed to be thought at his time and place. So it is for us. Heidegger was not a token of his culture, and neither are we. We are where we think, and our thinking is provoked by the history of the Americas (including the United States) and the Caribbean since the sixteenth century, when the very inception of modern/colonial patterns (i.e., coloniality) began to emerge. Yet our thinking … [does] not end – nor [is it] only located – here. (Mignolo and Walsh 2018: 2)

Note that while they claim that Heidegger's thought is a product of its place, that does not mean that it is limited to Germany, any more than decolonial theory is limited to parts of Latin America. It is a typical move of thinkers in Euro-Western thought to locate some kinds of theory in "other" places while refusing to recognize that Euro-Western thought is also located and that thought from elsewhere can also travel. Thought from anywhere can travel without making pretenses about universality. Thus, decolonial theory is place specific, relational, and always also practice. Theory is practice and practice is theory.

> For us, theory is doing and doing is thinking. Are you not doing something when you theorize or analyze concepts? Isn't doing something praxis? And from praxis – understood as thought-reflection-action, and thought-reflection on this action – do we not also construct theory and theorize thought? By disobeying the long-held belief that you first theorize and then apply, or that you engage in blind praxis without theoretical analysis and vision, we locate our thinking/doing in a different terrain. (Mignolo and Walsh 2018: 7)

The decolonial insistence on collapsing the theory/practice binary reflects the insistence on collapsing binaries in general that are seen as the product of Euro-Western thought since around 1500 that broke from previous ways of thinking that were more holistic. Other binaries such as those based on race and caste and human/nonhuman were also, according to Mignolo and Walsh, created around that time. The refusal to separate doing and thinking thus reflects the entanglement of living things and the Land in Place-Thought or in Aboriginal accounts of Country. Decolonial theory does not consider the human as created in relation to the less than human – a binary that, decolonial theorists insist, led to racism, sexism, and the subjugation of nature.

Decoloniality, for Mignolo and Walsh, has to be an ongoing practice of escaping the colonial matrix of power. They recognize that there is no easy endpoint when the work of decolonizing will be complete. Rather, they insist on the work of a decolonial pluriverse where other ways of thinking/acting become possible alongside coloniality.

> Decoloniality … does not imply the absence of coloniality but rather the ongoing serpentine movement towards possibilities of other modes of being, thinking, knowing, sensing, and living; that is, an otherwise in plural. In this sense, decoloniality is not a condition to be achieved in a linear sense, since coloniality as we know it will probably never disappear. (Mignolo and Walsh 2018: 81)

Decolonial and Anticolonial Geographies

One of the first texts I read while educating myself about decoloniality and decolonization was Eve Tuck and K. Wayne Yang's "Decolonization is not a metaphor" – a text written from an Indigenous perspective and reflecting Indigenous lived experience of how decolonization has been used in the Canadian academic context. They describe the way decolonization has been used to describe all kinds of theorization that has little or no impact on the brute facts of settler colonialism in a setter-colonial context.

> When metaphor invades decolonization, it kills the very possibility of decolonization; it recenters whiteness, it resettles theory, it extends innocence to the settler, it entertains a settler future. Decolonize (a verb) and decolonization (a noun) cannot easily be grafted onto pre-existing discourses/frameworks, even if they are critical, even if they are anti-racist, even if they are justice frameworks. The easy absorption, adoption, and transposing of decolonization is yet another form of settler appropriation. When we write about decolonization, we are not offering it as a metaphor; it is not an approximation of other experiences of oppression. Decolonization is not a swappable term for other things we want to do to improve our societies and schools. Decolonization doesn't have a synonym. (Tuck and Yang 2012: 3)

Decolonization for Tuck and Wang means something quite specific – the repatriation of Indigenous land and life. The challenge of Tuck and Wang is taken up by the geographers (one Indigenous, one settler) Sarah Hunt and Sarah de Leeuw:

> What are the limits of a project seeking to decolonize geography but absent of Indigenous peoples as experts or theorists? If Indigenous peoples and places continue to be *subjects* within scholarly contributions of settler geographers seeking to decolonize, is any decolonization really being done? What does it mean to read, write, and teach about decolonization absent of significant relationships with Indigenous peoples on whose land our universities are situated? (De Leeuw and Hunt 2018: 10)

Tuck and Yang's arguments are powerful ones, but certainly not the only ones regarding what decolonizing involves. Many disciplines and departments outside of settler-colonial contexts (such as in the UK) are striving to decolonize knowledge in, for instance, publications such as this one, and, most often, in syllabi that have for too long focused on White men from colonial and neocolonial powers. This decolonizing activity is not about giving Land back to Indigenous people but about recognizing that there are different places to look for knowledge than those we have become accustomed to. Sarah Radcliffe and Isabella Radhuber write that decoloniality provides a critique of how knowledge is produced in such a way that the norms of western knowledge are reproduced as if they were universal. And, perhaps more importantly:

> By delinking from Eurocentric knowledge and power relations, decoloniality seeks to re-theorize and re-make the world by learning from multiple and varied spaces, times and ontologies that provide non-universal and non-Eurocentric perspectives, and dismantle world-making colonial-modern relations.

> Analyzing a particular spatial-power configuration through decolonizing lenses highlights the scope for – and pathways to – recalibrating core sub-disciplinary concepts, frameworks and objectives in less exclusionary directions. (Radcliffe and Radhuber 2020: 2)

So, while decoloniality continues post-colonial critiques of dominant forms of colonial knowledge, it does more than postcolonialism by looking to other places for forms of geographical knowledge, and centering those ways of knowing in analysis. In this way, the seemingly all powerful forms of colonial knowledge become radically de-centered and thoroughly relativized. Decolonial thought, in this sense, is radically promiscuous in where it looks for theory including Black scholarship, Queer theory, feminism and Indigenous forms of knowledge. Importantly, "the decolonial turn encourages re-thinking the world from Latin America, from Africa, from Indigenous places and from the marginalised academia in the global South, and so on" (Radcliffe 2017: 329).

As Tariq Jazeel has argued, there is a tension in decolonial thought's desire to de-link from the processes of formal academia that have led to the production of disciplinary canons – a process that a book like the one you are currently reading is inevitably a part of (Jazeel 2017). The paradox, of course, is that decolonial thought is embraced by canon-producing academic disciplines and becomes part of the very thing it sought to be outside of. Decolonial thought becomes "main-streamed." Jazeel's paper was part of a special forum in the *Transactions of the Institute of British Geographers* on decolonizing geographies in which all of the papers reflected on the theme of the annual conference of the Royal Geographical Society with the Institute of British Geographers in 2017. In other words, a body of thought promoting knowledge produced outside of the norms of western colonial thought was being discussed in just about the most central bastion of disciplinary colonial thought imaginable. This clearly made Jazeel more than a little uneasy.

> In the face of a more planetary disciplinary canon, we might content ourselves that 'we now practice a worldly geography, we listen to those we did not listen to before, and therefore we can now feel better about ourselves'. Job done. (Jazeel 2017: 336)

An example of decolonial thought in action is Farhana Sultana's use of coloniality and the decolonial to think about climate change and climate justice. Reflecting on the COP26 meeting in Glasgow in 2021, she notes how the meetings were notably contentious with a dispiriting continuation of business-as-usual among the corporate and government elite (including armies of fossil fuel lobbyists) and, at the same time a loud and hopeful resistance by those who were excluded including Indigenous groups, the young and people of color. She uses this experience to frame the ongoing ways in which climate change is systematically unevenly experienced and accounted for within a global system of ongoing coloniality – the ongoing structures of colonialism that continue to exist even after colonialism, in a formal sense, has ended.

> Coloniality relies on racial domination and hierarchical power relations established during active colonialism and ongoing in post-colonial spacetimes, where the colonial matrix of power persists. Thus, climate coloniality occurs where Eurocentric hegemony, neocolonialism, racial capitalism, uneven consumption, and military domination are co-constitutive of climate impacts experienced by variously racialized populations who are disproportionately made vulnerable and disposable (Sultana 2022: 4)

Sultana lists the overwhelming number of ways in which climate coloniality continues to reproduce itself as Black and Brown people find themselves on the receiving end of both capitalist environmental ruination and the ongoing colonial logics of mainstream efforts to combat climate change, which ignore the knowledges and experiences of the majority of the world's population. In addition to listing material changes that are needed to decolonize climate change, Sultana asks us to focus on epistemological changes that need to happen. A process that in her words is, "not just about having a seat at the table (e.g. participation at the COP26) but determining what the table is, i.e. the terms of the debate or framing of the conversation and having decision-making power" (Sultana 2022: 8). Decolonizing climate, Sultana argues, involves providing alternatives to the epistemological violence of only seeing climate through the lens of coloniality. Her embrace of decolonial knowledge and practice is broad and, importantly includes art and folk knowledges.

> To celebrate resurgence in cultural practices of art, literature, oral traditions, poetry, and dance is to claim agency, desire, futurity, and spirit. Traditional folk songs and dances, plays and street theater, poetry and literature recitals, arts and handicrafts fairs, seasonal festivities and flower ceremonies, puppet shows and oratory recitals, collective cooking and sharing food, giving alms to wandering minstrals and holy folk, prayer ceremonies and rain dances – for many these are simultaneously coping mechanisms, refusals, resistance movements, and decolonial actions, where recollections of collective memories and practices as well as enactments for liberation remain the goal. However, it is vital to not fetishize pre-histories as frozen time or culture as magical solutions to systemic oppression, but recognize how they further propel decolonization and revolutionary resistance. They also constitute oppositional counterbalances to the coloniality of cannibalization of cultural artifacts, practices, and linguistics. It is a refusal of poverty porn and only-victim narratives, but cultivates fleshing out theories and grounding concepts. It is an affirmation of the humanity of the oppressed, and of fostering radical equality. At the same time, it is an understanding of our complicities in perpetuating harms and actively working to redress it through everyday praxis and re-education. (Sultana 2022: 9)

The problem with mainstream discussions of climate change, Sultana argues, is that they can become quite technocratic and physical science-based on the one hand, and obsessed with questions of finance and the economic on the other. What is not discussed is the need to rethink how people relate to the physical ecosystems – how, in other words, we might mobilize a notion of Land that reorganizes human relationships to the nonhuman world. "To decolonize," she argues "is to reveal, reassess, and dismantle colonial structures and discourses, to make them non-universal and demonstrate the hegemony deployed historically and through particular racialized colonial practices, and expose the everyday tactics of oppression and empire-building" (Sultana 2022: 10).

One place to look for the kind of work that Sultana is asking for is in the Civic Laboratory for Environmental Action Research (CLEAR) in Newfoundland, Canada associated with geographer, Max Liboiron. The work of Liboiron and CLEAR shows how decolonial and anticolonial thinking and practice do not have to be primarily focused on colonialism and postcolonialism. They could, for instance, be focused on the plastic pollution to be found in the guts of Atlantic cod. The problem of plastic in fish guts seems to be a problem we might designate to science. Is there plastic in the guts of Atlantic cod? How much plastic is in the guts of cod? Is it harmful? Questions such as these are answered through systematic laboratory science and laboratory science which, we might think, is neither colonial, decolonial, or anti-colonial. We would be wrong to think this. In Liboiron's book *Pollution is Colonialism* as well as the wider work of CLEAR we can see anticolonial work in action. Liboiron is a Métis/Michif geographer working on plastic pollution. Liboiron and their collaborators refer to CLEAR as an **anticolonial** rather than decolonial lab. Liboiron acknowledges a debt to Tuck and Yang's work noting that their understanding of decolonization is the specific context of Newfoundland, "comes largely from being an academic, where the verb *decolonize* is frequently invoked as something that you do to university course, syllabi, panels and other academic nouns. Yet in the face of all this 'decolonization', colonial Land relations remain securely in place. Appropriating terms of Indigenous survivance and resurgence, like decolonization, is colonial" (Liboiron 2021: 16). Rather than refer to CLEAR as a decolonial lab, they refer to it as anticolonial:

> … where anti-colonial methods in science are characterized by how they do not reproduce settler and colonial entitlement to Land and Indigenous cultures, concepts, knowledges (including Traditional Knowledge), and lifeworlds, an anticolonial lab does not foreground settler and colonial goals. There are many ways to do anti-colonial science: in addition to Indigenous sciences, there are, for example, also queer, feminist, Afro-futurist, and spiritual Land relations that are anticolonial. (Liboiron 2021: 27)

The anticolonial nature of CLEAR works right down to the minutia of the processes for calculating how much plastic is found in the guts of a fish. Liboiron provides an account of how their team would collect fish guts from fishermen for testing before realizing that some people used the guts to make food. They learned not to assume that what they saw as waste is, in fact, waste, and to make sure they had permission to use materials. The only way to find out how much plastic is inside the shellfish is to dissolve them in Potassium Hydroxide (KOH) – a process that produces hazardous waste that has to be disposed of. Liboiron recounts how easy it was to dispose of hazardous waste "just fill out a form and call a guy. Bam! Gone! (Somewhere!)" (Liboiron 2021: 66). This led to a Lab policy to not use hazardous substances, and therefore, to cease their work with shellfish. This policy was not simply because the substances were polluting but because the "somewhere" that the polluting substances ended up, in the context of Newfoundland and Labrador, was land colonized by settlers that was assumed to exist as a "standing reserve" ready for whatever uses the settlers saw fit. It is a small illustration of the provocation in the book's title – pollution is colonialism. But even this is not that simple. Anticolonial and decolonial work is always in process as shown by a footnote describing this decision.

> Based on the KOH issue, I made a new guideline for the lab: no processes that necessitate hazardous waste. It means we cannot study bivalves, crustaceans, and other invertebrates for plastic ingestion since KOH is the only way to "dissect" them. I gladly take up the restriction. Though now that's complicated as my Inuit colleagues want to study plastics in bivalves in their traditional food webs. Colonial technologies used for Inuit goals are …? What? Not colonialism, but we still have a problem. To be continued … (Liboiron 2021: 66)

Liboiron's book argues that pollution is colonialism (particularly in a settler-colonial context such as Canada) because it assumes the availability of land that is actually a much richer concept of Land (with capital L) for the Indigenous people who live, on, and belong to that Land. The rich sense of Land is reduced to a resource for the settlers to use and make decisions about. Even the seemingly micro scale processes of running a lab can make particular assumptions about very big questions of land and resources. "You can just order KOH whenever you want, because the infrastructure anticipates it use and disposal as hazardous waste … In this way, a seemingly simple and certainly common scientific research method that produces hazardous waste is involved in colonial Land relations, even though its users are also likely invested in environmental goods and perhaps see themselves as Indigenous allies – or are Indigenous scientists themselves" (Liboiron 2021: 66). Settler colonialism is based on the ongoing access to land that has traditionally been, and continues to be, a source of identity, wellbeing and livelihood to Indigenous people. Part of that equation is the expectation of land being available for purposes of pollution and thus pollution is colonialism. There are many illustrations such as the KOH account in Liboiron's book that illustrate the complexities of running a science lab that is anti-colonial. It is difficult and ongoing work. Doing anticolonial science is not the same as doing other kinds of counter-hegemonic science. Science can be feminist, for instance, but still be colonial. Using the term anticolonial specifies the commitment to good relations with L/land. Land remains at the center of calculations. Another important aspect of the anti-colonial science that Liboiron and CLEAR practice is that it is "place-based" and makes no pretence to being universal. It shares this with feminist standpoint theory but is not the same. It is connected to Indigenous ways of knowing the world and Place-Thought. Liboiron recounts the words of a CLEAR advisor and Kumeyaay Elder, Rick Chavolla.

> When you discover something scientifically, it applies to anything, anywhere. You can go anywhere in the world and say, "Yes! This works! This is what truth is! Truth was here in this place, and truth will be the same someplace else." For us, that's so far from the truth, so far from our knowledge as Indigenous people. We know that for knowledge you *must* understand where you are. (Rick Chavolla quoted in Liboiron 2021: 151–152)

Conclusions

When I arrived at the University of Edinburgh in 2019, I became involved with an informal group whose aim was to "decolonize" the building that geographers within the School of GeoSciences work in – the Institute of Geography at Drummond Street (Figure 14.1). The work of this group exists within a wider move to decolonize the curriculum in geography and to decolonize the University of Edinburgh in general. While I had a passing knowledge of decolonial thinking at the time, I took this as an opportunity to learn more about this process I had volunteered to be involved in. For the building to be decolonized, I thought, it was first necessary to consider how, specifically, the building was "colonized." The building has an interesting history. The building, and a building on the site before it, were originally Edinburgh's hospital – given a Royal Charter in 1725 by King George II and largely designed and built in its present form between 1848 and 1853. You can still see the drains that were sluices for blood in the building's basement. It served as the "Royal Infirmary" until 1879 and had over 200 beds. The building continued as a more specialized City Hospital for Infectious Diseases until 1903 when it was bought by the University of Edinburgh and converted into the Department of Natural Philosophy (what we now call Physics). The building is a listed building. It is seen as notable due to both its architectural features and because of its architect, David Bryce, who designed many significant buildings in Scotland. Our group, led largely by interested post-graduate students, conducted an audit of the current building noting how the walls were used for various maps and, in the Old Library in particular, pictures of (mostly) dead White men associated with the history of the

Figure 14.1 The Institute of Geography at the University of Edinburgh. Source: kim traynor / Wikimedia Commans / CC-BY-SA 2.0.

Department of Geography. Most of the walls were, and still are, bare white/cream walls with minimal decoration. Some examples of the work include making at least one gender-neutral bathroom, producing a publicly visible diverse archive of people and work from the building, increasing the visible representation of the diversity of the current inhabitants of the building, and properly representing some elements of the Department's past that are not currently represented – particularly the array of feminist scholars who have passed through the building in the last 40 years. All of this is important work, but it is hard to see how any of it can be called decolonial. It is equality, diversity, and inclusion work that had little, if anything, to do with colonialism. The building is in Edinburgh, if not the heart, then at least an important organ, of Empire. The city is associated with the Scottish Enlightenment and the hospital can be seen as part of that late-eighteenth and early-nineteenth-century process that has been widely celebrated for its role in the abolition of slavery movement but more recently critiqued with belated recognition of its complicity in aspects of the slave trade. There is no doubt that the building was built within an economy that benefitted from colonialism in general and the trade in enslaved African people in particular. Recently, NHS (National Health Service) Lothian, my local health body, conducted research into its historical links with the Drummond Street Building when it was the Royal Infirmary of Edinburgh. Their research confirmed that there were direct links between the Royal Infirmary and the institution of slavery. Benefactors of the original infirmary from 1729 were either themselves enslavers or benefitted directly from enslavement. The Royal Infirmary even owned an estate (including its enslaved and indentured laborers) in Jamaica called Red Hill pen. They owned this until 1892. The research concluded that income from slavery was used to provide healthcare for 1,390 people between 1808 and 1817 (Buck 2022). While there is a clearly a dark past to the building I work in, there is still the question of whether the process we are engaged in can or should be referred to as "decolonizing" rather than simply making the building more welcoming to the diversity of people who inhabit it and might visit in the future. The "decolonising the building" working group in Drummond Street were clearly never going to be involved in work that, in Tuck and Yang's terms, would repatriate land and life to Indigenous people. It is not situated on Indigenous land and there were and are no Indigenous people involved in the work. We have to ask, therefore, if "decolonizing" is the correct word to describe the important work we need to do.

This question sits alongside a wider question about how scholars in the Euro-Western world might appropriately engage with decoloniality. In the preface to her book *Decolonizing Geography: An Introduction*, the White, cis-gendered, UK-based geographer Sarah Radcliffe situates herself in order to contextualize her book on decolonization within the discipline (Radcliffe 2022). Radcliffe acknowledges her identity and the privileges that come with that and notes that "geographers of colour have argued rightly that geography's urgent task of decolonizing must not rest solely on racialized minorities" and that "as a white ally" she recognizes the importance of "white geographers informing themselves about de-colonizing and anti-racism." Using the language of decolonial thought she stresses the importance of constructing a "decolonial pluri-geo-graphy – or a world of many worlds" that requires White geographers to "overturn racialized exclusions and assumptions" (Radcliffe 2022: ix–x). Radcliffe has worked with scholars and others from the Global South for many years. I have not. So, my need to educate myself is much greater than Radcliffe's. There is a danger in a cis-gendered, White, male scholar from the UK – me – writing about decoloniality. It is perhaps a bigger danger to not engage. As with other critical theoretical perspectives, there is the possibility that decoloniality itself might become colonized as geographers who look a lot like me accumulate academic capital through citing decolonial work without doing the work of decolonization – or mistaking equity, diversity, and inclusion work for decolonizing. In the spirit of constant forward movement, scholars are already seeking to move "beyond the decolonial" (Sidaway 2023). As Tuck and Yang argued, decolonization "is not a swappable term for other things we want to do to improve our societies and schools" (Tuck and Yang 2012: 3) even when those things are offered in a radical spirit. Postcolonial, decolonial, and anticolonial theory in geography all locate a specific kind of matrix of power in the history of colonialism and in its ongoing lives and afterlives. All of them, in different ways, seek to both critique and undo these histories and geographies while recognizing that they are unlikely to disappear any time soon. They are theoretical interventions in the world that seek to create and maintain a theoretical and political pluriverse to challenge the universe that coloniality has sought to impose.

References

Bhabha, H. K. (1994) *The Location of Culture*, Routledge, London; New York.

Blunt, A. and McEwan, C. (2002) *Postcolonial Geographies*, Continuum, New York, N.Y.; London.

Blunt, A. and Wills, J. (2000) *Dissident Geographies: An Introduction to Radical Ideas and Practice*, Longman, Harlow.

Buck, S. (2022) *Uncovering Origins of Hospital Philanthropy: Report on Slavery and Royal Infirmary of Edinburgh*, Lothian Health Services Archive/Centre for Research Collections | University of Edinburgh, Edinburgh.

*De Leeuw, S. and Hunt, S. (2018) Unsettling decolonizing geographies. *Geography Compass*, 12, e12376.

Durkheim, E. (2008) *The Elementary Forms of Religious Life*, Oxford University Press, Oxford.

Gelder, K. and Jacobs, J. M. (1998) *Uncanny Australia: Sacredness and Identity in a Postcolonial Nation*, Melbourne University Press, Carlton South, Vic., Australia.

Gregory, D. (2004) *The Colonial Present: Afghanistan, Palestine, and Iraq*, Blackwell Publication, Malden, MA; Oxford.

*Hunt, S. (2014) Ontologies of indigeneity: the politics of embodying a concept. *Cultural Geographies*, 21, 27–32.

Jazeel, T. (2017) Mainstreaming geography's decolonial imperative. *Transactions of the Institute of British Geographers*, 42, 334–337.

Lai, C. Y. D. (1974) Feng-Shui model as a location index. *Annals of the Association of American Geographers*, 64, 506–513.

*Liboiron, M. (2021) *Pollution is Colonialism*, Duke University Press, Durham.

Mignolo, W. and Walsh, C. E. (2018) *On Decoloniality: Concepts, Analytics, Praxis*, Duke University Press, Durham.

Milroy, J. and Revell, G. (2013) Aboriginal story systems: re-mapping the west, knowing country, sharing space. *Occasion: Interdisciplinary Studies in the Humanities*, 5, 1–24.

Mohawk, J. (2005). *Iroquois Creation Story*. Mohawk Publications.

Muecke, S. (2004) *Ancient & Modern: Time, Culture and Indigenous Philosophy*, UNSW Press, Sydney.

Quijano, A. (2000) Coloniality of power and eurocentrism in Latin America. *International Sociology*, 15, 215–232.

*Radcliffe, S. A. (2017) Decolonising geographical knowledges. *Transactions of the Institute of British Geographers*, 42, 329–333.

Radcliffe, S. A. (2022) *Decolonizing Geography: An Introduction*, Polity Press, Medford.

Radcliffe, S. A. and Radhuber, I. M. (2020) The political geographies of D/decolonization: variegation and decolonial challenges of/in geography. *Political Geography*, 78, 102128.

Rose, D. B. (1996) *Nourishing Terrains: Australian Aboriginal Views of Landscape and Wilderness*, Australian Heritage Commission, Canberra, ACT.

Said, E. W. (1979) *Orientalism*, Vintage Books, New York.

Schiwy, F. and Ennis, M. (eds) (2002) *Special Dossier: knowledges and the known: Andean perspectives on capitalism and epistemology: introduction. Nepantla: Views from South* 3.1 1–14

*Sidaway, J. D. (2000) Postcolonial geographies: an exploratory essay. *Progress in Human Geography*, 24, 591–612.

Sidaway, J. D. (2023) Beyond the decolonial: critical Muslim geographies. *Dialogues in Human Geography*, 0, 1–22.

Singh, R. P. B. (2009) *Geographical Thoughts in India: Snapshots and Visions for the 21st Century*, Cambridge Scholars, Newcastle.

Spivak, G. C. (1999) *A Critique of Postcolonial Reason: Toward a History of the Vanishing Present*, Harvard University Press, Cambridge, MA.

Sultana, F. (2022) The unbearable heaviness of climate coloniality. *Political Geography*, 99, 102638.

Takeuchi, K. (1980) Some remarks on the history of regional description and the tradition of regionalism in modern Japan. *Progress in Human Geography*, 4, 238–248.

* Tuck, E. and Yang, K. W. (2012) Decolonization is not a metaphor. *Decolonization: Indigeneity, Education & Society*, 1, 1–40.

* Watts, V. (2013) Indigenous place-thought and agency amongst humans and non-humans (first woman and sky woman go on a European World Tour!). *Decolonization: Indigeneity, Education & Society*, 2, 20–34.

Whatmore, S. (2002) *Hybrid Geographies: Natures, Cultures, Spaces*, SAGE, London; Thousand Oaks, CA.

Chapter 15

Black Geographies

In the first edition of *Geographic Thought*, I included a final chapter called "Geography's Exclusions." Alongside brief accounts of postcolonialism and de-colonialism, I included a section on "Black[1] Geographies." I confessed to a degree of unease about including such a section at the end of 120,000 words about geographic thought, noting that I might be guilty of making a gesture that was inadequate. Now that a decade has passed since writing that edition, I have less leeway for ignorance. It is for this reason that in this second edition, I include a chapter on Black Geographies and have attempted to weave the work of Black geographers (and other geographers from structurally marginalized groups) throughout this book. I do so with an awareness of the politics of citation practices where White senior scholars such as myself have been guilty of listing a few Black scholars in order to fulfill some kind of diversity quota in their books and papers. On the other hand, they might cite such scholars but not really engage with their work beyond the act of citing it (Mott and Cockayne 2017; McKittrick 2021). Katherine McKittrick differentiates a practice of citation that enacts forms of ownership and colonialism – showing to the reader how much the writer knows (or thinks they know) – with a practice of referencing which enters into genuine collaboration and dialogue with the authors being referenced. In a textbook such as this, which gestures towards the possibility of knowing and reporting a whole discipline's worth of ideas, some citation is inevitable. I have not engaged with everything I have referred to. I believe, however, that Black scholarship by geographers, and by others on geographical themes, offers fundamental challenges and opportunities to our discipline and I hope, here, to do justice to this scholarship and these scholars. I would urge readers of this book to read accounts of Black Geographies by Black geographers who have informed my explorations, in addition to the chapter that follows (Bledsoe and Wright 2019; Hawthorne 2019; Noxolo 2020, 2022).

[1] Here I follow Noxolo (2022) and others in capitalizing "Black" to signify a political identity. I also capitalize "Brown" and "White" as keeping them lower case would signify that they are in some way "normal" in relation to Blackness.

Geographic Thought: A Critical Introduction, Second Edition. Tim Cresswell.
© 2024 John Wiley & Sons Ltd. Published 2024 by John Wiley & Sons Ltd.

A lot has happened in the world of geography, and Black Geographies in particular, since 2012, when the first edition was published. The intellectual work of Black Geographies, however, needs to be understood, like all theory, as formed through and influenced by the context that surrounds it. When the first edition of *Geographic Thought* was published, Black Lives Matter did not exist. Michael Brown, Eric Garner, George Floyd, and countless others had not yet been killed by the police. There had been no protests in Ferguson, Missouri. Black Lives Matter, a loose and decentralized organization founded by three Black women – Ayọ Tometi, Patrisse Cullors, and Alicia Garza – was formed in reaction to the killing of Black people by mostly White police forces. While it was killing and murder that were the immediate prompts for Black Lives Matter, the issues these killings represent – racism and **White supremacy** – had a much longer history. The killings of recent years have brought the problems of White supremacy and institutional racism to the attention of the White population. This was hardly news to Black and Brown people in the United States and elsewhere who would have directly experienced the everyday injuries that these structures of thought and action incur.

The immediate focus on the deaths of Black people at the hands of the police should not divert our attention from the fact that these deaths are part of a long history that most obviously includes the history of slavery, lynching, and Jim Crow in the United States. Nevertheless, the facts of disparity between the opportunities for White and Black and Brown people in the USA, UK, and elsewhere is still overwhelming. In the United Kingdom, 46% of Black households were in poverty compared to less than 20% for White families in 2019.[2] In 2021, Black families were diagnosed with Covid at rates that were roughly twice as high as the White population. Risks of hospitalization and death were also significantly higher for Black people.[3] Twenty thousand young Black men were stopped and searched by police during the first coronavirus lockdown in the UK. This represents more than 25% of all the Black men between 15 and 24 in the capital. The likelihood of a Black person being stopped and searched in the UK in 2018–2019 were close to 10 times the rate for White people.[4] If arrested for possession of drugs, Black and minority ethnic men were far more likely to be jailed than White people.[5] In the United States in 1966, a Black man earned, on average, 59% of what a White man earned. In 2018, the figure was 62%. The overall household poverty rate for Black households in the United States in 2018 was 20.7% compared to a rate of 8.1% for White households. The median household wealth for White people in 2016 was $41,800 while for Black households it was $3550. During the coronavirus pandemic in the USA, the Black population (representing 18% of the total) accounted for 33% of hospitalizations. Black men in the USA are five times more likely to be in prison than their White counterparts.[6]

All of these are contemporary brute facts with historical roots. Take the statistic of mean household wealth in the USA for instance. To understand why White people in 2016 had average wealth of $41,800 while Black people had, on average, $3550, we need to know that there is also a 30% gap in

---

[2] Social Metrics Commission, https://socialmetricscommission.org.uk/wp-content/uploads/2019/07/SMC_measuring-poverty-201908_full-report.pdf.

[3] Public Health England, https://assets.publishing.service.gov.uk/government/uploads/system/uploads/attachment_data/file/908434/Disparities_in_the_risk_and_outcomes_of_COVID_August_2020_update.pdf.

[4] Jamie Grierson, 'Met carried out 22,000 searches on young Black men during lockdown' *The Guardian* 8 July 2020. https://www.theguardian.com/law/2020/jul/08/one-in-10-of-londons-young-Black-males-stopped-by-police-in-may.

[5] Owen Bowcott and Frances Perraudin, BAME offenders 'far more likely than others' to be jailed for drug offences *The Guardian* 15 January 2020. https://www.theguardian.com/uk-news/2020/jan/15/bame-offenders-most-likely-to-be-jailed-for-drug-offences-research-reveals.

[6] 26 simple charts to show friends and family who aren't convinced racism is still a problem in America. *Business Insider*. https://www.businessinsider.com/us-systemic-racism-in-charts-graphs-data-2020-6?r=US&IR=T.

home ownership rates between White people and Black people in the United States. Property is the largest source of wealth for most people. The gap can be traced to the deliberate exclusion of Black people from the mortgage market. In the United States, before the 1968 Fair Housing Act, it was legal for mortgage providers and real estate agents to refuse to provide services to Black people. This locked Black people out of the housing market and thus denied them the opportunity to produce intergenerational wealth in the way that many White families have been able to. In Te-Nehisi Coates' important essay, "The Case for Reparations" (Coates 2014), he provides a detailed case for how historical injustice fueled by White supremacy in the housing market led to much of the inequity and injustice in present-day United States. When White people "red-lined" whole areas of cities so that the majority Black inhabitants of those areas could not buy houses, they set in place (literally) generations of relative poverty for Black people along with all the social and cultural baggage that inevitably accompany poverty.

One term that needs explaining when considering Black Geographies is "White supremacy." White supremacy refers to the belief that White people are inherently superior to other racial groups. This belief in inherent superiority is then used to justify inequality and inequity. The idea of White supremacy explains domination and exploitation as well as the disparity in life opportunities experienced by White people and others. While the term White supremacy might lead some people to think of members of the KKK or racist skinhead gangs, it actually refers to a more deeply engrained structure of feeling that, in Ruth Wilson Gilmore's words, undergirds racism defined as "the state-sanctioned and/or extra-legal production and exploitation of group-differentiated vulnerabilities to premature death, in distinct yet densely interconnected political geographies" (Gilmore 2007: 247). The key here is "premature death" – a fact seen in everything from generalized life expectancy to very specific cases of police brutality. As Gilmore reminds us, White supremacy has its own geographies. Gilmore makes the connections of geography to the projects of White supremacy clear in her work, noting the connection between the discipline, the project of colonialism and the emphasis on the visual in modernity.

> What counts as difference to the eye transparently embodies explanation for other kinds of differences, and exceptions to such embodied explanation reinforce rather than undermine dominant epistemologies of inequality (Gilroy 2000). Geographers from Linnaeus forward have figured centrally in the production of race as an object to be known, in part because historically one of the discipline's motive forces has been to describe the visible world (…). (Gilmore 2022: 108–109)

Gilmore traces the race/geography nexus through the "long, murderous twentieth century" (Gilmore 2022: 108) noting the links between the logics of visuality and White supremacy in obvious candidates such as environmental determinism and less obvious ones such as areal differentiation and social constructionism. She describes how race became a "thing" that could be quantified and mapped, rather than a sociospatial process and how other kinds of counting and mapping inevitably simultaneously map forms of racism.

> Any map of modernity's fundamental features – growth, industrialization, articulation, urbanization, and inequality - as measured by wealth, will also map historical-geographical racisms. Such a map is the product of rounds and rounds of globalization, five centuries' movement of people, commodities, and people *as* commodities, along with ideologies and political forms, forever commingled by terror syncretism, truce and sometimes love. (Gilmore 2022: 113)

The project of Black Geographies, in Gilmore's terms, is one that is not limited to a particular fenced in area of geographies of and about race – rather it is undergirded by a logic that is fundamental to understanding the historical processes of globalization and much else besides. While work we might retroactively call "Black Geographies" has existed for a long time, it was most notably named as such in the book *Black Geographies and the Politics of Place* edited by McKittrick and Woods (2007). In their introduction, the editors note the long history of geographies of Black people.

> Many geographic investigations of black cultures bring into focus empirical evidence based on ethnographic, demographic, or quantitative research. These studies locate where black people live. They bring to light labour-market discrimination, housing patterns, ethnic migrations, and how racialized ghettos contribute to (or defile) the urban environment. While that kind of work does importantly situate the materiality of race and racism, it can also be read as naturalizing racial difference in place. (McKittrick and Woods 2007: 6)

Classic work on the "where" of Blackness, McKittrick and Woods continue, can reduce Blackness to an observable fact within a positivist framework rather than seeing Black people as active agents in their own lives, "communities that have struggled, resisted, and significantly contributed to the production of space" (McKittrick and Woods 2007: 6). By centering Blackness in Black studies, Black Geographies opens up the possibility of exploring the Black production of place in a way that "brings into focus networks and relations of power, resistance, histories, and the everyday, rather than locations that are simply subjugated, perpetually ghettoized, or ungeographic" (McKittrick and Woods 2007: 7). Eleven years after the publication of *Black Geographies and the Politics of Place*, the American Association of Geographers chose Black Geographies as a key theme for the Annual Meeting in New Orleans. A Black Geographies Specialty Group of the Association had been formed in 2016. Since then, geography programmes have started advertising for lecturing positions and post-doctoral fellowships with a Black Geographies focus (including positions at Dartmouth College, Rutgers University, and the University of Edinburgh among others). The decision to designate Black Geographies a key theme was the culmination of long histories of entanglement between race and mobility.

Geography, Blackness, and Black Geographies

Before the rise of a body of work labeled Black Geographies, there was the long history of (mostly) White geographers working on geographies of race referred to by McKittrick and Woods. In social geography, for instance, the issue of racial segregation has been right at the heart of enquiry (Ley 1974; Peach *et al.* 1981; Western 1981; Jackson 1987). White geographers have kept themselves busy telling us about the various ways that race is not based on biology but is socially constructed (Jackson 1987; Mitchell 2000). White geographers have explored the various spaces and places associated with "race" – the reservation, the inner city, the borderlands. Delaney has argued that race is also always there in the "outer city," the "heartlands," and "unreserved space" (Delaney 2002). White geographers have carefully shown how ideas about race inform wider constructions of space and place in legal discourse (Delaney 1998). White geographers have critically assessed how landscape has been selectively formed to make Black lives in/visible (Schein 2009; Alderman and Inwood 2013). White geographers have even looked at themselves and critically analyzed Whiteness as race (Bonnett 1997). Linking this race work is an interest both in how race is reflected in arrangements of space and in how space is central to the construction of race. But, as McKittrick and Woods remind us, the focus of Black Geographies is not on the geographies of race. Black Geographies names a theoretical and activist approach within geography that is rooted in Black political and everyday experience as well as theorizing by Black scholars that cannot be reduced to reaction to either the empirical experience of White Supremacy or, epistemologically, to White theorization.

While there has a been a rich history of mostly White geographers writing about race, there has also clearly been a history of Black geographers within the discipline, however submerged or marginalized they may have been (Kobayashi 2014). We have seen, for instance, how the discipline was kept alive by the ground-breaking work of geographers such as Ibn Battuta and Ibn Khaldun during the Middle Ages (see Chapter 2). When they were exploring and writing, there was no conception of "race" as we know it and racial categories such as "Black" and "White" would have had little relevance. Nevertheless, it is important to note the long lineage of what we might now refer to as Black (and Muslim) geographers in the history of the discipline. It is also obviously the case that comments on the relative invisibility of Black geographers in the tradition of geography in the West obscures the obvious fact that there have been plenty of Black geographers in Africa, Brazil, and elsewhere, studying aspects of the lives of Black people but without a focus on Blackness as such. Prominent examples include the Nigerian geographer Akin Mabogunji and the Brazilian geographer Milton Santos. Akin Mabogunji was based at University College Ibadan (later University of Ibadan) in Nigeria. The University, like many in Africa, was started during the colonial period (in 1948) and was largely run under a British model and headed up by White British academics (Craggs and Neate 2020). This included the geography department. This colonial inheritance was not lost on Mabogunji who wrote in reference to "Third World" universities that "one thing common to most of these countries is that the spring-head of their present educational system lies in Western Europe, notably in France and Britain" (Mabogunji 1975: 288). Mabogunji traced the links between geography in France and Britain and the working of commerce and colonialism to show how the discipline was right at the heart of the colonial project. Where geography happens, he argued, is important, as the colonial geographies inherited in places like Nigeria was based on the experience of the colonizer – focusing on the perceived need to extract resources from the colonies. These colonial viewpoints, he argued, also privileged urban life in ways that made little sense in an Africa that still had largely rural populations leading to a profound ignorance of pressing rural issues such as food supply and water provision. Craggs and Neate recount the experience of another Nigerian geographer, Olusegun Areola who was forced to study soils in Wales when it would clearly be more useful to study soils in a Nigerian context.

> When Areola wanted to study soils in Nigeria for his Cambridge PhD under A. T. Grove, unfortunately Professor Barbour [Ibadan Department Head] … said, "No money, no money for you to be coming back to Nigeria and roaming about the place and doing field work. Whatever you want to do, do it in Britain."
> Areola ended up studying soils in Wales, with repercussions for his future career:
> When you do your PhD abroad, there is a period of adjustment, trying to establish new study areas, and it also affects your writing. Not many people are interested in soils in Montgomeryshire, temperate land, over here. So it was like starting all over, tropical soils. (Interview 2017) (Craggs and Neate 2020: 904)

At the core of Mabogunji's argument was the acknowledgement that the place of knowledge production matters to what knowledge is produced. Colonial forms of geography are unlikely to be the most appropriate for addressing the needs of colonized peoples.

Milton Santos was the leading Brazilian geographer during the late 1970s right up to his death in 2001. His work was published in English sporadically in the radical journal *Antipode* (Santos 1975) but his major works were only recently translated twenty years after his death (Santos 2021a, b). Santos would not have recognized the label "Black geographies" any more than Mabogunji would have. Nevertheless, his work was rooted in his identity as a Black Brazilian from the poorest part of Brazil – a fact that makes his success even more remarkable. Santos' work mirrored work from the Global North in the late 1970s that critiqued the quantitative revolution and sought to develop a geography with the object of space as its core. Geography he insisted, in line with many who have written since Santos, was part of a colonial project.

> Coming late into the world as an official science, geography has struggled since its infancy to disconnect itself from powerful interests. One of the great conceptual feats of geography was to hide the role of the state – and of class – in the organization of society and space. Its other was to justify the colonial project. (Santos 2021a: 12)

Santos wanted his new geography to shrug off the mantle of both capitalist and colonial inheritances and become a critical discipline. He objected to the "new geography" of spatial science and proposed, in its place, a "new geography" focused on the production and politics of space.

> I propose the objective of this renewed geography to be the study of human societies in their process of permanently reconstructing space inherited from preceding generations, across the many instances of production. (Santos 2021a: 150)

Reading Santos now, from the perspective of an Anglo-American geographer who does not read any language but English competently, it is enlightening to see how much of Santos' work ran parallel to, and, in many cases foreshadowed, ideas that were only widely discussed from the late 1980s onwards. His book *Por uma Geografia Nova: Da Critica da Geografia a uma Geogrfia Critica* was published in Portuguese in 1978 in Brazil and published as *For a New Geography* in English in 2021. In 1978, there were few people writing at length about the production of space in English. Lefebvre's book *The Production of Space* would not be translated into English until 1991.

> It is more than thirty years since the Second World War, a great number of geographers, consciously or otherwise, have collaborated perniciously in the expansion of capitalism and all forms of inequality and oppression in the Third World.
> We have to equip ourselves to move in exactly the opposite direction. In contemporary condition this requires courage, as much in study as in action, aiming to lay the foundations for the reconstruction of a geographical space that would truly be the space of man, the space of all people, and not space at the service of some, and of capital. (Santos 2021a: 168)

In 1978, Santos was critiquing the role of geography in capitalism and colonialism and proposing a forward-looking geography focused on the problem of space. As others have noted, he only infrequently reflected on the role of race in geography, happy as he was to locate his critique in what was then called "The Third World." He did, however, quite clearly critique the twin forces of capitalism and colonialism from a distinct Global South standpoint and his work has been adopted within Brazil as a precursor to what can now be called Black Geographies (Prudencio Ratts 2022).

In Chapter 8, we saw how feminist geography grew out of a recognition that women were underrepresented in the discipline of geography. Since the 1970s, there has been some success in increasing the number and visibility of women in geography. While progress has been slow and the current situation is still far from equitable, most students taking degrees in geography will encounter women as instructors, lecturers, and professors. If we turn our attention to race, the same is not true.

Geography in the Western world is far more White than the general population. David Delaney has put it this way:

> Geography is a field of study. Geography is also a social institution and a workplace. I have the impression that, as an institution, geography is nearly as White an enterprise as Country and Western music, professional golf, or the Supreme Court of the United States. It seems to me that, even in comparison with other academic disciplines, geography and people of color are not particularly interested in each other. (Delaney 2002: 12)

Delaney's observations ignore the long history of Black and Brown scholars in geography who have often been overlooked (Kobayashi 2014), not to mention the subsequent changes to the make up of the Supreme Court, that is now probably less White than the discipline of geography. Nevertheless, his observations are surely still broadly accurate. The relative absence of Black geographers leads Patricia Noxolo, more recently, to underline the link between the relative absence of Black geographers and the need for Black Geographies to be largely written by Black geographers:

Why? As I have set out, the main critique that Black Geographies mounts is that geographical knowledge has historically been exclusionary of Black spatial thought and agency: in response, Black Geographies centres Black spatial thought and agency. Therefore, unless the discipline actually recruits and retains enough Black geographers to sustain a field named 'Black Geographies', the name itself will become a risible fig leaf for a determinedly White discipline that continues the colonial business as usual of both surveilling Black lives and erasing Black spatial agency. (Noxolo 2022: 6)

This theme is foregrounded, in a radical way, by Black geographer Katherine McKittrick in her book *Demonic Grounds* (McKittrick 2006). In this book, McKittrick accounts for the disconnection between human geography and Black studies, arguing that:

> … if black people and communities are left out of, or are simply objects in, geographic studies, they are inevitably cast as unavailable or unreliable geographers and geographic subjects; black knowledges, experiences, and maps remain subordinate to and outside other traditional geographic investigations. (McKittrick 2006: 11)

McKittrick points to another process of exclusion in geographical theory. White geographers, particularly radical White geographers, often enroll Black writers and theorists into their work in order to flag an awareness of issues of race. This tends to work through waving at the presence of a limited group of internationally recognized non-White scholars. McKittrick mentions the feminist Black cultural theorist bell hooks, the Black British cultural theorist Stuart Hall, the theorist of the "Black Atlantic," Paul Gilroy, and the Black psychiatrist and philosopher Frantz Fanon. These are names that have frequently appeared on the pages of human geographers, partly in recognition of the rich geographies that appear in their writing. McKittrick has no wish to diminish their importance but argues that the use of their work is often gestural or works to suggest "hey look – see how we know some work by Black scholars!" and then proceed as normal. She reflects on the use of the work of bell hooks in particular:

> The discussion of the margin, homeplace, whiteness, and oppositional politics by hooks is particularly popular, as is hooks herself (as geography and as a body scale). And I have argued elsewhere that several of these geographic investigations of blackness and black culture stop at bell hooks: this conceptual end-game is detrimental to geographic investigations in that it reduces Black Geographies, black feminist geographies, and arguably hooks herself, to a transparent visual illusion – the black female body, knowable and knowing, and unaccompanied, answers the question of difference while allowing theorists to disregard heterogeneous ways of being black. (McKittrick 2006: 19)

This treatment of hooks, McKittrick suggests, is emblematic of a wider problem with human geography and its treatment of "the Black subject" and Black women in particular. Geographers, she argues, use Black bodies as part of arguments about difference and otherness in geography without fully taking on board the significance of Blackness for how geography might be theorized more generally:

> Through symbolic-conceptual positioning, the black subject (often, but not always, a black woman) is theorized as a concept (rather than a human or geographic subject) and is consequently cast as momentary evidence of the violence of abstract space, an interruption in transparent space, a different (all-body) answer to otherwise undifferentiated geographies. Spatially and conceptually, the black female subject is briefly granted one or two sentences to support 'arguments about difference and diversity' and raise some 'painful questions' about traditional geographic patterns. (McKittrick 2006: 19–20)

Throughout her book, she traces a much wider world of Black Geographies that exist in less likely sources than the books academic geographers normally read. Some of the places she looks are close to the normal ground of theoretically inclined geographers. The writing of Édouard Glissant, a Martinican writer and poet who brings a distinctly Caribbean perspective to cultural and literary theory, is particularly important. His work is clearly postcolonial in its intent but brings a lively sense of mobile hybridity to questions of belonging and identity. But there are many other places in which McKittrick finds overlooked Black Geographies. These include the world of creative writing in its various forms, including the novels of the Black Canadian writer Dionne Brand, the Jamaican poet and critic Sylvia Wynter, and poet and theorist Marlene Nourbese Philip.

In the work of Philip, for example, McKittrick explores how her work connects the space and place of the body to wider geographies of Black and female oppression and resistance. McKittrick provides an account of Philip's essay "Dis Place – The Space Between" (Philip 1994). Here, Philip connects the bodily "space between her legs" to a long history of material geographies of the plantation, the nation, the street, and other more conventional elements of human geography. The facts of sexual desire and abuse underline the conditions of Black women in the new world and thus link body space to other spaces in powerful ways:

> She emphasizes the different ways gender is lived via the experiential scale of the body to further denaturalize essentialist black places and spaces. While black women are positioned as objectified sexual technologies, they also continually make, remake, and articulate geography across practices of domination:
>
> …
>
> bodydeadbodiesmurderedbodiesimportedbredmutilatedbodiessoldbodiesboughttheEuropantrafficin bodiesthattellingsomuchaboutthemanandwhichhelpingfueltheindustrializationofthemetropolisesbodi escreatingwealththecapitalfeedingtheindustrialrevolutionsmany timesoverandoverandoverthebodies …
>
> Between the legs   thespace
> /within the womb   thespace
> colonized like place and space
> thesilenceof
> thespacebetween
>   the legs
> thesilenceof
> thespacewithin
>   thewomb … (Philip quoted in McKittrick 2006: 48–49)

McKittrick shows how Philip, through her essays and poetry, uses "the space between the legs" to chart a geography of moving Black bodies through landscapes of urbanization, industrialization, and colonization.

McKittrick has been a leading voice in the growing visibility of Black Geographies within the discipline and in the academic world beyond it. At the heart of this enterprise has been the entanglement of the idea of Blackness with broadly spatial themes. It asks what happens to ideas about Blackness (and, more broadly, "race") when geographical imaginations are taken seriously and, conversely, what happens to geography, when Blackness is taken seriously. These themes are clearly articulated by Camilla Hawthorne in a review of the field.

> … the use of "Black" in the phrase "Black Geographies" is intended, at least in part, to serve as a radical provocation to the discipline of geography. It is a call to center those subjects, voices, and experiences that have been systematically excluded from the mainstream spaces of geographical inquiry. It is also an invitation to consider how an analysis of space, place, and power can be fundamentally transformed by foregrounding questions of Blackness and racism. To put it another way, Black Geographies asks how the analytical tools of critical human geography can be used to engage with spatial politics and practices of Blackness, and how an engagement with questions of Blackness can in turn complicate foundational

geographical categories such as capital, scale, nation, and empire. By revealing the colonial and racist assumptions that undergird so many key concepts in geographical inquiry, Black Geographies can then point the way to their eventual undoing. (Hawthorne 2019: 8–9)

Global Black Geographies

As with other theoretical approaches in this book, Black Geographies are not a singular theory or approach. Their plurality should be evident from the phrase "Black Geographies." Even in the case of the United States, Adam Bledsoe and Willie Jamaal Wright have traced what they call "the pluralities of Black Geographies" (Bledsoe and Wright 2019). They consider three political/theoretical moments in the United States where Black thinkers and activists called for, and partially enacted, forms of Black territoriality and place-making outside of White majority society – Marcus Garvey's Back to Africa movement, which proposed a sovereign state of afro-descended people within Africa, the Black Panther Party for Self-Defense, which mobilized for autonomous Black communities with the United States, and the "Provisional Government of the Republic of New Afrika" that sought a Black sovereign nation-state within the United States. They chart these related but different examples of Black struggle "to foreground the fact that Black people—and thus Black political movements—have different diagnoses regarding the nature of anti-Blackness as well as different prescriptions for what Black liberation will look like and how this might be achieved" (Bledsoe and Wright 2019: 421).

There is no doubt that much of the impetus for Black Geographies comes from North America with its very specific history of slavery. In addition to Bledsoe and Wright's call for attentiveness to differences between Black Geographies in the American context, we need to think about Black Geographies in other places. In her popular book, *Why I'm no Longer Talking to White People About Race*, Reni Eddo-Lodge reflects on how issues of race and racism as well as the Black experience in the United Kingdom often look to the United States and the civil rights movement for inspiration and education (Eddo-Lodge 2017).

> I'd only ever encountered black history through American-centric educational displays and lesson plans in primary and secondary school. With a heavy focus on Rosa Parks, Harriet Tubman's Underground Railroad and Martin Luther King, Jr, the household names of America's civil rights movement felt important to me, but also a million miles away from my life as a young black girl growing up in north London. (Eddo-Lodge 2017: 1)

Throughout the book, Eddo-Lodge reminds the readers of the very long history of the United Kingdom's history of slavery as well as the very long list of injustices faced by Black people as they have sought to go about their everyday lives. Black Geographies in the UK are certainly connected to the American experience but are also clearly different.

Recognizing the geo-historical distinctiveness of Blackness in the heart of empire, and particularly the differences between Black experience in the UK and the USA, geographers have called for a Black British Geography (Noxolo 2020). Patricia Noxolo draws the links between Black British culture and the specificities of the British city as nodes in empire with their connections to both the Caribbean and Africa.

> A Black British Geography not only recognizes the British city as a geographical context for much of Black British culture, but (without forgetting the ongoing transnational salience of Kingston, Lagos, and other globally influential cities) also appreciates that Black British culture and the British city are to a large extent mutually constitutive. It becomes our task to demonstrate not only how the British city shapes Black British bodies, but also how Black British bodies articulate and produce the British city (Noxolo 2020: 510).

More recently, in a review of Black Geographies, Noxolo outlines the emergence of a Global Black Geographies that recognizes that the majority of Black people, and therefore surely the majority of Black Geographies exist outside of the United States and the United Kingdom (Noxolo 2022). While much is shared by Black geographies everywhere, including the deadening logic of White Supremacy, it is also the case that Black geographies are contextual. The characteristic space and places of Black life, for instance, vary depending on where they are located: "Where in the US intimate sites of Black existence like the stoop and the Black-owned business are key spatial sites and symbols of Black discourse and resistance (…), in the Caribbean, Canada and Australia studies of other intimate spaces – the lakou, the street corner, the mall, the cafe´ – reveal other political sites of surveillance and resistance through dance and playfulness" (Noxolo 2022: 1236). Despite this world of Black geographies, Noxolo notes how "Black Geographies so named are rooted in the powerful epicentre of US-focused Black knowledge production and are routing outwards from there to other centres of Black community" (Noxolo 2022: 1233). And yet, at the same time, the centering of Black experience in Black Geographies necessarily resists this US focus "re-reading Black Geographies as rooted in/routed through a more globalised and diverse vision of Blackness" (Noxolo 2022: 2).

> By centring Black spatial thought, rather than just noting its erasure, one of the most radical effects of Black Geographies might be to finally move Geography away from its postcolonial liberal impulse to try to include Black geographers in dialogue only on its own exclusionary terms (Hawthorne and Heitz 2018). Instead, Black Geographies might push the discipline towards a historical moment in which 'an ethical analytics of race [is] based not on suffering, but on human life." (McKittrick 2011: 948)

Camilla Hawthorne expands this plea at the end of her review of work in Black Geographies by noting how "A sizeable proportion of the Black Geographies scholarship has thus far been carried out in North America and, to a lesser extent, the Caribbean. There is an urgent need for Black Geographies research that considers the spatial politics of race and Blackness in other geographical contexts, and draws attention to the 'inherent pluralities' of Black Geographies" (Hawthorne 2019: 8). She suggests the need to recognize the global circulation of Black Geographies, the need for engagement with the African continent, and a focus on the role of seas and oceans in the constitution of Black Geographies – including her own work on the Mediterranean (Hawthorne 2017). In her commentary on the murder of Emmanuel Chidi Nnamdi, in addition to a sad litany of other Black people killed in Italy, she reflects on the Black Geographies emerging from increasingly desperate migration across the Mediterranean from Africa and the specificities of the newly acknowledged identity of afro-Italian. Part of the debate about what it means to be afro or Black Italian is how it is different from African-American. She links the experience of Blackness in Italy to the country's sense of itself as being a soft colonial power in Africa and the recent history of immigration where Blackness inevitably gets entangled with the label "immigrant" and its attendant images of overcrowded unseaworthy boats attempting the perilous journey across the Mediterranean.

> What, then, are the new and sometimes contradictory meanings being ascribed to blackness in Italy at this moment? How do the emergent struggles of black Italians relate to, for instance, African American histories and mobilizations from the other end of the diaspora? What are the geographical and historical constituents of blackness in Italy – as a category of racial subjection, as a form of self-identification, and even as the basis of a radical politics of liberation? They are admittedly enormous questions, but here I simply want to begin a reflection on the trials and travails of diasporic unity amid the specificities of different experiences of racialization, and about the possibilities of resistance to anti-Blackness in a violent and precarious European present. (Hawthorne 2017: 162)

Hawthorne's work provides another site of Black diasporic identity not just in Italy but in the fluid space of the "Black Mediterranean." Drawing on Paul Gilroy's well-known discussion of the "Black

Atlantic" (Gilroy 1993), Hawthorne traces how Italy has long been a crossroads and a key part of the larger Mediterranean world (Hawthorne 2022). This history makes any claim to racial or ethnic homogeneity spurious. Most obviously Italy is across the sea from Africa and has long been a site of migration for people from Sub-Saharan Africa as well as the colonization of parts of Spain and Sicily in the Moorish empire. And yet the Mediterranean is precisely the site where an imagined White European world was allegedly born in the classical world of Ancient Rome and Greece through a delineation between itself and the Africa over the sea. Despite this ongoing border work, Hawthorne enrolls Cedric Robinson to make the case that the Black Mediterranean was the precondition for racial capitalism – a kind of test case for emerging forms of racism and the use of enslaved labor that existed before the advent of the Atlantic trade in enslaved Africans.

Black Spatial Thought

Key to the project of Black Geographies is a focus on the creative and critical capacities of Black spatial thought. There are, of course, many instances of Black spatial thought in action. In the world beyond geography, the work of Paul Gilroy, Stuart Hall, bell hooks, and Frantz Fanon are comparatively well known (Fanon 1969; Hall 1978; Gilroy 1993; hooks 2014). Recent work by McKittrick and others has expanded the canon of Black thinkers referenced by geographers to include Sylvia Wynter and Édouard Glissant (Wynter 1971; Glissant 1997). Cedric Robinson's formulation of racial capitalism is also foundational (see Chapter 7) (Robinson 1983). Black spatial thought does not just exist in the pages of Black theorists and writers though; an important component of work in Black Geographies is the embrace of spatial thought in folk and popular culture and in the arts. In the remainder of this chapter, we will explore a selection of Black spatial thought from geography and beyond.

In *In the Wake,* the literary scholar, Christina Sharpe, provides us with a distinctly geographical way of thinking through Blackness in relation to specific spatial forms of the wake, the hold and weather (Sharpe 2016: 42). All of these are derived from the slave ship in the middle passage. The wake is the central word to think with, referring as it does to the wake of the ship – the trace it makes in the water that comes after its passage. A wake also refers to a funeral procession, the sense of being awake, the line of recoil of a gun and more besides. Sharpe uses this polysemous word to outline a kind of manifesto for "wake work" – the work that needs to be done in the wake of slavery. The hold refers to the underdeck dark space of the ship in which slaves were crammed to be shipped as cargo. Sharpe uses the hold as an analogy to think though, among other things, contemporary prisons in the United States where Black people are disproportionately imprisoned. The weather is the context for the ship – the total antiblack climate that continues to provide the context for Black death. The wake of slavery includes the "premature death" that is central to Gilmore's definition of slavery – premature death that is often state sanctioned or extralegal. To be in the wake, Sharpe tells us, "is to occupy and to be occupied by the continuous and changing present of slavery's as yet unresolved unfolding" (Sharpe 2016: 13–14).

There are other ways in which a geographical refocusing on the slave ship can fundamentally alter the production of theory while placing race generally, and Blackness specifically, at the center of our thinking. Recall the theorization of surveillance in contemporary society. The work of Michel Foucault has been central to critical theories of surveillance. In his book *Discipline and Punish*, he delineated his idea of discipline as something encoded in individual bodies through a sense of always being watched (Foucault 1979). His archetypal space for the construction of this concept was the *panopticon*. The panopticon, in Foucault's book, is described as the invention of the British utilitarian philosopher, Jeremy Bentham. It was a design for a prison that Bentham formulated in 1786, consisting of a circle of cells surrounding a watch tower. Each cell would be divided from the others by a solid wall but the front would be a metal grate, allowing the prisoner to be seen from the watch tower in the center of the circle. While a person in the watch tower could theoretically see every prisoner, the watcher was themselves invisible to the prisoners. In theory, the watchtower could be empty, and the prisoners would still feel as though they were being watched and act accordingly. The panopticon was an architectural seeing machine that allowed certain kinds of lines of sight and pre-vented others. Central to Foucault's argument was his claim that *discipline*, of the kind exercised and internalized in the panopticon, was replacing *punishment* – the exercise of spectacular torture on the individual body. While very few prisons were ever built on this model, the idea of a panopticon became a kind of spatial metaphor for the growth of surveillance in contemporary society using CCTV, biometrics, face recognition, and a host of other surveillance technologies. It is now com-monplace to talk of a metaphorical/theoretical panopticon. Foucault's account of discipline and the panopticon does not grapple with the importance of race to regimes of both discipline and punish-ment. Such an account is provided by the Black feminist scholar, Simone Browne, in her book *Dark Matters: On the Surveillance of Blackness* (Browne 2015).

Browne starts her argument by noting that the decline in public torture noted by Foucault is, at the very least, called into question when we take Blackness seriously, noting the grip of terror in both slavery and post-slavery lynchings as well as White mob violence. Like Robinson and Woods, she also looks to a very different place on which to base a discussion of surveillance – the hold of the slave ship rather than the panopticon. As Browne writes, "Drawing a line through panopticism by way of the slave ship is another means of interrupting Foucault's reading of discipline, punishment, and the birth of the prison, because, as Marcus Rediker put it, the slave ship was 'a mobile seagoing prison at a time when the modern prison had not been established on land'" (Browne 2015: 42). Just as Cedric Robinson revealed how a focus on the space of the plantation could transform accounts of capitalism that started with the Marxist focus on the English mill (see Chapter 7), so Sharpe and Browne both reconfigure spatial imaginaries by starting with the slave ship.

In her book *Black Faces, White Spaces*, Carolyn Finney reimagines the relationship between African-Americans outdoor spaces of "nature." She begins with a segment of the poem *Mercy* by the Black poet, Lucille Clifton.

> Surely i am able to write poems
> celebrating grass and how the blue
> in the sky can flow green or red
> and the waters lean against the
> chesapeake shore like a familiar,
> poems about nature and landscape
> surely but whenever i begin
> "the trees wave their knotted branches
> and..." why
> is there under that poem always
> an other poem?
>
> (Lucille Clifton in Finney 2014: 1)

The idea of the wilderness plays an important role in North American culture as a sacred space pro-viding evidence for God's blessing on a new country. Many of the leading figures in eco-philosophy and environmental preservation were Americans. One example was John Muir, the founder of the Sierra Club, who was instrumental in the formation of National Parks such as the one at Yellowstone. Despite their status as national spaces, however, Finney shows how Black people were excluded from the mobility necessary to access these wild spaces through both law and the prevalence of violent racist acts. This exclusion, she argues, was somewhat paradoxical as to some degree, Black people, like Native Americans, were made to signify nature and wild in racist ways – treated with a mixture of contempt and fake reverence. They were (and sometimes still are) described in ways that located them in nature – as not fully human, or primitive. For a long time Black people were exhibited in natural history museums and even zoos. Finney recounts how Ota Benga – a Congo Mbuti (Pygmy) man – was exhibited in an exposition in St. Louis in 1904 and in the Bronx Zoo in 1906. He was toured around as part of a traveling exhibit playing the role of both a scientific curiosity and a spectacle for the mostly White onlookers. The Bronx Zoo referred to him as a "missing link" between primates and humans. The very idea of race, Finney explains, is rooted in (false) ideas about nature and nature is often given as a justification for social differences and inequities between social groups. The National Park movement was happening at the same time as Whiteness was being linked to a new national identity and Blackness as a threat to the nation – as exemplified in the film *Birth of a Nation* (1915), which represented Black people in racist ways and celebrated the rise of the Ku Klux Klan.

> From the very beginning, the creation of "wilderness" and public lands (parks and forests) was the centerpiece of the nation-building project of defining who we are as Americans. These lands were our cathedrals, our representations of the world of, supposedly, the best of who we are and who we can be. From the beginning, African Americans as well as other nonwhite peoples were not allowed to participate on their own terms in this project. And when they were, the how, when, and where of their participation was determined by the dominant culture through legislation, rhetoric, science, and popular perception. (Finney 2014: 50)

Finney points to this history as a complicating factor in present-day Black attitudes to the environment and environmentalism. As Lucille Clifton's poem makes clear, there should be a way for Black people to engage with the natural world in traditional lyrical ways, but behind such an attempt there is the history that Finney had described where, in the words of *Strange Fruit*, the song made famous by Billie Holiday:

> Pastoral scene of the gallant South
> The bulging eyes and the twisted mouth
> Scent of Magnolias, sweet and fresh
> Then the sudden smell of burning flesh

The seeming innocence of the term "pastoral" and the presence of the natural beauty of the Magnolias (typical poetic subjects) are undercut by the dead Black bodies hanging from the poplar trees. Trees, Finney tells us, do not mean the same thing for Black people and White people.

Geographies of Black Life

The spaces of the plantation and slave ship both force a focus on the histories and ongoing geographies of premature Black death. An important component of Black Geographies, and Black spatial thought in general, however, is the need to focus on Black life – on the ongoing creativity of Black experience despite the general "weather" of antiblackness. The Great Dismal Swamp in North Carolina has historically been an inaccessible area. Before parts of it were drained it formed over 2000 square miles where there were no established paths across the swamp. It was a home to mosquitos, poisonous cottonmouth, rattler, and copperhead snakes, bears, bobcats, and wolves which prowled the islands between the gum trees and cypress that grew in the stagnant water. The swamp, in other words, was a significant barrier to White colonists. It was in this space that a free Black population lived during the era of slavery – a society of runaway enslaved people known as "maroons." Finney uses the space of the swamp as an example of an affirmative Black place living according to its own forms of governance (Finney 2014). Finney notes that the swamp was not entirely autonomous or divorced from the larger landscape of White supremacy. "Black people building a life in the swamp," she writes, "were also living under the continuous fear of being discovered. Slave owners did not relinquish their hold on their human property that easily" (Finney 2014: 122). Slave owners would train dogs to hunt runaway enslaved people and the swamp was a dangerous place to live. Nevertheless, this segment of wilderness did provide a kind of sanctuary for Black people and "[d]espite the challenges of day-to-day living in the swamp, any fear of the swamp was tempered by a more enduring fear of the white-population, slavery, and negative notions of Blackness that were embedded in the dominant culture. The wilderness, as defined in contemporary terms, became a place of refuge and possibility for black people" (Finney 2014: 122). Despite all the provisos, then, Finney described the Great Dismal Swamp as a Black landscape of relative autonomy where Black creative culture was alive and where Black people could thrive. She links this to other spaces of Black creative expression to affirm Black life that cannot be reduced to reactions or responses to White supremacy.

> Adaptiveness, resilience, fearlessness, and courage wasn't the anomaly, but was the reality. While fear as a by-product of white supremacy and oppression was/is certainly part of the lived reality for many African Americans, focusing solely on the fear denies the malleability of the black imagination to create and construct a reality that is not grounded primarily in fear, but in human ingenuity and the rhythms and flows of life. (Finney 2014: 123)

Finney's observations about the Great Dismal Swamp resonate with a significant and important strand in Black Geographies – the affirmation of Black life. In this book *Black Marxism*, Cedric Robinson made this affirmation a lynchpin of his thesis of racial capitalism. Black revolutionary cultures, he insisted were

> …formed through the meanings that Africans brought to the New World as their cultural possession: meanings sufficiently distinct from the foundation of Western ideas as to be remarked upon over and over by the European witnesses of their manifestations; meanings enduring and powerful enough to survive slavery to become the basis of an opposition to it. (Robinson 1983: 5)

While, in Gilmore's terms, racism hinges on the fact of premature death, it would be wrong to think of Blackness as defined only in these terms. The affirmation of Black life, often rooted in Africa, is equally important.

Another space associated with affirmative Black culture is the small plot of land given to enslaved people to grow their own vegetables within plantations. The Jamaican novelist and theorist, Sylvia Wynter reflects on the importance of the plot in her essay "Novel and History, Plot and Plantation" (Wynter 1971). She contrasts the plantation, a monoculture system designed to produce cash crops for export such as tobacco and sugar and marked by European models of social and racial hierarchy, with the plot, a small patch of ground marked by multiple crops with origins in Africa such as yams and cassava grown as food for subsistence by enslaved African people. Wynter describes the plot as a site of guerilla resistance to the exchange value-based system of the plantation.

> For African peasants transplanted to the plot all the structure of values that had been created by traditional societies of Africa, the land remained the Earth — and the Earth was a goddess; man used the land to feed himself; and to offer first fruits to the Earth; his funeral was the mystical reunion with the earth. Because of this traditional concept the social order remained primary. Around the growing of yam, of food for survival, he created on the plot a folk culture — the basis of a social order — in three hundred years. (Wynter 1971: 99)

While Wynter locates the plot as a site of traditional use-values based in Africa, the geographer Clyde Woods creates a wider sense of affirmative forms of knowledge in opposition to the plantation in what he calls **blues epistemology**. In *Development Arrested: The Blues and Plantation Power in the Mississippi Delta* (Woods 2017), Woods explores the regionalization of the area known as the Mississippi Delta – the poorest area of the United States where the population is 60% Black. Woods charts the interplay between two epistemologies in the region, one derived from the murderous logic of the plantation and other derived from the lives of the Black population – the blues epistemology. Woods takes Cedric Robinson's arguments about **racial capitalism** further (see Chapter 7). Unlike Robinson, Woods argues that slavery and the plantation should not be thought of as remnants of feudalism. Rather, he insists that the spatial reality and imagination of the plantation is the basis for capitalism in the USA. The social conditions of the Delta, Woods argues, are not the result of too little capitalism, but *too much* capitalism. Like Robinson, he places the plantation/slavery complex right at the center of explanations of capitalism, and not at its periphery. Similarly, he charts the emergence of a counter hegemonic collective within the logic of the plantation – the emergence of the blues epistemology among Black people in the Delta.

While the spatiality of the plantation was one of constant surveillance and hierarchical order, the blues epistemology is rooted in a critique of plantation epistemology based on an oral tradition, rhythm, and African routes. The blues epistemology might best be expressed through the idea of the "blue note." Woods quotes Mary Ellison's account of the blue note:

> The blue or flattened note, sung just under the note as it should have been sung on the Western musical scale, has become almost the hallmark of the blues.… Blue notes are not notes played out of tune but notes played in a specific way. It was created when slaves tried to fit African scales to European scales (Ellison 1989: 4)

Note the suggestion of a blue note being "out of tune" because it does not quite fit into the Western musical scale. This illustrates how an epistemology born out of Black life, with its different learned rhythms born out of different histories and geographies cannot be squeezed into dominant forms of understanding with their own spatial imaginaries. The Western musical scale can be thought of as a spatial arrangement of evenly divided intervals that organizes sound. Blue notes use this scale but do not quite fit.

> Born in a new era of censorship, suppression, and persecution, the blues conveyed the sorrow of the individual and collective tragedy that has befallen African Americans. It also operated to instill pride in a people facing daily denigration, as well as channeling folk wisdom, descriptions of life and labor, travelogues, hoodoo, and critiques of individuals and institutions (Woods 2017: 17)

Woods' account of the Blues as an epistemology that exists beyond the logic of the plantation is similar to both Finney's account of the Great Dismal Swamp and Wynter's account of the plot. Linking African cultural elements that preceded the logic of slavery and the plantation with the specific context of the plantation economy and White supremacy. The Blues, to Woods, is a form of knowledge that arises out of daily life and stands outside of the logic of dominant ways of knowing.

> Like other blocs in the region, working-class African Americans in the Delta and the Black Belt South have constructed a system of explanation that informs their daily life, organizational activity, culture, religion, and social movements. They have created their own ethno-regional epistemology. Like other traditions of interpretation, it is not a monolith; there are branches, roots, and a trunk. (Woods 2017: 16)

When Woods looks to the Blues as a site of theory rooted in daily life, he is finding affirmation in everyday creative culture – a site that is repeatedly sought out by scholars in the Black Geographies tradition. Finney, for example, ends her book with a call to recognize the creative ways in which Black artists and musicians are creating forms of Black environmentalism.

> ... I would like to see geography open up to engaging nonacademic spaces of knowledge production with the same respect given to more traditional spaces, such as academia. In particular, when doing work about race and difference, I find it is imperative to engage culture work – be it popular culture, art, or music in all its myriad forms – because it is in these spaces that people who are "different" are able to produce work about themselves without the boundaries and rules that can inhibit their voices that more traditional ways of knowing are unable to communicate. (Finney 2014: 133)

McKittrick, in her book *Dear Science and other Stories*, is a kind of academic embodiment of the affirmation of Black life and, particularly, Black place (McKittrick 2021). It not only draws on the kinds of "nonacademic spaces of knowledge production" that Finney and Woods write about, but exemplifies work that transgresses the boundaries and rules that the codified system of academic writing tends to privilege. It is, itself, a work of creative geography as well as being a work that uses creative non-disciplinary geographies as its source material. In *Dear Science*, McKittrick tells interdisciplinary stories that escape the reduction of Blackness to stories of death without denying the premature death that haunts the experience of Blackness in North America and elsewhere. She provides escape routes from discussions of Blackness that are rooted in biological determinism. In her words:

> I imperfectly draw attention to how seeking liberation, and reinventing the terms of black life outside normatively negative conceptions of Blackness, is onerous, joyful, and difficult, yet unmeasured and unmeasurable, Mnemonic black livingness. My heart makes my head swim. (McKittrick 2021: 3)

Among other spaces, McKittrick draws our attention to how archives, even when Blackness is made visible, tend to repeat stories of racism and violence and rarely Black joy. They thus become part of the "dehumanizing logics of white supremacy" (McKittrick 2021: 105). In contrast, McKittrick asks "what happens to our understanding of black humanity when our analytical frames do not emerge from a broad swathe of numbing racial violence but, instead, from multiple and untracked enunciations of black life?" (McKittrick 2021: 105). Her strategies for doing this involve looking to sites of knowledge production outside of academia, including poetry, art, and music which she theorizes as kinds of theory as well as kinds of practice. These sites of Black life are often set against forms of measure and mapping that, she argues, even when conducted with anti-racist intent, tend to replicate Blackness as an object marked with death.

> Black studies and anticolonial thought offer methodological practices wherein we read, live, hear, groove, create, and write across a range of temporalities, places, texts, and ideas that build on existing liberatory practices and pursue ways of living the world that are uncomfortably generous and provisional and practical and, as well, imprecise and unrealized. (McKittrick 2021: 5)

It is not only the sources she draws on that makes McKittrick's work stand out. It is the way the book itself is creative in its representational practices. It includes transparent pages, music playlists, texts arranged in columns that speak to and against each other, poetic glossaries, and more. The footnotes are not simply citations of work, but conversations with other scholars. In some places the footnotes are longer than the main body of the text. *Dear Science* is still an academic book but it punches through the margins of form in creative ways that formally match the liveliness that is her central theme. Perhaps the central strategy uniting all of these is her insistence that her chapters are *stories*. Stories, she points out, are an important part of Black life. Like much of Black life, they are distrusted by science. McKittrick uses stories as theory.

Stories and storytelling signal the fictive work of theory. I hope this move, at least momentarily, exposes the intricacies of academic work where fact-finding, experimentation, analysis, study, are recognized as narrative, plot, tale, and incomplete interventions, As story, theory is cast as fictive knowledge and insists that the black imagination is necessary to analytical curiosity and study. Story is theoretical, dance, poem, sound, song, geography, affect, photograph, painting, sculpture, and more. Maybe the story is one way to express and fall in love with Black life. Maybe the story disguises our fall. (McKittrick 2021: 8)

McKittrick tells her stories in often experimental ways and with reference to the diverse array of creative moments in Black life. To give one example, she looks to Black music forms as sites of practical theory. In particular, she focusses on the rhythm and sound (rather than lyrics) in Black music. This, of course, mirrors Woods' discussion of the blues epistemology, a work that McKittrick frequently engages with, as well as with the work of Black theorists and novelist Sylvia Wynter. To McKittrick, following Wynter:

Enthusiasm radically challenges and overturns the dominant order: the feeling of exaltation, emerging as a form of knowledge that is necessarily collaborative praxis, cites and sites Black joy and love. For it is in the waveforms of music – beats, rhythms, acoustics, notational moods, and frequencies that intersect with racial economies and histories – that rebelliousness is enunciated as an energetic (neurological, physiological) *affection for* black culture *as* life. (McKittrick 2021: 166–167)

McKittrick's discussion of rhythm and groove in particular, and Black life in general, resonates with Finney's discussion of the Great Dismal Swamp and Woods Blues epistemology, connecting affirmative sites of knowledge production beyond academia with insurgent forms of creative expression that exist alongside, but are not defined by, the spaces and times of White Supremacy.

Conclusions

Black Geographies provide a rich body of theory for thinking about the world we live in today and how we got here historically. At their heart, they center the production and experience of Blackness through Black spatial thought. Sites such as the plantation, the slave-ship, the plot, and the dance-floor, among others, are used to both account for the effects of White Supremacy and to affirm the creative potential of Black life. There is a danger, of course, in constraining the contribution of Black Geographies to the discipline of geography within this final chapter and I have tried to mitigate this by including the contributions of both Black geographers and the ideas associated with Black Geographies throughout the book. Examples include Clyde Woods' contributions to arguments about the region (in Chapter 3), and Ruth Wilson Gilmore's and Cedric Robinson's contributions to Marxism (in Chapter 7). Conversely, much of the work referred to in this chapter could equally well be in chapters on feminism (Chapter 8) or decoloniality (Chapter 14). The fact of the matter is that geographic thought does not fit neatly into paradigms that can be adequately surveyed in separate chapters. The theory and concepts in this chapter and others are always interconnected. Nevertheless, the centering of Blackness in Black geographies is an important contribution to geographic thought and, more importantly, to our understanding of how, in McKittrick's words, we "work out where and how black people imagine and practice liberation as they are weighed down by biocentrically induced accumulation by dispossession" (McKittrick 2021: 74). Black Geographies help us to understand both how White Supremacy is spatialized and how Black life has its own geographies beyond the logic of the plantation and slave ship. Importantly, they also point to the ways the discipline of geography might escape the prison house of its own racist past.

References

Alderman, D. H. and Inwood, J. (2013) Street naming and the politics of belonging: spatial injustices in the toponymic commemoration of Martin Luther King Jr. *Social & Cultural Geography*, 14, 211–233.

* Bledsoe, A. and Wright, W. J. (2019) The pluralities of black geographies. *Antipode*, 51, 419–437.

Bonnett, A. (1997) Geography, 'race' and Whiteness: invisible traditions and current challenges. *Area*, 29, 193–199.

Browne, S. (2015) *Dark Matters: On the Surveillance of Blackness*, Duke University Press, Durham.

Coates, T.-N. 2014 *The Case for Reparations*, The Atlantic

Craggs, R. and Neate, H. (2020) What happens if we start from Nigeria? Diversifying histories of geography. *Annals of the American Association of Geographers*, 110, 899–916.

Delaney, D. (1998) *Race, Place and the Law*, University of Texas Press, Austin, TX.

Delaney, D. (2002) The space that race makes. *Professional Geographer*, 54, 6–14.

Eddo-Lodge, R. (2017) *Why I'm No Longer Talking to White People About Race*, Bloomsbury Circus, London.

Ellison, M. (1989) *Extensions of the Blues*, Riverrun Press, New York.

Fanon, F. (1969) *The Wretched of the Earth*, Penguin Books, Harmondsworth.

Finney, C. (2014) *Black Faces, White Spaces: Reimagining the Relationship of African Americans to the Great Outdoors*, The University of North Carolina Press, Chapel Hill, NC.

Foucault, M. (1979) *Discipline and Punish: The Birth of the Prison*, Vintage Books, New York.

Gilmore, R. W. (2007) *Golden Gulag: Prisons, Surplus, Crisis, and Opposition in Globalizing California*, University of California Press, Berkeley, CA.

Gilmore, R. W. (2022) *Abolition Geography: Essays Towards Liberation*, Verso, London; New York.

Gilroy, P. (1993) *The Black Atlantic: Modernity and Double Consciousness*, Harvard University Press, Cambridge, MA.

Glissant, E. D. (1997) *Poetics of Relation*, University of Michigan Press, Ann Arbor, MI.

Hall, S. (1978) *Policing the Crisis: Mugging, The State, and Law and Order*, Macmillan, London.

Hawthorne, C. (2017) In search of black Italia. *Transition*, 123, 152–174.

* Hawthorne, C. (2019) Black matters are spatial matters: black geographies for the twenty-first century. *Geography Compass*, 13, e12468.

Hawthorne, C. A. (2022) *Contesting Race and Citizenship: Youth Politics in the Black Mediterranean*, Cornell University Press, Ithaca, NY.

hooks, b. (2014) *Black Looks: Race and Representation*, Routledge, New York.

Jackson, P. (ed.) (1987) *Race and Racism: Essays in Social Geography*, Allen & Unwin, London.

Kobayashi, A. (2014) The dialectic of race and the discipline of geography. *Annals of the Association of American Geographers*, 104, 1101–1115.

Ley, D. (1974) *The Black Inner City as Frontier Outpost: Images and Behavior of a Philadelphia Neighborhood*, Association of American Geographers, Washington, DC.

Mabogunji, A. (1975) Geography and the problems of the third world. *International Social Science Journal*, 27, 288–302.

McKittrick, K. (2006) *Demonic Grounds: Black Women and the Cartographies of Struggle*, University of Minnesota Press, Minneapolis, MN; London.

McKittrick K (2011) On plantations, prisons, and a black sense of place. *Social & Cultural Geography* 12(8): 947–963.

* McKittrick, K. (2021) *Dear Science and Other Stories*, Duke University Press, Durham.

McKittrick, K. and Woods, C. A. (2007) *Black Geographies and the Politics of Place*, South End Press, Cambridge, MA.

Mitchell, D. (2000) *Cultural Geography: A Critical Introduction*, Blackwell Pub., Malden, MA.

Mott, C. and Cockayne, D. (2017) Citation matters: mobilizing the politics of citation toward a practice of 'conscientious engagement'. *Gender, Place & Culture*, 24, 954–973.

* Noxolo, P. (2020) Introduction: towards a black British geography? *Transactions of the Institute of British Geographers*, 45, 509–511.

Noxolo, P. (2022) Geographies of race and ethnicity 1: black geographies. *Progress in Human Geography*, 46 (5), 1232–1240.

Peach, C., Robinson, V. and Smith, S. (1981) *Ethnic Segregation in Cities*, Croom Helm, London.

Philip, M. N. (1994) Dis place- the space between, in *Feminist Measures: Soundings in Poetry and Theory* (eds L. Keller and C. Miller), University of Michigan Press, Ann Arbor, MI, pp. 287–316.

Prudencio Ratts, A. J. (2022) Milton Santos: from new geography to black geography. *Dialogues in Human Geography*, 12, 470–472.

Robinson, C. J. (1983) *Black Marxism: The Making of the Black Radical Tradition*, Zed, London.

Santos, M. (1975) Geography, Marxism and underdevelopment. *Antipode*, 6, 1–9.

Santos, M. L. (2021a) *For a New Geography*, University of Minnesota Press, Minneapolis, MN.

Santos, M. L. (2021b) *The Nature of Space*, Duke University Press, Durham; London.

Schein, R. H. (2009) Belonging through land/scape. *Environment and Planning A: Economy and Space*, 41, 811–826.

Sharpe, C. E. (2016) *In the Wake: On Blackness and Being*, Duke University Press, Durham.

Western, J. (1981) *Outcast Cape Town*, University of Minnesota Press, Minneapolis, MN.

*Woods, C. A. (2017) *Development Arrested: The Blues and Plantation Power in the Mississippi Delta*, Verso, London; New York.

Wynter, S. (1971) Novel and history, plot and plantation. *Savacou*, 5, 95–102.

Glossary

Abolition geography   Associated with Ruth Wilson Gilmore. An argument for the kind of world that is necessary for prisons to not exist – a world based on spaces and places of hope that already exist. An example of radical place-making.

Absolute space   A view of space as independent of what occupies it. A potentially unlimited expanse within which everything else exists. Associated with Isaac Newton.

Actor-network theory   A set of ideas associated with Bruno Latour, John Law, and others that accounts for the way agency is produced through the interrelations between things in networks. Best known for its insistence on the agency of nonhumans.

Affect   A pre-cognitive sensation resulting from an encounter with a person or thing. Often translated, in retrospect, as feeling or emotion.

Agency   The capacity of an agent to act in the world. Often interpreted as the capacity to act freely.

Anarchism   A social and political philosophy opposed to institutionalized hierarchy and particularly the state. Associated with the writings of geographers Elisée Reclus and Peter Kropotkin.

Anthropocene   The current geological age in which humans are the dominant agents of change in the physical environment.

Areal differentiation   The study of how human and physical phenomena vary over the earth's surface.

Anti-foundational   An attribute of philosophies and theories that reject approaches that assert deep (foundational) causes for surface events. Often associated with postmodernism.

Assemblage theory   A set of ideas associated with Manuel DeLanda which explores how distinct wholes (assemblages) are made from the relations between contingent heterogenous parts.

Base–superstructure model   A Marxist model that asserts that an economic base composed of relations and forces of production determines superstructural phenomena such as "culture" and "beliefs."

Being-in-the-world   An idea from phenomenology (particularly Heidegger and then Merleau-Ponty) that human existence (being) is always located. "Being" is always "being there."

Biopolitics   An idea associated with Michel Foucault that refers to the way in which ideas from the biosciences are applied to human-political behavior. These ideas are not just "about" people but, in some sense, make them up and regulate them.

Bioregionalism   A belief, most often associated with the green movement, that natural regions (often defined by watersheds of rivers) exist. This is often accompanied by a belief that human life should be organized to correspond to these natural regions.

Blues epistemology   A way of knowing rooted in the experience of enslaved Africans on plantations who produced their own creative forms of Black life, understanding, and expression rooted in the blues. Associated with Black geographer, Clyde Woods.

Capitalocene   An alternative to Anthropocene that posits capitalism as the main agent of change in current physical environments, rather than all humans.

Central place theory   A theory that seeks to explain and predict the number, size, and arrangement of human settlements. Associated with German geographer Walter Christaller.

Geographic Thought: A Critical Introduction, Second Edition. Tim Cresswell.
© 2024 John Wiley & Sons Ltd. Published 2024 by John Wiley & Sons Ltd.

Chora   A philosophical term derived from Plato which refers to a region or area in which things come into being. Has since come to mean region. Also used by feminists to refer to an undifferentiated space associated with the maternal body.

Chorography   The geographical description of regions.

Chorology   The study of relations between things in a particular region. Associated with American geographer Richard Hartshorne.

Class   In Marxism, this refers to a social group with a particular position in the relations of production. More generally, it refers to people with the same social, economic, and educational status.

Colonial matrix of power   Also known as the coloniality of power. A concept accounting for the ongoing colonial, Western process of domination through domains of authority, economy, gender/sexuality and knowledge. Associated with decolonial scholar Anibal Quijano.

Colonialism   The establishment and maintenance of territory in one location by people from another. A process whereby the metropole claims sovereignty over the colony.

Coloniality   Arrangements of power originating in colonialism that continue to define knowledge, culture, economics, and practice despite the formal end of colonialism.

Constructionism/constructivism   The belief that scientific knowledge is constructed by scientists rather than originating in the natural world. It is opposed to objectivist beliefs that are central to positivism.

Contextual explanation   A form of explanation that relies on the particular context for its validity. This can often be the context of a particular place or location.

Critical phenomenology   A form of phenomenology that foregrounds the experiences of difference, marginalization, and power rather than assuming universal human experience.

Critical physical geography (CPG)   A field of study that applies critical accounts of social power to our understanding of physical systems. A theoretical approach that combines deep knowledge of physical landscapes with awareness of social systems and unequal power relations.

Critical rationalism   The claim that scientific theories should be rationally criticized and that empirical content of theories should be tested in ways that might falsify them. Associated with the philosopher Karl Popper.

Critical realism   A position that accepts the existence of a real, objective world but insists that this is not always empirically observable. There are, in other words, "real" structural generative mechanisms that can be actualized to produce particular observable phenomena. Associated with Roy Bhaskar and Andrew Sayer.

Critical regionalism   A focus on the ways in which regions are articulated and assembled in relation to the local and global. Here, regions are understood as constructs rather than essences. Associated with an influential essay by architectural critic Kenneth Frampton.

Critical theory   A school of thought that seeks to examine and critique culture and society. Generally seeks to transform and liberate society rather than simply explain, describe, or understand it. Associated with the Marxist and neo-Marxist Frankfurt School (Theodor Adorno, Walter Benjamin, etc.).

Decoloniality   A school of thought and body of theory, arising out of Indigenous and Latin American thinkers, that seeks to decouple from Euro-Western and colonial modes of thinking and being to create other modes of thinking and being that are not determined by coloniality.

Deductive reasoning   A process of reasoning that starts from a general hypothesis or statement and works toward specific conclusions.

Discourse   A set of meanings, assumptions, and practices that create a commonsense view of the world which we use to communicate and negotiate versions of truth. Produces the possibility for particular statements and practices to be valid or acceptable. Associated with the work of Michel Foucault.

Dwelling   A philosophical concept associated with Martin Heidegger. Refers to the process and practices by which people make themselves at home in the world. This is an important part of the humanistic conception of place.

Ecofeminism   A branch of feminism that considers both the links between women and nature and their mutual devaluation in patriarchal societies.

Ecumene (oikoumene)   A term used in the Greek/Roman world to refer to the inhabited earth.

Empiricism   An approach that privileges observation and experimentation over theory.

Environmental determinism   A set of ideas, dominant in late-nineteenth- and early twentieth-century geography, that asserts that environmental (natural, climatic) conditions determine and delimit human life. Associated with Ellen Semple, Ellsworth Huntingdon, and Griffith Taylor.

Epistemology   The theory of knowledge – or how we can know what we know.

Essentialism   The philosophical belief that entities have characteristics that are necessary for their identity – these are essences. These essences are permanent and unalterable. They are most often given as attributes of "nature."

Ethnogeomorphology   An approach to geomorphology that emphasizes the knowledge that communities have about the physical landscape and explores how social groups relate to and manage the land.

Feminism   A set of political and social philosophies and practices that center on the role of gender in the constitution of social life. The main focus is on the critique of hierarchical construction of binaries associated with masculinity and femininity.

Feminist empiricism   An approach that uses empiricist and positivist methodologies, including quantitative methods, to further the aims and objectives of feminism.

Fetishism   The attribution of powers to objects that do not, in fact, have them. Some geographers have been accused of "spatial fetishism" – attributing to space powers that actually originate in "society."

Flat ontology   An ontology that has no hierarchical system of levels. Any supposed "higher" level can always be reduced to the levels below it. Existence is exhausted by the sum of entities at the "lowest" level and their first-order properties. Hence, "global" is nothing other than the sum of entities that are "local."

Gender   What society makes out of sexual difference – ideas and practices of masculinity and femininity.

Genres de vie   Ways of life as defined by the French geographer Paul Vidal de la Blache. These tended to be specific to particular regions or *pays*.

Geopolitics   A set of ideas that accounts for the relationships between politics and space – most often, territory – at scales ranging from the local to the global. Often used in the actual practice of politics in order to develop strategy in an international context. Critical geopolitics analyzes and critiques these practices.

Geosophy   The study of geographical knowledge and imagination. Connected to the writing of American geographer John Kirkland Wright.

GIScience   The academic theory that supports and informs the use of Geographic Information Systems (GIS).

Golden spike   A unique reference point or marker chosen to mark the start of the geological era of the Anthropocene.

Governmentality   The art of government – the way governments attempt to produce a governable population. Derived from the work of Michel Foucault.

Gravity model   A model that attempts to predict the degree of interaction (migration, trade, etc.) between two places. Based on Newton's laws of gravity, gravity models generally focus on the size of places and the distance between them in order to predict levels of interaction.

Habitus   The way in which the norms of wider society are incorporated into the habits and movements of the body. An idea associated with French sociologist Pierre Bourdieu.

Hegemony   A way of ruling through the definition of what counts as common sense rather than through force or direct oppression. Associated with the Italian Communist writer Antonio Gramsci.

Hermeneutics   The theory and practice of the interpretation of all manner of texts.

Historical materialism   The Marxist theory of history that insists that history proceeds inevitably on a path from feudalism to communism owing to the development of the forces of production.

Humanism   A human-centered view of the world that bases understanding of our world on humans rather than God or some other external agent. This view was born in Renaissance Europe.

Humanistic geography   An approach to geography drawing on philosophies of meaning, phenomenology, and existentialism, which places human experience and subjectivity at the center of understanding our world.

Hybrid geographies   An approach to geography which draws on actor-network theory to insist on the networked mixing of human and nonhuman elements in the creation of worlds.

Idealism   The philosophical belief that reality is fundamentally or primarily a mental product. Associated with Immanuel Kant among others.

Ideology   Meanings that tend to serve to reproduce hierarchies of power and serve the interests of those in positions of power.

Idiographic   Approaches and methods which focus on the particular and specific rather than the general and universal.

Imperialism   The creation and maintenance of unequal power relations (in economic, social, and cultural terms) between states where one state dominates others in the form of empire.

Inductive reasoning   A style of reasoning that starts with particular instances and builds up to general theories and hypotheses.

Intentionality   The way in which consciousness is always consciousness directed at something. An idea central to phenomenology.

Landscape   The material topography of a segment of land as seen from one point in space. This definition, which centers on materiality and vision, has been challenged by geographers who insist on the lived and experienced qualities of the material world.

Least effort   The idea that forms of movement attempt to expend as little effort (time and space) as possible in achieving their outcomes. This idea is at the center of much of spatial science.

Lebensraum   Literally "living room." The idea, associated with German geographer Friedrich Ratzel, that strong states need more space and naturally tend to expand into the territory of weak states. Most famously linked to Nazi ideology.

Locale   The local material context for social relations to unfold. An aspect of place.

Location   An objective point in space marked by longitude and latitude and with specific distances from other locations. An aspect of place.

Logical Positivism   A philosophical position that only statements verifiable through direct observation or through logical proof are meaningful in relation to truth. Associated with the Vienna Circle of philosophers.

Mappa Mundi   A medieval map of the world.

Masculinist   Beliefs, values, and practices that tend to reproduce and reinforce the dominance of men in a hierarchical arrangement of genders. These are often deeply engrained and unconscious.

Marxism   A set of theories associated with Karl Marx that seeks to understand the world of capitalism and then transform it.

Metanarratives   A big idea that purports to explain large segments of history and experience. Often, these are ideologies used to support present arrangements of society.

Mobility   The movement of people, objects, or ideas. As a social product, mobility has become central to work in the "new mobilities paradigm" or "mobility turn."

Modernism   A general term for modern thought and practice. Often associated with the application of rationality and abstraction that is applicable anywhere.

Mutual aid   An idea derived from the work of anarchist geographer Peter Kropotkin, which asserts that there is a natural tendency toward cooperation between humans. This idea challenges conventional readings of Darwin which emphasize competition.

Nation   A community of people who identify themselves as sharing a culture in common. Often, this community also shares a territory.

Nation-state   A term used to describe a state that bases its legitimacy on the existence of a nation with which it territorially coincides.

Natural kinds   Kinds or classes of things that can be said to exist in nature rather than being the product of human interests.

Neighborhood effect   A concept that suggests that neighborhoods have direct and indirect effects on individual behaviors, beliefs, and social outcomes.

Nomothetic   A general kind of explanation that tends toward universal explanations and the production of explanatory laws.

Nonrepresentational theory   A style of work that emphasizes practice over representation. The emphasis is on the processes of becoming in life rather than the already achieved. Associated with the work on British geographer Nigel Thrift.

Ontology   The philosophical study of the nature of being and existence. Asks the question, "what exists?"

Patriarchy   A social order based on male privilege and female subordination.

Phase space   A term derived from physics to describe how possible futures exist as potentials in present forms of space. Translated into human geography by British geographer Martin Jones.

Phenomenology   A philosophical methodology that aims to account for the essential characteristic of things as accessed through human experience. Associated with philosophers Edmund Husserl and Maurice Merleau-Ponty and imported into geography by humanistic geographers such as Yi-Fu Tuan and David Seamon.

Place   A meaningful lived space combining location, locale, and sense of place.

Placelessness   A term used to describe places that appear to have a lack of identity and where it is impossible or difficult to become an existential insider. Associated with the Canadian geographer Edward Relph.

Plantation bloc   A place-based development alliance centered on the continuation of beliefs and practices associated with the plantation economy and its aftermath in the American South. A concept originating in the work of Clyde Woods.

Plantationocene   An alternative version of Anthropocene that posits the advent of slavery-based plantation economies as the starting point for the current geological era. This centers colonialism, capitalism, and ongoing racism in our understanding of global environmental change.

Pluriverse   A concept emerging from decolonial theory that recognizes the coexistence of multiple place-based ways of thinking and acting beyond the supposed universality of Eurocentric and colonial systems of thought and action.

Positionality   The recognition that the way in which we know things is based on who we are, our particular biography, and our locatedness within social structures of class, gender, race, etc.

Postcolonialism   Refers to both the period following the end of colonialism and a theoretical body of work that takes a critical stance toward colonialism in all its forms.

Posthumanism   A philosophical position that decenters the human and human agency in our understanding of the world. Posthumanism moves away from an overvaluing of individual subjectivity and focuses instead on how agency is distributed between humans and non-human animals and objects.

Postmodernism   Refers to both the thought of the period following (or at the end of) modernism and to theory that is opposed to thought associated with modernity – particularly the rigid certainties of forms of science, planning, and structuralism.

Postphenomenology   A revision and critique of phenomenology that considers experience in a relational way that includes both humans and their technologies. Postphenomenology moves away from a knowing human subject and sees experience as a relational achievement of human and non-human objects.

Poststructuralism   A wide range of philosophical and theoretical approaches and ideas that are opposed to structuralism, and particularly to the binary logic that tends to define structuralist approaches. Most often associated with twentieth-century French thinkers such as Michel Foucault, Jacques Derrida, Gilles Deleuze, and Julia Kristeva.

Political ecology   A body of work that focuses on the connections between environmental change on the one hand, and social, economic, and political systems on the other.

Positivism   A philosophical approach which emphasizes the role of direct sensory experience and mathematical logic in the construction of authentic knowledge. Associated with the work of Auguste Comte and widely held to inform much of spatial science.

Possibilism   A response to environmental determinism that insisted on the possibility of many different ways of life in the same environmental conditions – thus rejecting environmental determinism. Associated with French geographer Paul Vidal de la Blache.

Practice   A theoretical term for human activity – the things people do. Often contrasted with representation.

Productive forces   A Marxist term for the sum of technologies of labor (knowledge, tools, arrangements of space, etc.) and labor power at a given moment in a society's development. Together with the relations of production, they form the mode of production, the development of which is seen as the prime driver of history.

Promiscuous realism   A claim that there are countless ways of classifying the world into kinds and that the way we classify depends on our theoretical interests.

Racial capitalism   A view of capitalism that sees it as arising from the extraction of value from people of marginalized racial identities – particularly Black people. Associated with the work of Cedric Robinson and Ruth Wilson Gilmore.

Reification   Treating an abstraction as a real thing.

Region   An area of land, usually at a mesoscale, between the local and the global.

Relational space   A view of space as the product of relations between objects. Space, in this view, arises at the same time as the objects in it and can thus be contrasted with absolute space (above). Associated with poststructuralist geographies and particularly the work of British geographer Doreen Massey.

Relations of production   The social arrangement of humans that is mobilized in order to produce things. These are the arrangements people have to enter into in order to produce and reproduce their means of life at a given point in history. Together with the productive forces, they form the mode of production.

Relative space　A view of space as something which exists relative to objects – such as the distance between A and B. Should A or B change position, then the space they produce would change too.

Representation　The use of signs to stand in for something else. The most obvious example is language – where the word stands for the thing it refers to.

Scientific realism　A belief that objects of scientific knowledge exist independently of the mind. Scientific realists believe that good theories approximate the truth about the real world.

Sense of place　The subjective meanings, either personal or shared, that are attached to place. An aspect of the definition of "place."

Situated knowledge　Knowledge that is a product of the place and time in which it was produced. To be contrasted with views of knowledge as objective "views from nowhere."

Social constructionism　A view that asserts that particular objects or categories are products of particular social arrangements rather than objects of nature with inherent qualities. Often opposed to "essentialism."

Social theory　Theories which are used to describe and explain social phenomena. Often, these take normative positions and seek to critique traditional social thought.

Spatial fix　An argument in Marxist geography that capitalism temporarily resolves its contradictions by abandoning some (no longer profitable) spaces and producing new ones.

Spatial interaction theory　A body of ideas in spatial science that seeks to provide law-like generalization about how given spaces interact with each other. Movement between spaces is taken to be a product of attributes of spaces.

Spatial science　A term used to describe geography in its quantitative and positivist guise that emerged in the 1960s.

Spatiality　The influence of space on existence. This term describes the total interweaving of space with other aspects of social life, insisting that space is always social and vice versa.

Standpoint theory　A theory that insists on the value of a person's particular knowledge, which is based on the position they hold within a context. Standpoint theory insists that recognition of a person's standpoint produces a more profound "truth" than the "objectivity" associated with traditional science. Most clearly developed in feminist theory.

State　An organized political community that is recognized as having sovereignty in international law. Also a term used to apply to the government or the public sector more widely within a society.

Structuralism　A theoretical approach to the world that sees the variety of surface phenomena as being caused or determined by some invisible, deep-generating structure such as the mode of production or the system of grammar.

Structuration theory　A theory which sees structure and agency as mutually constitutive, with agency being authorized by structures and structures being produced through the repeated actions (agency) of individuals. Associated with British sociologist Anthony Giddens and French sociologist Pierre Bourdieu.

Structure　A deep generative arrangement or mechanism (such as the relations of production, or patriarchy, for instance) that produces or determines everyday actions.

Teleology　A circular mode of argument where the end point prefigures the beginning. An argument where the end point is logically inherent to the argument.

Territory　A clearly defined and bounded space which defines particular actions both within it and between the inside and outside. A controlled space.

The Event　An idea in poststructuralism and nonrepresentational theory that describes the potential for surprise and excess in life that is not reducible to any overarching structure or determination.

Time-space (space-time)　A concept which insists on the necessary interweaving of time and space. Space is always temporal, and time is always spatial. This defines the four dimensions within which we live.

Topology　A term derived from mathematics and geometry. In relational geographical theory, topology and the topological are used to refer to spaces defined by connections and networks (what connects to what) rather than accurate shape or form.

Topography　A term used to describe the shape or form of the land.

Topos　From the Greek for "place." Used by Aristotle to describe the place that emerges from space through a process of becoming.

Transnationalism　A social and geographical phenomenon resulting from the increasing mobility and connections between people across national boundaries. Transnationalism questions the continuing centrality of national borders in understanding social and cultural life.

Uneven development   The idea that processes of economic development occur unevenly across space and that there is a capitalist logic that connects more-developed spaces with less-developed spaces. This uneven process is necessary for the survival of capitalism. This idea is associated with the geographer Neil Smith.

Vincularidad   An idea originating from Andean decolonial thinkers that refers to the interrelational interconnectedness of all living organisms with the Land and the cosmos.

White supremacy   A set of beliefs and actions based on the idea that white people constitute a superior race and should rightfully dominate society to the detriment of other racialized groups.

Index